Socioscientific Issues-Based Instruction for Scientific Literacy Development

Wardell A. Powell
Framingham State University, USA

A volume in the Advances in Educational Technologies and Instructional Design (AETID) Book Series

IGI Global
PUBLISHER of TIMELY KNOWLEDGE

Published in the United States of America by
IGI Global
Information Science Reference (an imprint of IGI Global)
701 E. Chocolate Avenue
Hershey PA, USA 17033
Tel: 717-533-8845
Fax: 717-533-8661
E-mail: cust@igi-global.com
Web site: http://www.igi-global.com

Library of Congress Cataloging-in-Publication Data

Names: Powell, Wardell Anthony, 1969- editor.
Title: Socioscientific issues-based instruction for scientific literacy development / Wardell Anthony Powell, editor.
Description: Hershey, PA : Information Science Reference, 2021. | Includes bibliographical references and index. | Summary: "This book contains research on the development of a research-based integrated socioscientific issues-based instruction for use in the K-12 system, teacher education preparation, and informal education centers"-- Provided by publisher.
Identifiers: LCCN 2020010887 (print) | LCCN 2020010888 (ebook) | ISBN 9781799845584 (hardcover) | ISBN 9781799856337 (paperback) | ISBN 9781799845591 (ebook)
Subjects: LCSH: Science--Study and teaching. | Science teachers--Training of.
Classification: LCC Q181 .S6677 2021 (print) | LCC Q181 (ebook) | DDC 507.1/2--dc23
LC record available at https://lccn.loc.gov/2020010887
LC ebook record available at https://lccn.loc.gov/2020010888

This book is published in the IGI Global book series Advances in Educational Technologies and Instructional Design (AETID) (ISSN: 2326-8905; eISSN: 2326-8913)

British Cataloguing in Publication Data
A Cataloguing in Publication record for this book is available from the British Library.

Advances in Educational Technologies and Instructional Design (AETID) Book Series

ISSN:2326-8905
EISSN:2326-8913

Editor-in-Chief: Lawrence A. Tomei Robert Morris University, USA

MISSION

Education has undergone, and continues to undergo, immense changes in the way it is enacted and distributed to both child and adult learners. In modern education, the traditional classroom learning experience has evolved to include technological resources and to provide online classroom opportunities to students of all ages regardless of their geographical locations. From distance education, Massive-Open-Online-Courses (MOOCs), and electronic tablets in the classroom, technology is now an integral part of learning and is also affecting the way educators communicate information to students.

The **Advances in Educational Technologies & Instructional Design (AETID) Book Series** explores new research and theories for facilitating learning and improving educational performance utilizing technological processes and resources. The series examines technologies that can be integrated into K-12 classrooms to improve skills and learning abilities in all subjects including STEM education and language learning. Additionally, it studies the emergence of fully online classrooms for young and adult learners alike, and the communication and accountability challenges that can arise. Trending topics that are covered include adaptive learning, game-based learning, virtual school environments, and social media effects. School administrators, educators, academicians, researchers, and students will find this series to be an excellent resource for the effective design and implementation of learning technologies in their classes.

COVERAGE

- Educational Telecommunications
- Web 2.0 and Education
- K-12 Educational Technologies
- Hybrid Learning
- Adaptive Learning
- Digital Divide in Education
- Game-Based Learning
- Instructional Design
- Curriculum Development
- Social Media Effects on Education

The Advances in Educational Technologies and Instructional Design (AETID) Book Series (ISSN 2326-8905) is published by IGI Global, 701 E. Chocolate Avenue, Hershey, PA 17033-1240, USA, www.igi-global.com. This series is composed of titles available for purchase individually; each title is edited to be contextually exclusive from any other title within the series. For pricing and ordering information please visit http://www.igi-global.com/book-series/advances-educational-technologies-instructional-design/73678. Postmaster: Send all address changes to above address.

Titles in this Series

For a list of additional titles in this series, please visit:
http://www.igi-global.com/book-series/advances-educational-technologies-instructional-design/73678

Ethical Research Approaches to Indigenous Knowledge Education
Ntokozo Mthembu (University of South Africa, South Africa)
Information Science Reference • © 2021 • 260pp • H/C (ISBN: 9781799812494) • US $185.00

Digital Learning Architectures of Participation
Nigel Ecclesfield (FRSA, UK) and Fred Garnett (FRSA, UK)
Information Science Reference • © 2021 • 289pp • H/C (ISBN: 9781799843337) • US $165.00

Participatory Pedagogy Emerging Research and Opportunities
Martha Ann Davis McGaw (McGaw Davis Foundation, USA) and Simone McGaw-Evans (Georgia State University, USA)
Information Science Reference • © 2021 • 150pp • H/C (ISBN: 9781522589648) • US $155.00

Visual Approaches to Instructional Design, Development, and Deployment
Shalin Hai-Jew (Kansas State University, USA)
Information Science Reference • © 2021 • 228pp • H/C (ISBN: 9781799839460) • US $185.00

Teaching Language and Literature On and Off-Canon
José Manuel Correoso-Rodenas (Universidad Complutense de Madrid, Spain)
Information Science Reference • © 2020 • 370pp • H/C (ISBN: 9781799833796) • US $195.00

Handbook of Research on Advancing Language Equity Practices With Immigrant Communities
Lourdes Cardozo-Gaibisso (Universidad ORT Uruguay, Uruguay) and Max Vazquez Dominguez (University of North Georgia, USA)

For an entire list of titles in this series, please visit:
http://www.igi-global.com/book-series/advances-educational-technologies-instructional-design/73678

IGI Global
PUBLISHER of TIMELY KNOWLEDGE

701 East Chocolate Avenue, Hershey, PA 17033, USA
Tel: 717-533-8845 x100 • Fax: 717-533-8661
E-Mail: cust@igi-global.com • www.igi-global.com

Table of Contents

Section 5
Integrating Modeling and STEM in Socioscientific Issues for Scientific Literacy

Section 6
Concluding Remarks

Detailed Table of Contents

Research on the Socioscientific Issues (SSI) framework provides compelling evidence of its ability to provide contextualized science learning as preparation for informed citizenship. However, few SSI studies provide insight as to whether the pedagogical demands of SSI warrant modification or refinement when implemented with young children. Through this exploratory teacher research study, SSI units were developed and implemented in the teacher-researcher's first, second, and fourth grade science classrooms over a one-year period. Results suggest that while SSI provided a critical real-world context for science learning and student discourse, several developmentally-influenced challenges were evident including students' difficulties in shifting from emotional to cognitive bases of argumentation, unintended activation of student fears, confusion between real and imaginary scenarios, and uneven emotion regulation. Findings and recommendations provide a foundation for future research on curricular and pedagogical supports that can facilitate successful elementary SSI implementation.

Section 3
Socioscientific Issues for Scientific Literacy in the Undergraduate Classroom

Fostering informed Socioscientific Reasoning (SSR) is an essential component of developing scientific literacy. In this chapter, the authors suggest that enhancing Nature of Technology (NOT) understanding can be one way to leverage students' informed socioscientific reasoning. The authors describe a proposed NOT conceptual framework with four dimensions and detailed components, and present an analysis of students' reasoning of various socioscientific issues using this framework. Finally, the authors present the finding that NOT components were present in student discussions with varying levels of understanding. The SSR analysis reveals that students with NOT informed understanding can appreciate the integrated characteristics of technology, so as to make sophisticated decisions about science and technology that will change society in fundamental ways, for both better and worse.

Section 4
Socioscientific Issues for Scientific Literacy in Preservice Teacher Training

This chapter explores the association between elementary preservice teachers' (PSTs) perceptions of teaching using socioscientific issues (SSI) and the context in which the PSTs initially engaged with SSI. One course engaged with SSI while learning pedagogical strategies during an elementary science methods (decontextualized) course. The second course engaged with SSI via an environmental sciences course (contextualized). The contextualized course examined gray wolf management in Northern California as part of an ecology unit, which was followed by a series of debriefings regarding the implementation of the SSI. The findings indicate that while PSTs from both courses generally held more positive perceptions of SSI instruction after engaging with SSI, the students in the contextualized course perceived a greater ability to answer SSI-related questions and expressed more positive attitudes towards science content in the post-course data. Additionally, students in both courses perceived a greater need for teacher training on SSI implementation post-course.

This study aims to assess how preservice teachers' supporting reasons vary in relation to six subject areas, namely, sociology/culture (S), environment (E), economy (E), science (S), ethics/morality (E), and policy (P), in different socioscientific issues (SSI) and how preservice teachers' decisions regarding these issues interact with the aspects of value, knowledge, and personal experience. Exploratory sequential design was utilized in this study and 47 third-grade preservice science teachers. Data were collected via written reports in two different SSI scenarios, namely, gene therapy and preimplantation genetic diagnosis. Preservice teachers' written reports were analyzed using the SEE-SEP model. The results showed that preservice teachers' supporting reasons mainly stemmed from the subject areas of ethics/morality (41%) and science (32%). The results also revealed little use of justifications from the subject areas of policy and sociology/culture (11% and 10%, respectively), and subject area of economy (6%) was the least referred.

Current policy documents across the world call for changes in K-12 science teaching to use integrated STEM strategies to provide a more authentic learning environment for students. Though the importance of integrated STEM education is established through national and international policy documents, there remains disagreement on focus, models, and effective approaches for integrated STEM instruction. A primary focus of STEM policies is addressing STEM workforce issues. However, other important foci for global STEM initiatives are more equitable education, poverty reduction, and increased STEM literacy and awareness. This chapter critiques current implementations of STEM as focused only on technical aspects of engineering design which will not meet any of the goals of integrated STEM. Rather, the authors propose that integration of SSI into STEM lessons will promote the social thinking necessary in engineering design and enhance work toward achieving not only STEM workforce preparation, but also increased STEM literacy and equity within STEM.

Preface

As more and more environmental practices and policies that advantage some, and disadvantage many, are instituted around the world, and in the United States in particular, I became interested in editing a volume that I believe would share ideas from leading proponents of the socioscientific issues movement on how educators can best situate their students in becoming scientifically literate citizens to make informed decisions on these egregious environmental practices.

The use of socioscientific issues as a key pedagogical strategy has the potential to develop scientific literacy among school-age children, enhance their ability to engage in discourse and other practices that promote environmental health and social justice activities. Studies have shown that socioscientific issues improve students' skills to explain phenomena scientifically (Kosterman & Sadler, 2009; Nielson, 2012; Sadler, Klosterman, & Topcu, 2011; Zeidler, Sadler, Simmons, & Howes, 2005). Additionally, studies have also shown that socioscientific issues instruction resulted in improvement students' ability to evaluate and design scientific inquiry (Zeidler, Applebaum, & Sadler, 2011; Bencze, Sperling, & Carter, 2011), interpret data and evidence scientifically (Powell, 2019 & 2020: Zeidler, Sadler, Applebaum, & Callahan, 2009). Studies have shown that socioscientific issues enhance students' moral and character development (Owens, Sadler, & Zeidler, 2017; Zeidler, 2014).

The findings from these studies have elevated the socioscientific issues framework to a point where it should play a more pivotal role in the school science curriculum. To this end, I wanted to spearhead a book project that highlighted the work of science educators at the national and international levels who were engaging in cutting edge research on the impact of socioscientific issues on scientific literacy. I wanted these researchers to share the work they were doing in a manner that would be meaningful to practitioners in the field. As a result, this volume shares the implementation and outcomes of socioscientific issues in research practices that were implemented with students at the elementary, secondary, and tertiary levels.

This volume is organized into eleven chapters to serve as great resources for practitioners and science education researchers who are interested in engaging students at all levels in socioscientific issues investigation.

Wardell A. Powell
Framingham State University, USA

REFERENCES

Bencze, L. L., Sperling, E., & Carter, L. (2011). Students' research informed socio-scientific activism: Re/Visions for a sustainable future. *Research in Science Education, 42*(1), 129–148. doi:10.100711165-011-9260-3

Kosterman, M. L., & Sadler, T. D. (2010). Multi-level assessment of scientific content knowledge gains associated with socioscientific issues-based instruction. *International Journal of Science Education, 32*(8), 1017–1043. doi:10.1080/09500690902894512

Nielson, J. A. (2012). Science in discussion: An analysis of the use of science content in socioscientific discussion. *Science Education, 96*(3), 428–456. doi:10.1002ce.21001

Owens, D. C., Sadler, T. D., & Zeidler, D. L. (2017). Controversial issues in the science classroom. *Phi Delta Kappan, 99*(4), 45–49. doi:10.1177/0031721717745544

Powell, W. A. (2020). Framework for Using Geospatial Technology to Promote Middle School Students' Critical Thinking on Socioscientific Issues. In *M, Grassetti & Zoino-Jeannetti. Next Generation Digital Tools and Applications for Teaching and Learning Enhancement* (pp. 47–78). IGI Global. doi:10.4018/978-1-7998-1770-3.ch004

Powell, W. A., & Fuchs, D. (2019). Using Socioscientific Issues to Enhance Evidence-Based Reasoning Among Middle School Students. In S. Robinson & V. Knight (Eds.), *Handbook of Research on Critical Thinking and Teacher Education Pedagogy* (pp. 150–176). IGI Global. doi:10.4018/978-1-5225-7829-1.ch009

Sadler, T., Klosterman, M., & Topcu, M. (2011). Learning science content and socio-scientific reasoning through classroom explorations of global climate change. In T. D. Sadler (Ed.), *Socio-scientific issues in the classroom: Teaching, learning and research* (pp. 45–77). Springer. doi:10.1007/978-94-007-1159-4_4

Zeidler, D. L. (2014). Socioscientific Issues as a Curriculum Emphasis: Theory, Research, and Practice. In N. G. Lederman & S. K. Abell (Eds.), *Handbook of Research on Science Education* (Vol. II, pp. 697–726). Routledge.

Zeidler, D. L., Applebaum, S. M., & Sadler, T. D. (2011). Enacting a socioscientific issues classroom: Transformative transformations. In T. D. Sadler (Ed.), *Socio-scientific issues in science classrooms: Teaching, learning and research* (pp. 277–306). Springer. doi:10.1007/978-94-007-1159-4_16

Zeidler, D. L., Sadler, T. D., Applebaum, S., & Callahan, B. E. (2009). Advancing reflective judgment through socioscientific issues. *Journal of Research in Science Teaching, 46*(1), 74–101. doi:10.1002/tea.20281

Zeidler, D. L., Sadler, T. D., Simmons, M. L., & Howes, E. V. (2005). Beyond STS: A research-based framework for socioscientific issues education. *Science education*, *89*(3), 357–377. doi:10.1002ce.20048

Acknowledgment

I am eternally grateful to all the contributing authors of this volume. Their desire to share their experiences, expertise, and creativity in science education through a national and international lens made this volume a reality. I am humbled by their desire to showcase the use of socioscientific issues as a key pedagogical strategy for enhancing students' development of scientific literacy. To each of these authors, I want to say thank you from the bottom of my heart.

I could not have completed this project without the guidance of "El Maestro," Dana L. Zeidler. His wisdom and guidance from the inception of this project and throughout made it possible.

Mrs. Valerie Hytholt, I owe you a debt of gratitude for all of your feedback.

Finally, I would like to thank my wonderful wife, Novene, and our children Caresse and Anthony, for their support, encouragement, and understanding of the time it took to make this project a reality.

Wardell A. Powell
Framingham State University, USA

Introduction

Serendipity of Socioscientific Issues:

An Introduction to Socioscientific Issues-Based Instruction for Scientific Literacy Development

Serendipity is a word that has endured throughout the centuries partly because of its rich, melodious tone, and because its definition or interpretation has varied slightly with its application by different individuals under different circumstances. Horace Walpole (1754) is credited with having coined the term in a letter to his uncle, Sir Horace Mann, explaining to him:

I once read a silly fairy tale, called "The Three Princes of Serendip;" as their Highnesses travelled, they were always making discoveries, by accidents and sagacity, of things which they were not in quest of: for instance, one of them discovered that a mule, blind of the right eye, had travelled the same road lately, because the grass was eaten only on the left side, where it was worse than on the right—now do you understand Serendipity?" (Rosenau,1935, p. 91)

Examining an array of English dictionaries, you will come to find definitions of serendipity that tend to center around something like "the happy knack of making unexpected and delightful discoveries by accident." I have a hankering for this conglomerate of descriptions because it implies the intercourse of two things: 1) a knack; and, 2) a discovery. Hence, serendipity may be thought of as the gift of finding, by the combination of chance and wisdom, certain valuable or agreeable things not sought, while in pursuit of something else. "Accidental discovery" is not a synonym for serendipity. Serendipity is, rather, a discovery made because the individual is mentally equipped not to disregard or overlook something new of strange while on the quest for another goal. The fact that such discoveries come with them a sense of happiness and delight both touch on and highlight the emotive nature of act.

I view my own research program on socioscientific issues (SSI) as a journey that has come about through a kind of serendipity (Zeidler, 1984; Zeidler, 2014; Zeidler Herman & Sadler, 2019; Zeidler & Schafer, 1984). I wasn't looking to develop a particular research program on SSI, and I surely was not any more intelligent than the other graduate students in the program at Syracuse University. However, in retrospect, I think I was simply mentally equipped for the excursion. I was pursuing a fundamental question of how to better educate in the sciences in a manner that developed more than the cognitive functions of the student, but in a manner that also developed the moral faculties of the individual. The journey took me from the domain of the "hard sciences" to other domains of inquiry that tapped psychology, sociology, and philosophy. So being mentally equipped was no more than having a kind of *predisposition to wonder* beyond my immediate comfort zone and taking advantage of the flexibility afforded to me by my Ph.D. program, which provided opportunities to explore beyond my immediate program of study.

This is precisely what SSI enable both science educators and their students to be able to do. SSI provide a context for teachers to afford the conditions for students to experience serendipity. As students search for answers to scientific issues, they find moral and ethical principles embedded in sociocultural concerns – not prescribed, but thrust upon them in surprisingly interesting ways. SSI is the arena where students find the crossroads of science, morality, care, discourse and action. SSI, by its very nature, favors those the most with the prepared mind, one that possess a predisposition to inquiry in unfamiliar territory. It allows even laggardly souls to find interest and purpose, because educators have encouraged "accidental" discoveries that are not merely fortuitous, but destined to be unveiled in different ways by different students with different perspectives about how to construct meaning, relevance, and shared understanding.

This is precisely what the authors who contributed chapters in this book have done by way of investigating or advancing related themes that connect to developing multiple perspectives related to SSI. It is important to understand that there is much interrelated scholarship surrounding SSI that draws from areas of epistemological maturation, sociomoral discourse, emotive reasoning, character/citizenship education, nature of science, argumentation, and socioscientific reasoning, among other areas of study (Bencze, et al., 2020; Herman, Zeidler, & Newton, 2018; Evagorou, Nielsen, & Dillon, 2020; Lederman, Antink, & Bartos, 2014; Newton & Zeidler, 2020). The "common core" for these themes is centered on the fact that SSI is uniquely positioned as a sociocultural process within a progressive science education framework.

SSI is, de facto, connected to the quality of personal choices about community and global issues. This is what makes SSI viewed as "Science-in-Context." It requires a blend of socioscientific reasoning, evoking dimensions of understanding and recognizing elements of complexity, inquiry, perspective taking, skepticism and affordances & limitations of science (Kinslow, Sadler, & Nguyen, 2019; Sadler et al., 2007), as well as the exercise of socioscientific perspective taking (SSPT) (Kahn & Zeidler, 2016; 2019). SSPT is not just a specialized case of perspective taking. SSTP requires three conditions and are necessary and sufficient in the context of SSI, which include engagement in a personally relevant issue, shifting between etic and emic orientations, and being embedded moral context that compels one to consider ethical principles and exercise moral reasoning. This latter point is salient and what distinguishes SSPT as something more than a specialized case of perspective taking: "One can connect with others and shift their viewpoint to appreciate others' perspectives for evil, or for no particular reason. What makes perspective taking appear to be a positive behavior is the assumption of a moral context condition" (Kahn & Zeidler, 2019, p. 621).

The chapters that follow in this book add to and extend our present contextually understanding of the primacy of perspectives in SSI research and teaching. As an overview, we find that Chapters 1 and 2 focus on aspects of teaching SSI at the elementary level. Chapters 3,4 and 5 examine SSI at the middle school level. Chapter 6 moves to the undergraduate level with respect to SSI and the Nature of Technology. (It is noted that no chapters focus at the high school level but a cursory review of the research from 2000 to 2020 reveals hundreds of studies, including seminal research, at that level). The next two chapters, 7 and 8, examine important aspects of SSI and teacher training. The last two chapters, 9 and 10, present ideas related to the integrations of SSI with scientific modeling and STEM initiatives. As an advanced organizer, I highlight the foci of each chapter and suggest some points of interest to see on along the journey:

Chapter 1: No Child Too Young – A Teacher Research Study of Socioscientific Issues Implementation at the Elementary Level by Sami Kahn

Sami draws from her rich experiences as an educator who believes, to paraphrase Jerome Bruner, that any child can be taught any subject in some intellectually honest form. She believes in children and they believe in her, and that makes all the difference in the world. She studies and highlights what refinements are needed to teach SSI to young children to help other educators not to underestimate their abilities in developing scientific literacy. Sights not to miss along the way include the exemplars of major themes for the teacher journey and student feedback.

Chapter 2: Integrating Scientific Modeling and Socio-Scientific Reasoning to Promote Scientific Literacy by Li Ke, Laura A. Zangori, Troy Sadler, and Patricia J. Friedrichsen

This group of educators offers the benefit of their years of rich classroom and research experiences to advance a Socioscientific Issue and Model Based Learning (SIMBL) to promote scientific literacy. In this work, they focus on the scientific practice of modeling as a way to negotiate complex SSI and employ socioscientific reasoning. Pay particular attention to the illustrative example of the SIMBL framework.

Chapter 3: Impact of Socioscientific Issues on Middle School Students' Character and Values for Global Citizenship by Wardell Powell, Mark Newton, and Dana Zeidler

This group of science educators have had the benefit of collaborating on past SSI projects, and focusing on aspects of scientific literacy that develop students' character and responsibility, and is an important part of engaging with others within the global community. While many details of the research are presented, it is worth noting the stops along the way that examine social and moral compassion, moral and ethical sensitivity, perspective taking, empathetic concerns, socioscientific accountability, and feelings of responsibility, because of their direct connections to the emotive conditions of moral reasoning.

Chapter 4: A Socioscientific Issue Approach to Understanding Middle School Students Beliefs and Intentions Toward Climate Change by Wardell Powell

Wardell combines his experiences of teaching using SSI and applies it toward a specific area of research about exposing middle school students' beliefs toward climate change. Part of the appeal here is providing a specific example of implementing an SSI program that is intricately connected to developing scientifically literate citizens. Some points of interest include how the curriculum and teaching was aimed at scaffolding questions that lead to students' deeper understanding of curriculum content and proposing solutions to the issues.

Chapter 5: Blending Creative Drama and Computer-Supported Collaborative Learning for Socioscientific Argumentation by Aysegul Namdar and Bahadir Namdar

Aysegul and Bahadir take us on a tour of two areas that one would tend to think were quite diverse areas of study: creative drama and computer-supported collaborative learning so as to advance socioscientific argumentation. This blending of learning opportunities provides opportunities for students to diverse modalities and formats to encourage deep processing of the elements of socioscientific argumentation. It is worth seeing how the authors also attend to scientific literacy by providing a space for using creative drama in flipped classrooms.

Chapter 6: Enhancing Socioscientific Reasoning Through Nature of Technology by Hyunok Lee and Hyunju Lee

Part of this journey takes the reader along a road less traveled to examine how students' socioscientific reasoning (SSR) may be fostered by an understanding of the nature of technology (NOT). Hyunok and Hyunju adventure into an important area that has been understudies in science education. The authors synthesize from other areas of research the salient elements of NOT and apply them to developing scientific literacy through SSI and NOT. Be on the lookout for a matrix containing descriptions of NOT components that connect categories, NOT components, and contents. This provides a rather nice conceptual framework for contextualizing NOT within SSI.

Chapter 7: Teacher Train – The Association Between Course Context and Preservice Teachers' Perceptions of SSI Instruction by Mark Newton and Melanie Kinskey

In this excursion, Mark and Melanie take us into the underexplored realm of elementary preservice teacher education as it relates to developing a skill set for the appreciation of how SSI may be leveraged to enhance scientific literacy. This chapter addresses the reluctance of elementary teachers to delve into rigorous science instruction with their students, and provided empirical evidence for pedagogical success in addressing this problem with preservice teachers. One would not want to miss how they smartly contextualized SSI course content and integrated it as a sort of capstone experience in what would otherwise have been a conventional methods course. The authors' collective years of experience with teaching at this level as well as working with preservice teachers grounds the research and suggestions in authentic understanding.

Chapter 8: Use of the SEE-SEP Model in Preservice Science Teacher Education – The Case of Genetics Dilemmas by Dilek Karisan and Umran Betul Cebesoy

Another interesting journey is led by Dilek and Umran, who argue that scientific literacy must necessarily include consideration of moral and ethical factors through the exploration of SSI. To address this, they use the "SEE-SEP" model as an analytic framework to examine how preservice teachers explore decisions that tap six subject areas (Science, Environment, Ethics/morality, Sociology, Economy and Policy) that trigger their values, knowledge and personal experiences related to the issues at-hand. Points of interest along the way include how the rich integration of subject areas in their research is seamlessly integrated with the extant scholarship in the field.

Chapter 9: Intersections of Integrated STEM and Socio-Scientific Issues by Gillian Roehrig, Khomson Keratithamkul, and Benny Hiwatig

The scientific literacy adventure continues in the work by Gillian, Khomson and Benny, who bring us the crossroads of STEM and SSI. The authors problematize the misguided tendency to focus on STEM as merely a pipeline for workforce needs, as well as the misrepresentation of engineering practices. They argue that STEM could be better positioned within an SSI framework to better attend to questions of inclusivity and authenticity through moral and ethical considerations. The voyage will take the reader through well-crafted arguments and classroom examples of integrating STEM and SSI in K-12 classrooms.

Chapter 10: Socio-Scientific Issues and Model-Based Learning by Benzegül Durak and Mustafa Topçu

In this journey, Benzegül and Mustafa present extensions of Sadler and colleagues work (note chapter 2) on model-based learning applied in an SSI context. Their discussion of meta-modeling knowledge provides a good point of departure to guide the reader into integrating scientific modeling within SSI. A scenic point to stop and walk about is the discussion of model-oriented issue-based teaching as it is applied to the socioscientific issue and model-based learning framework (SIMBL).

Chapter 11: Unifying Themes in Socioscientific Issues-Based Instruction for Scientific Literacy Development by Wardell Powell

Wardell brings us safely to port, concluding our present journey with a vivid reminder of the immediacy for scientific literacy in today's exponentially changing world, and how the socioscientific issues framework may help us to navigate future uncertainties through carefully crafted scientific inquiry. It will be worth remembering, like a postcard sent to oneself during a trip, to remind oneself of something important experienced and not to be forgotten upon return from that excursion, that the quest for good thinking in the pursuit of knowledge, is as ageless as the ages. The thoughtful placement of SSI in science education helps us to remember those principles and concepts that ought not to be forgotten, and will serves us well in our future endeavors.

At the beginning of this introduction, I mentioned that serendipity is a discovery made because the individual is mentally equipped not to disregard or overlook something new or strange while on the quest for another goal. The extant research base in this field of study is filled with voluminous examples of teachers and students surprising and engaging discoveries while implementing SSI (e.g., Zeidler, Applebaum & Sadler, 2011). For teachers, their journey typically is a quest to improve their pedagogical science education practices. For students, their journey is one that leads them to make personally relevant connections to subject matter that might otherwise seem distant and remote. The chapters in this volume provide concrete examples of where the voyage of SSI instruction may take us on our quest to secure a sense of scientific literacy for all. SSI is a conceptual framework that favors and promotes an open mind through the exploration of perspectives, positions and orientations where moral and ethical issues intersect science (Kahn & Zeidler, 2017). This *is* the serendipity of socioscientific issues. This is also the enactment of a *predisposition to wonder*. In this regard, I am reminded of a famous quote by Louis Pasteur (1854):

"Dans les champs de l'observation le hazard ne favorise que les esprits prepares" which translates to something close to "Where observation is concerned, chance favors only the prepared mind"

Dana L. Zeidler
University of South Florida, USA

REFERENCES

Bencze, J. L., Pouliot, C., Pedretti, E., Simonneaux, L., Simonneaux, J., & Zeidler, D. L. (2020). SAQ, SSI & STSE education: Defending and extending 'Science-in- Context.' *Cultural Studies of Science Education*, 1–27. doi:10.100711422-019-09962-7

Evagorou, E., Nielsen, J. A., & Dillon, J. (Eds.). (n.d.). *Science Teacher Education for Responsible Citizenship: Towards a pedagogy for relevance through socio-scientific issues*. Springer. doi:10.1007/978-3-030-40229-7

Herman, B.C., Zeidler, D.L., & Newton, M. (2018). Students' emotive reasoning through place-based environmental socioscientific issues. *Research in Science Education*, 1-29. . doi:10.100711165-018-9764-1

Kahn, S., & Zeidler, D. L. (2016). Using our heads and HARTSS*: Developing perspective-taking skills for socioscientific reasoning (*humanities, arts, and social sciences). *Journal of Science Teacher Education*, *27*(3), 261–281. doi:10.100710972-016-9458-3

Kahn, S., & Zeidler, D. L. (2017). A case for the use of conceptual analysis in science education research. *Journal of Research in Science Teaching*, *54*(4), 538–551. doi:10.1002/tea.21376

Kahn, S., & Zeidler, D. L. (2019). A conceptual analysis of perspective taking: Positioning a tangled construct within science education and beyond. *Science & Education*, *28*, 605–638. doi:10.100711191-019-00044-2

Kinslow, A. T., Sadler, T. D., & Nguyen, H. T. (2019). Socio-scientific reasoning and environmental literacy in a field-based ecology class. *Environmental Education Research*, *25*(3), 388–410. doi:10.1080/13504622.2018.1442418

Lederman, N. G., Antink, A., & Bartos, S. (2014). Nature of science, scientific inquiry, and socioscientific issues arising from genetics: A pathway to developing a scientifically literate citizenry. *Science & Education*, *23*(2), 285–302. doi:10.100711191-012-9503-3

Newton, M., & Zeidler, D. L. (2020). Developing socioscientific perspective taking. *International Journal of Science Education*, 1–18. doi:10.1080/09500693.2020.1756515

Pasteur, L. (1854). *Formal inauguration of the Faculty of Arts and Douai and of the Faculty of Science at Lille, University of Lille Nord de France*. https://en.wikiquote.org/wiki/Louis_Pasteur

Rosenau, M. J. (1935). Serendipity (Presidential address delivered before the Society of American Bacteriologists at its Thirty-sixth Annual Meeting, Chicago, December 28, 1934). *Journal of Bacteriology,29*(2), 91-98.

Sadler, T. D., Barab, S., & Scott, B. (2007). What do students gain by engaging in socioscientific inquiry? *Research in Science Education*, *37*(4), 371–391. doi:10.100711165-006-9030-9

Zeidler, D. L. (1984). Moral issues and social policy in science education: Closing the literacy gap. *Science Education*, *68*(4), 411–419. doi:10.1002ce.3730680406

Zeidler, D. L. (2014). Socioscientific Issues as a Curriculum Emphasis: Theory, Research and Practice. In N. G. Lederman & S. K. Abell (Eds.), *Handbook of Research on Science Education* (Vol. 2, pp. 697–726). Routledge.

Zeidler, D. L., Applebaum, S. M., & Sadler, T. D. (2011). Enacting a socioscientific issues classroom: Transformative transformations. In T. D. Sadler (Ed.), *Socio-scientific issues in science classrooms: Teaching, learning and research* (pp. 277–306). Springer. doi:10.1007/978-94-007-1159-4_16

Zeidler, D. L., Herman, B. C., & Sadler, T. D. (2019). New directions in socioscientific issues research. *Disciplinary and Interdisciplinary Science Education Research*, *1*(11), 1–9. doi:10.118643031-019-0008-7

Zeidler, D. L., & Schafer, L. E. (1984). Identifying mediating factors of moral reasoning in science education. *Journal of Research in Science Teaching*, *21*(1), 1–15. doi:10.1002/tea.3660210102

Section 1

Socioscientific Issues for Scientific Literacy in the Elementary Grades

Chapter 1
No Child Too Young:
A Teacher Research Study of Socioscientific Issues Implementation at the Elementary Level

Sami Kahn
Princeton University, USA

ABSTRACT

Research on the Socioscientific Issues (SSI) framework provides compelling evidence of its ability to provide contextualized science learning as preparation for informed citizenship. However, few SSI studies provide insight as to whether the pedagogical demands of SSI warrant modification or refinement when implemented with young children. Through this exploratory teacher research study, SSI units were developed and implemented in the teacher-researcher's first, second, and fourth grade science classrooms over a one-year period. Results suggest that while SSI provided a critical real-world context for science learning and student discourse, several developmentally-influenced challenges were evident including students' difficulties in shifting from emotional to cognitive bases of argumentation, unintended activation of student fears, confusion between real and imaginary scenarios, and uneven emotion regulation. Findings and recommendations provide a foundation for future research on curricular and pedagogical supports that can facilitate successful elementary SSI implementation.

DOI: 10.4018/978-1-7998-4558-4.ch001

INTRODUCTION

I wait for it each day...that magical moment when my students forget that I'm there! At that instant, I envision my classroom as a rainforest and I, an intrepid explorer listening intently to its sounds. Quiet and tentative at first, they rise to a delicious cacophony of voices giggling, arguing, squealing, and buzzing, with only an occasional "Dr. Kahn!" squawking like a raucous bird piercing the discursive mist. For me, these are the happiest times in elementary science teaching; when young children are given the tools, opportunities, and encouragement to share their ideas, they can be magnificently unguarded, incredibly insightful, and at times oblivious to your presence. They have so much to say! That's only true, I think, if you earn their trust and create an environment that is safe for their deepest thoughts. The beauty of teaching science to young children is that they are only mildly indoctrinated into societal norms; elementary students say what's on their minds, providing what could be an ideal context for enacting SSI. ~Teacher/Researcher Journal Entry, 2016

The *Next Generation Science Standards* (NGSS Lead States, 2013) call for engaging students' interests in science-related issues while providing them with the content knowledge and practical skills to apply science in their everyday lives. Arguably, the Socioscientific Issues (SSI) framework addresses this goal as it utilizes the negotiation of real-world, ill-structured, socially-related scientific issues as the context for science learning while mediating argumentation (Chinn & Brewer, 1993; Lin & Mintzes; 2010), moral and character development (Zeidler, 2014), moral sensitivity (Fowler, Zeidler, & Sadler, 2009), informal reasoning (Karpudewan & Roth; 2018), Nature of Science (Abd-El-Khalick, 2006; Khishfe, Alshaya, BouJaoude, Mansour, & Alrudiyan, 2017), perspective taking (Kahn & Zeidler, 2019), and the cultural perspectives necessary for responsible global citizenship (Lee, Chang, Choi, Kim, & Zeidler, 2012). These hallmarks of SSI, which differentiate it from earlier science and technology frameworks, would seem to bolster the hypothesis that SSI should be particularly effective at the elementary level as developing the whole child, including socioemotional and ethical aspects of their character is a key goal for teachers of young children (Denham, 2015). In addition, literacy and numeracy, which form the foundation of the *Common Core State Standards* (National Governors Association Center for Best Practices, 2010) and comprise the bulk of elementary teaching time (Blank, 2013), are easily embedded within SSI curriculum as SSI is an interdisciplinary approach to science teaching that challenges students to utilize and evaluate a variety of texts and media sources and analyze data in real-world contexts (Zeidler & Kahn, 2014). This would seem to position SSI as a particularly *efficient* framework for elementary teachers who often struggle with limited time to "cover" all that needs to get done each day. While advocates do indeed describe

SSI as a framework that is sensitive to the developmental and epistemological needs of the learner and offer it as an appropriate framework for K-12 science learning, most of the existing research has been gleaned from work in secondary classrooms or within teacher education programs (Linn & Mintzes, 2010). Very few studies have actually been conducted within elementary classrooms and those that have focused on particular outcome variables such as environmental awareness (Burek & Zeidler, 2015), argumentation through an online environment (Evagorou, 2011; Evagorou & Osborne, 2013), critical thinking (Vieira & Tenreiro-Vieira; 2016), and informal reasoning (Karpudewan & Roth, 2018), or have identified particular lessons that have been successfully implemented (Dolan, Nichols, & Zeidler, 2009) rather than exploring modifications that might be required to ensure that curriculum and pedagogy were age-appropriate for elementary students. In addition, currently no empirical study has documented implementation of SSI with children as young as first and second grade. Developmentally-appropriate curriculum design and practice is essential for students at every level, but is particularly salient for students at an early age (Copple, Bredekamp, Koralek, & Charner, 2014) especially in science (Yoon & Onchwari, 2006) as it sets the course for all future learning. While teachers' content knowledge, lack of time spent on science, and low administrative support are frequently cited as barriers to science excellence at the elementary level (Roth, 2014), the question of whether there are challenges inherent to the SSI framework that present themselves at the elementary level remains unanswered. To address this question, this chapter describes the results of a one-year teacher research study implemented by an elementary science teacher/researcher to address the following questions:

Research Question #1 – What key themes characterize implementation of the Socioscientific Issues (SSI) framework at the elementary level?

Research Question #2 – What, if any, modifications or refinements are warranted when the Socioscientific Issues (SSI) framework is implemented with young children?

Answering these questions can provide teachers, teacher educators, and researchers with a more robust foundation for age-appropriate SSI curriculum development, pedagogy, and ongoing research.

BACKGROUND

Socioscientific Issues (SSI) as a Vehicle for Scientific Literacy

The Socioscientific Issues (SSI) framework utilizes the negotiation of real-world, socially-related scientific issues as the context for science learning (Zeidler, 2014). Common examples of SSI studied in the classroom include genetic engineering, the use of animals in research, fluoridation of tap water, and nuclear energy, among many others. SSI are, by definition, complex and ill-structured because they are related to, but not fully answered, by science; in other words, science informs but doesn't dictate their resolution. Advocates of SSI assert that providing students with contextualized opportunities for argumentation about scientific issues prepares students for participatory citizenship as adults. Unlike other pedagogical approaches that connected science to societal issues, such as STS (Science, Technology, and Society), SSI considers students' psychological and epistemological development, as well as the development of character and virtue (Zeidler, Sadler, Simmons, & Howes, 2005). Premised on a notion of Vision II scientific literacy (Roberts, 2007; Roberts & Bybee, 2014) that emphasizes contextualized, student-centered science learning as preparation for informed citizenship in an ever-changing society, SSI extends its framework to include the promotion of moral development and reflective reasoning as students grapple with the ethical implications of their decisions. This "functional scientific literacy" (Zeidler & Sadler, 2011) is "contextually and culturally sensitive to the needs of the learner" (p. 180) and compels students toward better moral choices that reflect "internal consistency, respect for multiple perspectives, and tolerance for dissenting voices" (p. 181). An array of empirical research suggests that SSI mediates argumentation (Chinn & Brewer, 1993), moral and character development (Zeidler & Sadler, 2008), moral sensitivity (Fowler, Zeidler, & Sadler, 2009), Nature of Science (Abd-El-Khalick, 2006), as well as the cultural perspectives necessary for responsible global citizenship (Lee, Chang, Choi, Kim, & Zeidler, 2012).

While SSI lessons can take many forms, a typical lesson sequence might entail introducing students to the issue, having them identify the constituencies, research the issues, reflect upon their values orientation, engage in discourse with others, and reach a decision or consensus (Kahn, 2019; Zeidler & Kahn, 2014). Throughout this experience, students are prompted to question their existing beliefs, collect and examine evidence through multiple research experiences, and develop arguments. Of paramount importance is the opportunity for students to experience dissenting voices that challenge their viewpoints and to exercise perspective taking (Kahn & Zeidler, 2016). While ethical issues are integrated into SSI, teachers do not instruct students in what to believe but rather, they help students to acquire flexibility, open-mindedness, and perspective taking abilities so that students can integrate content knowledge with emotive, rational, and intuitive influences (Sadler & Zeidler, 2005). Presley and colleagues (2013) suggest a framework for SSI implementation that features certain curriculum design elements (such as utilizing a compelling issue and providing a culminating experience), learner experiences (such as engaging in higher-order practices and negotiating the social dimensions of issues), and teacher attributes (such as knowledge of the science content and a willingness to learn with the students). The authors also highlight the importance of creating a collaborative and respectful classroom environment to ensure that students feel safe sharing their ideas. In sum, a substantial body of research on SSI instruction suggests that thoughtfully-designed curriculum and pedagogy implemented within SSI-friendly classroom environments can imbue students with critical scientific, intellectual, and interpersonal skills that can prepare them for science-related decision making in an ever-changing, global society.

What does this vision of SSI look like at the elementary level? While advocates of SSI describe it as a framework that is sensitive to the developmental and epistemological needs of the learner (Zeidler, Sadler, Simmons, & Howes, 2005) and offer it as an appropriate framework for K-12 science learning (Zeidler & Kahn, 2014), most of the existing research has been gleaned from work in secondary classrooms or within teacher education programs (Linn & Mintzes, 2010). Very few studies have actually been conducted within elementary classrooms and those that have focused on particular outcome variables such as environmental awareness (Burek & Zeidler, 2015), argumentation through an online environment (Evagorou, 2011), or have identified lessons that have been successfully implemented (Dolan, Nichols, & Zeidler, 2009) rather than exploring modifications that might be required to ensure that curriculum and pedagogy were age-appropriate for elementary students. This leads naturally to the question, can a framework like SSI that focuses on complex, real-world, and even thorny subjects be appropriate for such young students?

Elementary Science Teaching Aims and Challenges

Young children's abilities in science have long been underestimated, as child developmental theories have traditionally focused on deficits rather than assets in children's thinking (Gelman & Baillargeon, 1983). For example, using simplistic interpretations of Piaget's stage theory to drive curriculum led to assumptions that young children are concrete, pre-operational, and unable to think in relational terms (National Research Council, 2000). More recent research has led cognitive developmental researchers to recognize the vast capacity of young learners insofar as relational, abstract, and experimental thinking (National Academies of Sciences, Engineering, and Medicine, 2018).

Yet while children's immense capabilities are being recognized, elementary teachers struggle to implement rigorous science curriculum due to several practical barriers including time, content knowledge, and administrative support (Roth, 2014). Arguably, emphases on accountability in recent years have led to inordinate amounts of time on test preparation and content that "counts" toward testing results. Time spent on science in elementary schools has been decreasing over the past decades, with schools allocating greater time to literacy and math (Blank, 2013). Although increased time on science alone wouldn't address gaps in scientific literacy, evidence suggests that greater time on science in elementary school can contribute to increased achievement (Curran & Kitchin, 2019). Therefore, identifying curricular frameworks that enable elementary teachers to incorporate science more readily within their day, perhaps through integration with literacy, math, and/or social studies, could provide benefits for students' science learning.

Interdependence between disciplines was the impetus behind the development of the Common Core Curriculum which provides motivation for interdisciplinary curricular approaches that capitalize on the way students learn (Pearson & Hiebert, 2013). In addition to reducing artificial siloing of subject areas, quality frameworks that emphasize common interdisciplinary skills such as arguing from evidence, communicating findings, evaluating media sources, and so on, provide teachers with efficient ways of meeting standards from a variety of disciplines.

Of course, elementary teaching is not solely (or perhaps even primarily) about academics. Developing the whole child, including socioemotional and ethical aspects of their character is key for teachers of young children. Social-Emotional Learning (SEL) is widely recognized as a key factor in future social and academic performance (Rimm-Kaufman & Hulleman, 2015; Taylor, Oberle, Durlak, & Weissberg, 2017), and many elementary schools integrate programs that emphasize SEL competencies including developing care and concern for others, making responsible decisions, negotiating conflict constructively, establishing positive relationships, and managing one's emotions (Collaborative for Academic, Social, and Emotional Learning, 2017). Relatedly, contemporary visions of moral education in elementary years are said to result from creating a sociomoral atmosphere that is based upon mutual respect and support of character development (DeVries & Zan, 2012) as evidenced by teachers helping students to resolve their conflicts and prompting them to reflect on sociomoral issues (Hildebrandt & Zan, 2014).

In theory, the research that supports SSI, particularly insofar as its ability to develop character and prepare students for global citizenship (Lee, et al., 2012) seems wholly aligned with contemporary goals of elementary education. By introducing young children into real-world discourse, they could gain an understanding of the consequences of their actions both locally and globally. In doing so, SSI may be positioned to empower children to become informed decision makers able to consider the ethical, humanistic aspects of scientific decisions, thus ensuring a more equitable tomorrow. Of course, such lofty motivations would be meaningless for a science program if science content were not being mastered. Fortunately, a strong research base supports SSI as a meaningful context for students to apply content in a manner that is consistent with US national standards, and in an interdisciplinary way. To that end, the following framework is advanced to guide the inquiry and interpretations in the present study.

Figure 1. Key elements of SSI framework that align with elementary teaching goals (Kahn, 2019)

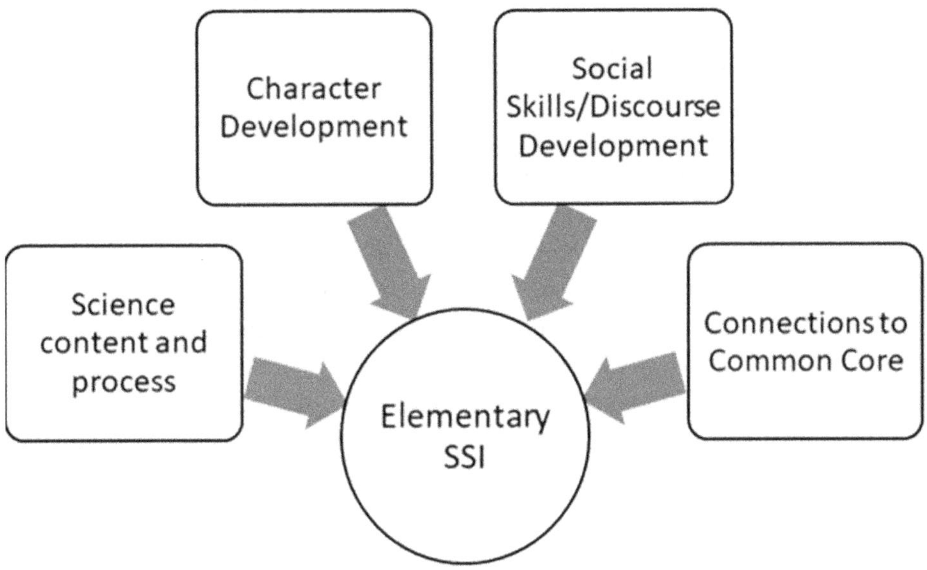

Theory alone, however, cannot achieve these goals. Therefore, this study examines the practical implications of SSI in the elementary classroom with an eye toward revealing any modifications or refinements that might be warranted in order to meet the abovementioned goals of elementary science education and scientific literacy.

METHODS

Teacher Research

I (the teacher/researcher) selected teacher research (Cochran-Smith & Lytle, 1999; Rust, 2009) as the appropriate methodology for this study as it aims to bridge the gap between research and practice by allowing the teacher to be an informant to scholars while reflecting on her/his own practice. Teacher research, rather than teacher action research, was invoked in this study as the emphasis was on teachers' written reflections and students' written artifacts as the key data, as opposed to outcomes from experimentation (Feldman, 1996). This approach allowed for a more descriptive picture of SSI implementation in an elementary classroom.

Study Context and Reflexivity Statement

During the 2015-2016 academic year, I designed and implemented a series of SSI units in my first, second, and fourth grade classes while serving in my twelfth year as an elementary science teacher at a single gender independent preparatory school in a major city in the northeast U.S. While the students at the school were ethnically and socioeconomically diverse, they could all reasonably be considered high achieving students due to admissions standards. Students had science three 50-minute periods over two- week cycle and were seen in half groups so that only half of a homeroom class came to science at any one time, thereby ensuring small class sizes (approximately 12 students/class). The total number of student participants included in the study were: 1st grade (48 students); 2nd grade (24 students); 4th grade (50 students).

Given that teachers' science content knowledge, lack of time spent on science, and low administrative support are frequently cited as barriers to science excellence at the elementary level (Roth, 2014), I recognized that my situation was not representative of most classrooms. I was fortunate to have small class sizes, supportive administrators, tremendous flexibility in terms of curriculum development, and freedom from standardized, high-stakes testing. In addition, I was a science specialist rather than a classroom teacher; in other words, I taught science classes all day, every day. I also have a master's degree in a science discipline, in addition to doctoral degrees in science education and law. This unusual combination of educational pursuits is particularly uncommon for elementary teachers who often have little content background in science and would not typically be expected to gain equal expertise in both the scientific and societal aspects inherent in SSI (Sadler, 2011). Unquestionably, my background and relatively flexible teaching environment afforded me the opportunity to test an array of different pedagogical approaches in general, and specifically, to hone my skills in enacting SSI.

In a sense, my teaching circumstances created a perfect testing environment for research in this area as I was able to address (i.e., control for) each of the limitations typically affecting elementary science teachers. While most classroom studies strive for generalizability by arguing that they are representative of a typical classroom ("If it happens there, it can happen anywhere") my context represented the antithesis of the "norm," thereby constituting a critical case sample (Patton, 1990) that controls for the typical variables such that we can say, "If it doesn't happen there, it won't happen anywhere (Patton, 1990, p. 175). Stated another way, any challenges I found in implementing SSI were likely not due to the barriers listed above but rather, could more readily be attributed to those inherent to SSI implementation at the elementary level itself. A summary of this argument is presented in Table 1.

Table 1. Factors contributing to the emergence of a critical case sample in present study

Common Barriers to Elementary Science Teaching	How Addressed in Current "Critical Case" Study
Lack of teacher content knowledge	Teacher-Researcher holds graduate degrees in a science discipline as well as science education
Lack of administrative support	Teacher-Researcher had full support of department head and principal to implement SSI
Lack of time spent on science	Teacher-Researcher taught in a self-contained science class with the opportunity to expand to in-class science support
Lack of resources	Teacher-Researcher was provided with an array of hands-on materials as well as library access and online tools.

In addition to the above factors influencing the likelihood of my successful SSI implementation, there are several additional points worthy of disclosing to the reader in order to provide context for the study. In line with my constructivist philosophy, I utilized Responsive Classroom (Responsive Classroom, 2020) which is a student-centered, socioemotional approach to teaching and discipline. In my science classroom, this manifested itself in a daily meeting at the start of class to build community, shared establishment of rules, logical consequences for misbehavior, and positive teacher language (Denton, 2007). Creating a classroom of mutual respect, trust, and a safe place for students to share ideas are conditions that have been identified as supportive of SSI implementation (Presley, et al., 2013). Also, I implemented Cooperative Learning techniques (Slavin, 1980) to help students to develop meaningful relationships and mutual interdependence with teammates, promote personal accountability, and support consensus-building. Finally, I also made Nature of Science (Lederman, 2007) tenets apparent by creating a bulletin board of photographs of my students engaging in practices and demonstrating habits of mind of scientists, and also explicitly pointing out (and asking students to point out) activities that emulate practices of scientists, such as demonstrating skepticism, engaging in argument using evidence, being creative, etc. During my twelve years of teaching at the school, I was recognized for high quality curriculum development and pedagogy, evidenced through the receipt of excellent teaching evaluations, teaching awards, and curriculum development grants.

Curricular Context

The units that were implemented in this study were as follows: Beaver Relocation (first grade), Mars Colonization (second grade), Rare Earth Elements (fourth grade) and Wetland Development (first, second, and fourth grade). Therefore, there were two SSI units completed per grade; one grade-specific and one (Wetlands Development) common to all three grades. With this design, I was able to experience lessons both within and across grade levels. These topics were selected in an effort to identify contexts that were both contemporary and personally relevant to students as these qualities have been strongly identified as necessary for interest in science learning (Stolz, Witteck, Marks, & Eilks, 2013; Zeidler, Sadler, Applebaum, & Callahan, 2009). The full units lasted between 4-6 weeks each, while the specific SSI lessons lasted 3-4 classes. The four SSI lessons that were included in this study are discussed more fully below.

First Grade (Six and Seven Year Olds): Beaver Relocation

This activity was designed to complement a unit on animal architects which embedded engineering design within the study of life science. After using various media to investigate beaver anatomy and behavior, students designed, built and tested beaver dams for their ability to hold back water (and create a pond) for a specified amount of time. Students were then introduced, through video, to a real-life controversy in which a beaver dam is causing flooding in a town. The townspeople in the video were considering the options of a) leaving the dam alone; b) moving the dam; or c) destroying the dam. After viewing the video, students engaged in a town hall meeting where they role-played town residents and discussed their positions on the issue. The activity culminated in a town hall meeting where students shared their opinions and a project of their choice in which students communicated facts about beavers' physical adaptations and ability to change their environment, and the student's position on the relocation controversy (Kahn, 2019).

Second Grade (Seven and Eight Year Olds): Mars Colonization

The second-grade unit on space consisted of a series of investigations that simulated the processes involved with space travel including planning a mission, determining what to bring, rocket design, rocket fuel testing, and exploring planetary surfaces. Students worked collaboratively with their "mission" teams to research and compare Earth vs. Mars environments and decide on a mission plan while considering weight, time, and economic constraints. For the SSI lesson, I placed two signs up at either end of my classroom; one said "Agree" with a thumbs-up icon, and the other said "Disagree" along with a thumbs-down icon. I instructed my students to listen to the statement I was about to make, and then stand somewhere along the imaginary spectrum line to reflect their opinion on the statement. I then read the statement, "The United States should build a colony on Mars." I asked students to share their ideas using the sentence frame, "I think that _______ because ________" which I had written on the whiteboard. Students moved to their desired position on the line and we engaged in discussion. Students were permitted to move along the line if they felt persuaded by arguments. After the discussion, teams of students worked collaboratively to develop planetary proposals for colonies on Mars. The proposals required students to describe how basic human needs would be met and what rules would organize the colony. This lesson was adapted from Zeider and Kahn (2014) and originally written for middle school.

Fourth Grade (Nine and Ten Year Olds): Rare Earth Elements

As the capstone activity for an extensive unit on environmental science, I adapted an activity on rare earth elements ("REEs") from one written for middle school (Zeidler & Kahn, 2014). The lesson began by showing students an infographic entitled, "Elements of a Smartphone" which depicted the various rare earth elements used in this device. All of my students either had or had used smartphones, iPads, and other electronics that use REEs, so bringing that to the forefront helped to situate REEs as an issue with personal significance. Students then viewed a "60 Minutes" report on REEs that gave a very basic overview of what they are and how they are used. It also introduced students to the controversy over the mining of REEs, which was halted in the US due to environmental concerns and caused the US to rely more on China, who provides the majority of REEs that the US uses. I randomly divided the class into groups representing a mining company, environmentalists, a Smartphone manufacturer, and a group of consumers, and gave students two classes to prepare for a congressional sub-committee meeting. Students voted on the outcome of the meeting.

First, Second, and Fourth Grades: Wetland Development

After several weeks of engaging in activities about the critical roles wetlands play in flood control, water filtration, providing nurseries for wildlife, food production, and so on, my students were introduced to Bullfrog Pond, a fictitious wetland area that was under consideration for development (for more information on the activity, see Kahn, 2019). Teams of students were provided with blank maps of the pond and were instructed to decide on the best locations for various paper game pieces including homes, stores, farms, factories, roads, and so on. They were also able to add features to their town, if time permitted. In addition, the teams were randomly assigned roles as homeowners, business owners, and farmers, and were challenged to try to make decisions on behalf of those constituencies. To ensure that students understood the potential consequences of their decisions, and to check that they understood the perspective taking that was being requested, we brainstormed a matrix that included the advantages and disadvantages to each group of developing a town near the pond. For example, homeowners might enjoy the benefits of a beautiful view, the opportunity to fish and go boating, and high home values near the pond, but trash from the homes and destruction of the shoreline could harm the wildlife and increase flooding, respectively. Once teams had completed their maps, they shared their plans at a town hall meeting where they presented their rationales, discussed potential environmental impacts of their plan, and proposed mitigation plans to decrease potential impacts. The townspeople were then charged with determining a final plan through debate, negotiation, and compromise.

Table 2. Summary of SSI Curricular Context

SSI Topic	Grade Level	Driving Science and Engineering Questions	Ethical Issues	Opportunity to Share Diverse Perspectives
Beaver Relocation	1st	How do animals change their environment to meet their needs? How can we use drawings and models to communicate our ideas for solving problems?	Human-Animal Conflicts, Animal Rights, Environmental Stewardship	Town Hall Meeting, Beaver Proposal and Gallery Walk
Mars Colonization	2nd	What do humans need to survive? What are the key planetary features of Mars? What do we need to consider in order to ensure habitability and sustainability?	Allocation of Resources, Planetary Protection, Pioneering/ Manifest Destiny	Yes/No spectrum line; United Nations "Planetary Colony Proposal" and vote
Rare Earth Elements (REEs)	4th	What are renewable and non-renewable resources? What properties of REEs make them desirable in manufacturing? How do environmental quality and "green technologies" factor into decisions about resource use?	Economic and Environmental Tradeoffs, Government Regulation of Public Safety	Congressional Sub-Committee Hearing and vote
Wetland Development	1st, 2nd, and 4th	What are wetlands, and what important functions do they serve? How does development affect wetlands, and what steps can mitigate those impacts?	Property Rights, Land Use, Environmental Sustainability	Town Hall Meeting and Presentation of Development and Mitigation Plans

Data Collection and Analysis

During the SSI implementation stages, I maintained a systematic reflective journal which was written in narrative form and influenced Linda Valli's (1997) work on teacher reflection. I also collected artifacts of student work and, at the end of each SSI activity, elicited feedback (oral from first and second grade, oral and/or written in fourth grade) from my students. I also met regularly for peer debriefing sessions with a colleague with whom I shared a room to discuss my impressions of the classes.

Table 3. Exemplars of Major Themes from Teacher Journal and Student Feedback

Theme	Teacher Journal Exemplar	Student Feedback Exemplar	Researcher Interpretation
Student Pride	[Fourth grade student] who was initially very unhappy about being on the 'environmentalist' team exclaimed, "Wow! I can't believe we won! It was really hard getting all of this information together and convincing them, but we did it!" (Wetland Development) When I asked the [first grade] 'homeowners' how they did on their development plan, the entire team smiled with their toothless grins with [student] exclaiming, 'We did awesome!" as he put his arms around his partners. (Wetland Development)	"Today was my best science day! I'm good at this kind of research and I got to do it!" (Fourth grade student during Rare Earth Elements) "We thought the dam would break but it was strong! We're pretty smart beavers!" (First grade student during Beaver Relocation) "I love our town. We disagreed at first but when we each gave an idea and put it all together, it worked!" (2nd grade student during Wetland Development)	These exemplars indicate that SSI can provide an opportunity for students to feel pride as they grapple with challenging curriculum and identify areas relative of strength.
Doing "Real Science"	N/A	"I liked planning our Mars colony because we were talking like real scientists." (2nd grade student during Mars Colonization) "Testifying in front of Congress was amazing...I felt like a real scientist." (Fourth grader during REEs) "I liked researching the REEs, getting all the facts, and arguing for it. I was like an expert...a real scientist!" (4th grade student during REEs)	These exemplars indicate that the scenarios and activities prompted students to feel that they were doing the work of, and feeling like, "real scientists." Interestingly, this theme did not show up in the teacher/researcher journal.
Doing the "Right Thing"	[First grade student] was set on finding a compromise for [his two friends]. After they argued for a while, he said, "C'mon guys, let's move the beavers to a different place... they'll be happy and your house won't get flooded. It's the right thing to do!" (Beaver Relocation)	"The government needs to protect people... if we mine REEs, we have to do it safely. It's just the right thing to do!" (Fourth grade student) "The beavers are 'natural' and they were here first, so they should be left alone... it's wrong to hurt them just for doing what beavers like to do." (First grade student during Beaver Relocation)	These exemplars demonstrate the manner in which SSI prompts students to consider issues of justice, fairness, consequences of actions and inactions, and invocation of moral decision making.
Student Frustration	[Fourth Grade Student] left the room in tears and sat outside crying. When I asked him what had happened, he said, 'They're not listening to me! I keep telling them not to keep developing [on the wetlands] because it will hurt the animals, but they keep doing it!" (Wetland Development)	"It was hard to talk because [student] kept arguing louder and louder." (Second grade student during Mars Colonization) "We couldn't find as much information about REEs as the other teams...that made me upset." (Fourth grade student)	These exemplars indicate that students' emergent social/emotional skills (particularly cooperation skills) may be a factor in incidences of frustration during SSI. In addition, uneven media resources (real or perceived) can also be a source of frustration.
Student Fear	One [second grade] student also mentioned that it [colonizing Mars] would be scary because, "if a terrorist was there, it would be hard to get away." And then it spread through the class. Several students became visibly upset and began talking about terrorists in space. [First grade student] was very upset after the video and just kept repeating, "I don't want anything to happen to the beavers. Don't let anything happen to the beavers." I think the video was just too real.	"If we make a colony with a machine that can give off oxygen and it breaks, we could all die!" (Second grade student during Mars Colonization) "What if we stop buying REEs from China and they get mad? They could attack us!" (Fourth grade student during REEs)	These exemplars indicate isolated but impactful incidents where students became visibly distressed by discussions or scenarios posed.
Teacher Self-Doubt	I feel like a failure today...more of a juggler than a teacher...I know what I want but I always come up short. I honestly thought I knew my students but today's discussion was eye-opening.	N/A	These exemplars convey the challenging nature of SSI for teachers insofar as the immensity of the societal issues and unpredictability of the student discourse on the issues.
Teacher Fulfillment	I wait for it each day...that magical moment when my students forget that I'm there... these are the happiest times in elementary science teaching... I'm so happy that [fourth grade student] was able to shine! I had no idea he would be so good at this. And everyone noticed! He's the new star of the fourth grade! Today was a very good day! I'm so pleased that this SSI activity encouraged my urban-dwelling students to envision themselves in more rural surroundings, as global citizens, and as future decision makers.	N/A	These exemplars convey the strong feelings of fulfillment that can be present for teachers implementing SSI due to the strong student engagement and opportunities for varied student strengths to be highlighted.

I conducted inductive thematic analysis of the data in three stages: (a) organizing and familiarizing, (b) coding and reducing (utilizing constant comparative method), and (c) interpreting and representing. (Ary, Jacobs, & Sorenson, 2010). I highlighted keywords and phrases and developed codes for each. Based on co-occurring codes (i.e., data coded by multiple codes), I consolidated several codes into broader themes such as "student pride" and "teacher self-doubt." I implemented analyst triangulation (Patton, 2002) by randomly selecting ten quotes from my journal and sending them to two colleagues with background in social science research but not specifically in education. I asked them to categorize the quotes within the themes in a manner that made sense to them. No discrepancies were noted in any of the categorizations, thus yielding high inter-rater reliability. I then compared my themes against student artifacts seeking to refute my journal findings. I found that the only discrepancy I could identify was that the artifacts reflected a much less critical stance than had my journal, with students expressing strong positive feelings about "doing real science." Although that phrase did not come up in my journal, I added it to the themes due to its prevalence in student evaluations.

RESULTS

The first research question of this study asked, "What key themes characterize implementation of the Socioscientific Issues (SSI) framework at the elementary level?" The inductive thematic analysis described resulted in the emergence of seven themes: 1) Student Pride; 2) Doing "Real Science;" 3) Doing the "Right Thing;" 4) Student Frustration; 5) Student Fear; 6) Teacher Self-Doubt; 7) Teacher Fulfillment.

Each of the themes identified are discussed more fully below.

Student Pride: Student pride was a theme across all grades and lessons, with several students noting positive feelings about coming to consensus with their teams, collaboratively researching and developing their arguments on particular topics, grappling with challenging questions or activities, and being able to persuade others to see their perspectives. Of particular note was that several students indicated that the SSI activities helped to students uncover and/or exercise certain academic or social strengths (e.g., online research, developing evidence-based arguments, negotiating, etc.) that might not be highlighted in more traditional science classes. Perhaps not surprisingly, these situations often led to feelings of fulfillment in the teacher as well (coded as "Teacher Fulfillment").

Doing "Real Science": A particularly interesting theme emerged regarding students' belief that engaging in SSI emulated authentic science practice and, moreover, that they viewed themselves as being "real scientists." Students cited specific activities including testifying before a Senate Sub-Committee (in the 4th grade REE lesson) and drafting and debating proposals for colonies on Mars. This theme did not emerge in the teacher reflective journal.

Doing the "Right Thing": This theme was particularly apparent in the first grade beaver relocation controversy and the fourth grade rare earth elements debate. Students invoked very strong stances on moral decision making using justifications of justice, fairness and compassion. An excerpt from a fourth- grade class discussing the question of REEs is below:

Student 1: "China should have to give the US REEs because we are dependent on them. It's not right to keep them when we've been buying them for so long."
Student 2: "But why should they give them up? It's their stuff!"
Student 1: "But we really, really need them. When someone needs something, you should give it to them because it's right."
Student 3: "Maybe we have something they need so we can trade!"
Student 1: "Yeah, maybe. But they should give'em to us anyway. It's just right."

A similar discussion emerged in a first-grade class as they discussed the beaver dilemma; in this case, doing the "right thing" is inferred from an admonition from doing the "wrong" thing:

Teacher: What if a beaver dam is causing a flood in a town…should it be moved or destroyed?

Student 1: The beavers are 'natural,' and they were here first, so they should be left alone.
Student 2: But I don't want houses to be flooded.
Student 1: It's wrong to hurt them just for doing what beavers like to do.

Student Frustration: This theme emerged at various points during the SSI units across all grades. The main trigger for students' frustration stemmed from social interactions with their teammates. Although cooperative learning techniques were implemented in all science classes throughout the year, with particular scaffolding of turn taking, active listening, and appropriate language for disagreements, the SSI activities proved to be fertile ground for students feeling "talked over" by their teammates, and at times, not feeling valued or listened to. Another source of student frustration, although far less common, was students' perceptions that other students or teams had easier positions to defend or more readily accessible resources from which to glean information.

Student Fear: This theme did not emerge in many classes, but when it did, it was rather extreme. In one situation, a first grader became visibly upset after viewing the video of the town grappling with whether to destroy a beaver dam. In a second-grade class about Mars colonization, one student expressed fears about an oxygen system failing while another student expressed fear about the possibility of a terrorist making their way to the Mars colony:

Teacher: (after students have moved to their position on an agree/disagree line in response to the statement, "The United States should build a colony on Mars.") Would anyone like to share why they don't think that the United States should build a colony on Mars?
Student 1: Because if a terrorist were there, it would be hard to get away.
Student 2: But there aren't terrorists in space!
Student 1: There could be, and they could have guns and then we'd be dead.
Student 3: But the guns wouldn't shoot in space, would they? (followed by several students worriedly hypothesizing about the functionality of guns in space)
Student 4: I don't want to get shot in space!

This brief interchange was followed by the teacher acknowledging student worries, admitting that she didn't know whether guns worked in space, and quickly refocusing the class on the key discussion points. Later in the period, the teacher shared the results of some quick online research on the topic with all students calm and engaged. However, the interlude had a profound impact on the teacher, who expressed self-doubt in her teacher reflection journal, a theme introduced more fully in the next section.

Teacher Self-Doubt: The unpredictable nature of the abovementioned interchange led the teacher to express doubts about her knowledge of her students:

I honestly thought I knew my students, but today's discussion was eye-opening. We went from talking about what we would need to live on Mars to terrorism! Should I have anticipated this? I feel badly that I didn't realize just how much world events are impacting them or how a discussion of space could lead us down such a dark path. I tried to turn it into a "teachable moment" but I still feel sad. Is this a sign of the times, where children are so sophisticated that they're paralyzed by fear? Where are our future explorers going to come from? How can I help my students to balance the realities of the world with an ability to dream? (Journal Entry, 2015)

Another source of teacher self-doubt stemmed from the challenge of trying to fit SSI lessons, which can be quite time consuming, into the relatively rigid time structures of the elementary science classroom:

I feel like a failure today…more of a juggler than a teacher. We had to rush through the lessons just to get everything done in time before the next unit. I know what I want, but I always come up short. My dream is to keep this unit going for the rest of the year so that my students can write to Congress, start a campaign, Skype with schools across the globe, change the world! Instead, we become immersed for a short time, and then we move on. I want to do so much more!!! (Journal Entry, 2016)

A third and final source of self-doubt emerged from the realization that students will often express positions and preferences that run counter to that of the teacher during the implementation of SSI. This teacher journal entry followed a fourth-grade class on wetland development:

I couldn't help but be reminded of Garrett Hardin's, The Tragedy of the Commons, which describes the tendency of individuals/groups to exploit shared resources in favor of their own self-interest; this situation ends with them depleting the resources, negatively impacting themselves and the common good. Having been trained as an ecologist myself, it was painful to watch my students quickly forget the interdependency and delicate balance that we had studied only minutes before. When given the opportunity to sprawl, they did so, completely unchecked. Perhaps it was because they weren't asked to take on the role of an environmentalist. Or perhaps it was because there were no economic or legal constraints on their building. The same students who only moments before had talked about the importance of preserving the food chain were now oblivious to its existence. I gently asked the class what they thought might be some of the consequences of their development, and they revisited the "downsides" of development for each of the constituencies, but I still couldn't shake the thought of how quickly they had decimated the land. Was the whole first half of this unit a loss?"

Teacher Fulfilment: At several points across all grades and lessons, the teacher expressed strong feelings of fulfillment from SSI implementation. The key sources of these positive feelings tended to come from two distinct situations - seeing a student who typically had challenges thriving, and feeling that students had gained meaningful knowledge or skills by successfully navigating challenging topics, as follows:

I was a little nervous about [fourth grade student's] ability to handle challenges from other students. He tends to be very rigid in his thinking and gets frustrated easily when other students don't see things his way. But he stood up in the town hall meeting, presented his team's plan beautifully, and was able to respond to the other teams' very pointed challenges without getting defensive. He actually listened to them and considered their ideas! I was so proud of him! After the meeting, I asked the class for their feedback and [same fourth grade student] said it was his best day in science and one of his best days in school! I'm so happy that he was able to shine! I had no idea he would be so good at this. And everyone noticed! He's the new star of the fourth grade! Today was a very good day!

I was so pleased that this SSI activity encouraged my urban-dwelling students to envision themselves in more rural surroundings, as global citizens, and of course, as future decision makers. With expansion of urban and suburban areas encroaching on animal territories, the challenges of urban wildlife management are very real and will be an issue that my first graders will undoubtedly confront in their adult lives. This activity helped my students to see the complexity of this controversial issue; while their initial responses were rather simplistic and very emotionally driven, more study of the issue led to richer intellectual debate. Moreover, we were able to address the critical tenets that science can't answer every question and technology can't solve every problem.

DISCUSSION

The results suggest that SSI implementation provided an engaging real-world context for science learning and student discourse and appeared to elicit strong feelings of pride and satisfaction in many students. Students embraced the feeling of being "real scientists" through research and discourse and were generally able to identify and put forth evidence to support their arguments. These results support prior research on SSI as an effective framework for supporting students' argumentation, content knowledge, and understanding of NOS (Zeidler, 2014). However, several developmentally-influenced challenges were evident, including students' difficulties in shifting from emotional to cognitive bases of argumentation, unintended activation of student fears, confusion between real and imaginary scenarios, and uneven emotion regulation, particularly in regard to cooperative skills. These findings, interpreted in light of literature on psychosocial development, can address the study's second research question which asked, "What, if any, modifications or refinements may be warranted when the Socioscientific Issues (SSI) framework is implemented with young children?"

Erikson's (1959; 1963) Psychosocial Development Theory would suggest that students of this age grapple with industry versus inferiority, two competing conditions whose positive resolution manifests as feelings of competence in adulthood. It was clear that many students experienced "industry" as they worked through collaborative research and debates and were able to successfully overcome social and learning challenges, ultimately resulting in feelings of pride and accomplishment. However, the emphasis that SSI places on discourse and debate can also prompt embarrassment and feelings of inferiority and frustration in students who are not as adept as others or are simply less likely to assert themselves. Research on community theory (McMillan & Chavis, 1986) might be useful in identifying ways that elementary teachers can create an environment of acceptance that honors diversity in the way people communicate their ideas, and then build it into SSI instruction explicitly. Dovetailing with this theory, intentional integration of SEL approaches (Collaborative for Academic, Social, and Emotional Learning, 2017) within SSI could reinforce the desired classroom values and attributes of acceptance and trust which, in turn, would also support critical goals of elementary teaching outside the scientific context. Additionally, providing digital opportunities for discourse, through apps such as VoiceThread, or online SSI environments such as WISE (Web-based Science Learning Environment) (Evagorou, 2011) adapted for early elementary students can provide opportunities for more private scaffolding of students' skills, thereby promoting feelings of competence. This approach might also provide a safe environment for mastery for ELL's and students with learning and language disabilities, and ultimately, reduce the overall taxing of emotion regulation in the class. And, although the teacher in the present study utilized cooperative learning techniques in her classes, additional emphasis on cooperative learning to enhance students' interpersonal and discourse skills may aid students SSI negotiations as it has in secondary science classes (Day & Bryce, 2013).

Some students in the study also seemed to experience "inferiority" in relation to the positions they were assigned to research or the unevenness (or *perceived* unevenness) of available resources to support those positions. Ensuring that selected topics have positions that can be strongly supported by evidence, and that the resources supporting those positions are readily and equitably available is important for SSI implementation at the elementary level, as it is in general (Presley, et al., 2013). Teachers can also assure students that they will have a chance to share their own positions, regardless of those to which they were assigned, in order to assuage such concerns.

The difficulties some students had in distinguishing between real and hypothetical scenarios, such as in the beaver relocation controversy and the Mars colonization debate, are in line with Piaget's (1926) description of the Concrete Operational Stage (age 7-11) which suggests that abstract thinking can be quite challenging for some children. This understanding typically prompts teachers to use "concrete" scaffolds such as manipulatives, demonstrations, and so on. The challenge with SSI, however, is that when more concrete tools were used to engage the students (such as photos or videos of real beavers), they became emotionally connected to the subject and many were unable to shift to cognitive spheres of argumentation. This finding suggests that SSI may wield a double-edged sword when implemented with very young children (such as first and second graders) as the emotional hook that is often used to engage students in the SSI lesson is precisely the piece that seemed to make argumentation and cooperative skill demands challenging. A possible source of guidance for science education researchers may be found in social studies education, specifically from the work on historical empathy (Foster, 2001) as it parallels SSI pedagogy in its use of role play and emotive engagement. This line of research may also shed light on managing SSI's emphasis on real-world societal issues which, in the present study, lent itself to students expressing concerns and fears that would ordinarily not be tapped in a science class. Recalling the vignette describing the fear of terrorists on a Mars colony, an emotional contagion (Schoenewolf, 1990) took root when one student's fear quickly spread throughout the class. This example again illuminates not only students' challenges with distinguishing between real/current and hypothetical/future threats, but also in regulating their emotions (i.e., managing the intensity of emotional displays in ways that are appropriate to the audience and the situation; Cole, Martin, & Dennis, 2004; Garner, 2010). SEL approaches integrated within SSI might again prove helpful in preparing students and teachers in anticipating and regulating fears. For example, having a Morning Meeting during which the teacher asks students to vote on whether they would like to travel in space (by entering a check mark on a chart or placing a chip in yes/no cups, for example) and then allowing students to express their thoughts (and fears) in the supportive meeting environment might help students to ease their concerns while also alerting the teacher to possible areas of sensitivity, all before the specific question of planetary colonization is even posed. Of course, ensuring that the issues and potential resolutions themselves don't unnecessarily tap fearful subjects seems advisable, particularly for the youngest students. For example, in the first grade beaver dam dilemma, limiting the options to keeping the beaver dam where it is or moving it, without including the option to destroy it, achieves roughly the same content and process objectives (albeit in a slightly less realistic way) without unnecessarily activating student fears (Kahn, 2019).

Although not the focus of this study, several aspects of teacher identity emerged, not unlike Zeidler, Applebaum, and Sadler's (2011) work which speaks of the "transformative transformations" that occurred as the teacher of focus implemented SSI at the secondary level. Many parallels were noted here, but specifically, the issues of SSI being time consuming, requiring teacher flexibility due to its unpredictable nature, and requiring teacher open-mindedness to student positions that differ from the teacher's own, were all present in this study. Continued work in preparing pre-service and in-service teachers for the specific nuances of SSI implementation is essential.

Challenges notwithstanding, SSI also provided highly rewarding teaching experiences in the present study. Particularly noteworthy was that SSI provided the teacher (and her students) the opportunity to recognize student talents that wouldn't otherwise have been exercised in more traditional science lessons. Science assessment has been criticized for focusing on a narrow range of strengths and characteristics in defining talent in science (Sumida, 2010), yet identifying students' strengths and reinforcing them through science practice is a valuable means of encouraging students' interest in and propensity towards future scientific pursuits (Kahn, 2018). Closer examination of the ways that SSI can uncover and enhance students' strengths, perhaps by incorporating tenets of Strength-Based Instruction (Armstrong, 2012) seems warranted.

To summarize, some practices that may prove helpful in implementing SSI at the elementary level are as follows:

- Seek balance insofar as the strength of positions on issues and the availability of resources on those positions
- Use real-world scenarios but be cognizant that realistic videos, books and other media may be "too real" for some young children
- Consider explicitly integrating SEL and Cooperative Learning approaches, particularly in regard to teambuilding, consensus building, and creating a safe and supportive classroom environment
- Pay attention to students' strengths which may be illuminated through SSI
- Respond to student worries or fears with understanding and reassurance, but also explore ways of helping students to regulate their emotions in positive ways

LIMITATIONS AND FUTURE RESEARCH DIRECTIONS

The findings of this exploratory study should be considered in light of its small scale, individual researcher, and unique setting, all of which may affect transferability to other contexts. In addition, teacher effects, such as the impact of the teacher's explicit emphasis on NOS in her class, may also limit transferability. That said, it should be remembered that this study was analyzed as a critical case sample involving ideal rather than typical circumstances, so that challenges in implementation could more readily be attributed to the pedagogical framework rather than to the teacher or school setting. Nonetheless, future studies would benefit from additional sources of data including student interviews, classroom videos, and more detailed analysis of student discourse, particularly in regard to conflict resolution.

This study prompted a series of questions for future research including:

- How can SSI researchers and curriculum developers leverage elementary teachers' understanding of child development to improve SSI outcomes?

- What other frameworks, perhaps from social studies or language arts, can help facilitate interdisciplinary implementation of SSI at the elementary level?
- How does teacher identity relate to implementation of SSI in the elementary classroom?
- How can SSI be implemented with non-readers/writers, students with disabilities, ELLs, etc.?
- What might a model of pedagogical content knowledge (PCK; Shulman, 1987) for various SSI topics at the elementary level involve?

CONCLUSION

Due to the paucity of research on SSI at the elementary level, this exploratory study lays critical groundwork for an expansive research agenda focused on the unique needs of elementary students and teachers in an SSI classroom. Specifically, identifying developmentally-appropriate curriculum and pedagogy, with particular attention to the psychosocial and cognitive needs of elementary-aged students, would solidify SSI as a K-12 framework and would ensure that students are better prepared in the areas of argumentation, NOS, informal and moral reasoning, and so on at the higher grade levels. Although some modifications may be warranted, particularly in the areas of issue and media selection, argumentation scaffolds, and classroom management, this work makes clear that even the youngest learners are worthy and capable of preparing for informed citizenship and socioscientific decision making.

REFERENCES

Abd-El-Khalick, F. (2006). Socioscientific issues in pre-college science classrooms. In D. L. Zeidler (Ed.), *The role of moral reasoning and discourse on socioscientific issues in science education* (pp. 41–61). Springer.

Armstrong, T. (2012). *Neurodiversity in the classroom*. Alexandria, VA: Association for Supervision and Curriculum Development.

Ary, D., Jacobs, L. C., & Sorensen, C. (2010). *Introduction to research in education*. Belmont, CA: Wadsworth.

Blank, R. K. (2013). Science instructional time is declining in elementary schools: What are the implications for student achievement and closing the gap? *Science Education*, *97*(6), 830–847. doi:10.1002ce.21078

Burek, K., & Zeidler, D. L. (2015). Seeing the forest for the trees! Conservation and activism through socioscientific issues for young students. In M. Mueller & D. J. Tippins (Eds.), *Ecojustice, citizen science & youth activism: Situated tensions for science education* (pp. 425–441). Springer.

Chinn, C. A., & Brewer, W. F. (1993). The role of anomalous data in knowledge acquisition: A theoretical framework and implications for science instruction. *Review of Educational Research*, *63*(1), 1–49. doi:10.3102/00346543063001001

Cochran-Smith, M., & Lytle, S. (1999). The teacher research movement: A decade later. *Educational Researcher*, *28*(7), 15–25. doi:10.3102/0013189X028007015

Cole, P. M., Martin, S. E., & Dennis, T. A. (2004). Emotion regulation as a scientific construct: Methodological challenges and directions for child development research. *Child Development*, *75*(2), 317–333. doi:10.1111/j.1467-8624.2004.00673.x PMID:15056186

Collaborative for Academic, Social, and Emotional Learning. (2017). *Social and emotional learning (SEL) competencies*. Retrieved March 2, 2020, from https://casel.org/wp-content/uploads/2019/12/CASEL-Competencies.pdf

Copple, C., Bredekamp, S., Koralek, D., & Charner, K. (2014). *Developmentally appropriate practice: Focus on children in first, second, and third grades*. National Association for the Education of Young Children.

Curran, F. C., & Kitchin, J. (2019). Early elementary science instruction: Does more time on science or science topics/skills predict science achievement in the early grades? *AERA Open*, *5*(3). Advance online publication. doi:10.1177/2332858419861081

Day, S. P., & Bryce, T. G. (2013). The benefits of cooperative learning to socio-scientific discussion in secondary school science. *International Journal of Science Education*, *35*(9), 1533–1560. doi:10.1080/09500693.2011.642324

Denham, S. A. (2015). Assessment of SEL in educational contexts. In J. A. Durlak, C. E. Domitrovich, R. P. Weissberg, & T. P. Gullotta (Eds.), *Handbook of social and emotional learning: Research and practice* (pp. 285–300). New York: Guilford.

Denton, P. (2007). *The power of our words: Teacher language that helps children learn*. Turners Fall, MA: Northeast Foundation for Children.

DeVries, R., & Zan, B. (2012). *Moral classrooms, moral children: Creating a constructivist atmosphere in early education*. Teachers College Press.

Dolan, T. J., Nichols, B. H., & Zeidler, D. L. (2009). Using socioscientific issues in primary classrooms. *Journal of Elementary Science Teacher Education, 21*(3), 1–12. doi:10.1007/BF03174719

Durlak, J. A., Weissberg, R. P., Dymnicki, A. B., Taylor, R. D., & Schellinger, K. B. (2011). The impact of enhancing students' social and emotional learning: A meta-analysis of school-based universal interventions. *Child Development, 82*(1), 405–432. doi:10.1111/j.1467-8624.2010.01564.x PMID:21291449

Erikson, E. H. (1959). Identity and the life cycle. *Psychological Issues, 1*, 1–71.

Erikson, E. H. (1963). *Childhood and society*. Norton.

Evagorou, M. (2011). Discussing a socioscientific issue in a primary school classroom: The case of using a technology-supported environment in formal and nonformal settings. In T. D. Sadler (Ed.), *Socio-scientific issues in science classrooms: Teaching, learning and research* (pp. 133–159). Springer. doi:10.1007/978-94-007-1159-4_8

Evagorou, M., & Osborne, J. (2013). Exploring young students' collaborative argumentation within a socioscientific issue. *Journal of Research in Science Teaching, 50*(2), 209–237. doi:10.1002/tea.21076

Feldman, A. (1996). Enhancing the practice of physics teachers: Mechanisms for the generation and sharing of knowledge and understanding in collaborative action research. *Journal of Research in Science Teaching, 33*(5), 513–540. doi:10.1002/(SICI)1098-2736(199605)33:5<513::AID-TEA4>3.0.CO;2-U

Foster, S. J. (2001). Historical empathy in theory and practice: Some final thoughts. In O. L. Davis, E. A. Yeager, & S. J. Foster (Eds.), *Historical empathy and perspective taking in the social studies*. Rowman and Littlefield.

Fowler, S. R., Zeidler, D. L., & Sadler, T. D. (2009). Moral sensitivity in the context of socioscientific issues in high school science students. *International Journal of Science Teacher Education, 31*(2), 279–296. doi:10.1080/09500690701787909

Garner, P. W. (2010). Emotional competence and its influences on teaching and learning. *Educational Psychology Review, 22*(3), 297–321. doi:10.100710648-010-9129-4

Gelman, R., & Baillargeon, R. (1983). A review of some Piagetian concepts. In Handbook of child psychology (4th ed., pp. 167–230). New York, NY: Wiley.

Hildebrandt, C., & Zan, B. (2014). Constructivist approaches to moral education in early childhood. In L. Nucci, T. Krettenauer, & D. Narvaez (Eds.), *Handbook of moral and character education* (pp. 180–197). Routledge.

Kahn, S. (2018). From access to assets: Strength-based visions for inclusive science education. In M. Koomen, S. Kahn, C. L. Atchison, & T. A. Wild (Eds.), *Towards inclusion of all learners through science teacher education* (pp. 105–114). Brill. doi:10.1163/9789004368422_012

Kahn, S. (2019). *It's still debatable! Using socioscientific issues to develop scientific literacy, K-5*. NSTA Press.

Kahn, S., & Zeidler, D. L. (2016). Using our heads and HARTSS*: Developing perspective-taking skills for socioscientific reasoning (*Humanities, ARTs, and Social Sciences). *Journal of Science Teacher Education*, *27*(3), 261–281. doi:10.100710972-016-9458-3

Kahn, S., & Zeidler, D. L. (2019). A conceptual analysis of perspective taking in support of socioscientific reasoning. *Science & Education*, *28*(6-7), 605–638. doi:10.100711191-019-00044-2

Karpudewan, M., & Roth, W. M. (2018). Changes in primary students' informal reasoning during an environment-related curriculum on socio-scientific issues. *International Journal of Science and Mathematics Education*, *16*(3), 401–419. doi:10.100710763-016-9787-x

Khishfe, R., Alshaya, F. S., BouJaoude, S., Mansour, N., & Alrudiyan, K. I. (2017). Students' understandings of nature of science and their arguments in the context of four socio-scientific issues. *International Journal of Science Education*, *39*(3), 299–334. doi:10.1080/09500693.2017.1280741

Lederman, N. G. (2007). Nature of science: Past, present, and future. In S. K. Abell & N. G. Lederman (Eds.), *Handbook of research on science education* (pp. 831–879). Lawrence Erlbaum Associates.

Lee, H., Chang, H., Choi, K., Kim, S. W., & Zeidler, D. L. (2012). Developing character and values for global citizens: Analysis of pre-service science teachers' moral reasoning on socioscientific issues. *International Journal of Science Education*, *34*(6), 925–953. doi:10.1080/09500693.2011.625505

Linn, S.-S., & Mitzes, J. J. (2010). Learning argumentation skills through instruction in socioscientific issues: The effect of ability level. *International Journal of Science and Mathematics Education*, *8*(6), 993–1007. doi:10.100710763-010-9215-6

McMillan, D. W., & Chavis, D. M. (1986). Sense of community: A definition and theory. *Journal of Community Psychology*, *14*(1), 6–23. doi:10.1002/1520-6629(198601)14:1<6::AID-JCOP2290140103>3.0.CO;2-I

National Academies of Sciences, Engineering, and Medicine. (2018). *How people learn II: Learners, contexts, and cultures*. National Academies Press.

National Governors Association. (2010). *Common core state standards.*

National Research Council. (2000). *How people learn: Brain, mind, experience, and school: Expanded edition*. The National Academies Press.

Next Generation Science Standards Lead States. (2013). *Next Generation Science Standards: For States, By States*. The National Academies Press.

Patton, M. Q. (1990). *Qualitative evaluation and research methods*. Sage.

Patton, M. Q. (2002). *Qualitative research and evaluation methods* (3rd ed.). Sage.

Pearson, P. D., & Hiebert, E. H. (2013). Understanding the common core state standards. *Teaching with the Common Core Standards for English language arts PreK–2*, 1-21.

Piaget, J. (1926). *The language and thought of the child*. Harcourt Brace.

Presley, M. L., Sickel, A. J., Muslu, N., Merle-Johnson, D., Witzig, S. B., Izci, K., & Sadler, T. D. (2013). A framework for socio-scientific issues based education. *Science Educator*, *22*, 26.

Responsive Classroom. (2020). *Responsive Classroom: Principals and Practices*. Retrieved from https://www.responsiveclassroom.org/about/principles-practices/

Rimm-Kaufman, S., & Hulleman, C. S. (2015). SEL in elementary school settings: Identifying mechanisms that matter. In J. A. Durlak, C. E. Domitrovich, R. P. Weissberg, & T. P. Gullotta (Eds.), *Handbook of social and emotional learning: Research and practice* (pp. 151–166). Guilford.

Roberts, D. A. (2011). Competing visions of scientific literacy: The influence of a science curriculum policy image. In C. Linder, L. Ostman, D. A. Roberts, P. Wickman, G. Erickson, & A. MacKinnon (Eds.), *Promoting scientific literacy: Science education research in transaction* (pp. 11–27). Routledge/Taylor & Francis Group.

Roberts, D. A., & Bybee, R. W. (2014). Scientific literacy, science literacy, and science education. In N. G. Lederman & S. K. Abell (Eds.), *Handbook of research in science education* (Vol. 2, pp. 545–558). Routledge.

Roth, K. (2014). Elementary science teaching. In N. Lederman & S. Abell (Eds.), *Handbook of research on science education* (Vol. 2, pp. 361–394). Routledge.

Rust, F. O'C. (2009). Teacher research and the problem of practice. *Teachers College Record, 111*(8), 1882–1893.

Rust, F., & Meyers, E. (2006). The bright side: Teacher research in the context of educational reform and policy-making. *Teachers and Teaching, 12*(1), 69–86. doi:10.1080/13450600500365452

Sadler, T. (2011). Situating socioscientific issues in classrooms as a means of achieving goals of science education. In T. Sadler (Ed.), *Socio-scientific issues in the classroom: Teaching, learning and research* (pp. 1–9). Springer. doi:10.1007/978-94-007-1159-4_1

Sadler, T., & Zeidler, D. (2005). The significance of content knowledge for informal reasoning regarding socioscientific issues: Applying genetics knowledge to genetic engineering issues. *Science Education, 89*(1), 71–93. doi:10.1002ce.20023

Schoenewolf, G. (1990). Emotional contagion: Behavioral induction in individuals and groups. *Modern Psychoanalysis, 15*(1), 49–61.

Shulman, L. (1987). Knowledge and teaching: Foundations of the new reform. *Harvard Educational Review, 57*(1), 1–22. doi:10.17763/haer.57.1.j463w79r56455411

Slavin, R. E. (1980). Cooperative learning. *Review of Educational Research, 50*(2), 315–342. doi:10.3102/00346543050002315 PMID:6698102

Stolz, M., Witteck, T., Marks, R., & Eilks, I. (2013). Reflecting socio-scientific issues for science education coming from the case of curriculum development on doping in chemistry education. *Eurasia Journal of Mathematics, Science and Technology Education, 9*(4), 361–370.

Sumida, M. (2010). Identifying twice-exceptional children and three gifted styles in the Japanese primary science classroom. *International Journal of Science Education, 32*(15), 2097–2111. doi:10.1080/09500690903402018

Taylor, R. D., Oberle, E., Durlak, J. A., & Weissberg, R. P. (2017). Promoting positive youth development through school-based social and emotional learning interventions: A meta-analysis of follow-up effects. *Child Development, 88*(4), 1156–1171. doi:10.1111/cdev.12864 PMID:28685826

Valli, L. (1997). Listening to other voices: A description of teacher reflection in the United States. *Peabody Journal of Education, 72*(1), 67–88. doi:10.120715327930pje7201_4

Vieira, R. M., & Tenreiro-Vieira, C. (2016). Fostering scientific literacy and critical thinking in elementary science education. *International Journal of Science and Mathematics Education, 14*(4), 659–680. doi:10.100710763-014-9605-2

Yoon, J., & Onchwari, J. A. (2006). Teaching young children science: Three key points. *Journal of Early Childhood Education, 33*(6), 419–423. doi:10.100710643-006-0064-4

Zeidler, D. L. (2014). Socioscientific issues as a curriculum emphasis: Theory, research, and practice. In N. G. Lederman & S. K. Abell (Eds.), *Handbook of research on science education* (Vol. 2, pp. 697–726). Routledge.

Zeidler, D. L., Applebaum, S. M., & Sadler, T. D. (2011). Enacting a socioscientific issues classroom: Transformative transformations. In T. D. Sadler (Ed.), *Socio-scientific issues in science classrooms: Teaching, learning and research* (pp. 277–306). Springer. doi:10.1007/978-94-007-1159-4_16

Zeidler, D. L., & Kahn, S. (2014). *It's debatable! Using socioscientific issues to develop scientific literacy, K-12*. NSTA Press.

Zeidler, D. L., & Sadler, D. L. (2011). An inclusive view of scientific literacy: Core issues and future directions of socioscientific reasoning. In C. Linder, L. Ostman, D. A. Roberts, P. Wickman, G. Erickson & A. MacKinnon (Eds.), Promoting scientific literacy: Science education research in transaction (pp. 176-192). New York, NY: Routledge / Taylor & Francis Group.

Zeidler, D. L., & Sadler, T. D. (2008). The role of moral reasoning in argumentation: Conscience, character and care. In S. Erduran & M. Pilar Jimenez-Aleixandre (Eds.), *Argumentation in science education: Perspectives from classroom-based research* (pp. 201–216). Springer Press.

Zeidler, D. L., Sadler, T. D., Applebaum, S., & Callahan, B. E. (2009). Advancing reflective judgment through socioscientific issues. *Journal of Research in Science Teaching, 46*(1), 74–101. doi:10.1002/tea.20281

Zeidler, D. L., Sadler, T. D., Simmons, M. L., & Howes, E. V. (2005). Beyond STS: A research-based framework for socioscientific issues education. *Science education, 89*(3), 357–377. doi:10.1002ce.20048

KEY TERMS AND DEFINITIONS

Psychosocial Development: A theoretical approach to understanding the manner in which humans change through life, both psychologically and socially, due to social and environmental influences.

Scientific Literacy: A goal of science education that prepares students to apply science in their everyday lives and in society.

Socioscientific Issues: A framework that uses ill-structured, scientifically-related societal issues as the context for science learning.

Teacher Research: An educational research method that values teachers as knowledge makers but does not require the introduction of a specific intervention to be tested.

Chapter 2
Integrating Scientific Modeling and Socio-Scientific Reasoning to Promote Scientific Literacy

Li Ke
University of North Carolina at Chapel Hill, USA

Laura A. Zangori
https://orcid.org/0000-0002-7512-5559
University of Missouri, USA

Troy D. Sadler
https://orcid.org/0000-0002-9401-0300
University of North Carolina at Chapel Hill, USA

Patricia J. Friedrichsen
University of Missouri, USA

ABSTRACT

Socio-scientific issues (SSI) are widely advocated as a productive context for promoting scientific literacy that aims to prepare responsible citizens who can use science in their daily lives. However, many teachers find it challenging to enact SSI and consider SSI and discipline-based instruction as mutually exclusive approaches to science teaching. In this chapter, the authors present their framework for SSI instruction, socio-scientific issue and model-based learning (SIMBL), that emphasizes both disciplinary knowledge and its social implications. In particular, the authors argue that the integration of scientific modeling and socio-scientific reasoning (SSR) can advance students' competencies in both areas, thus promoting students' scientific literacy. The authors use an illustrative example from their work with elementary students to demonstrate the connection between students' modeling practice and their SSR. The authors conclude the chapter by introducing the epistemic tools developed to support students' modeling practice and SSR as well as implications for classroom enactments.

DOI: 10.4018/978-1-7998-4558-4.ch002

INTRODUCTION

Over the past several decades different reform efforts have advocated that scientific literacy is the primary goal of school science. The term scientific literacy has been used to describe a general and functional understanding of science on the part of the general public (DeBoer, 2000). Yet, like other constructs such as inquiry, scientific literacy has been broadly used and variedly interpreted so that its meaning and purpose has become unclear. Roberts (2007) and Roberts and Bybee (2014) synthesized the extant literature on the construct and proposed a heuristic framework useful for understanding different perspectives on scientific literacy (SL) as a way of framing the purposes for science education. The SL framework is a continuum between two extremes, which Roberts (2007) called Vision I and Vision II. At one extreme, Vision I tends to prioritize a well-defined body of academic knowledge and practices associated with the discipline of science. That is, achievement of Vision I requires that students develop understandings of science content and how science works with an aim to prepare future scientists. At the other extreme, Vision II highlights scientific knowledge and perspectives relevant in society that go beyond the boundaries of scientific disciplines. Under Vision II, the goal is for students to become responsible citizens who can use science and make informed decisions in their lives.

Previous research on socio-scientific issues (SSI) has shown that engaging students in the negotiation of complex societal issues is an effective way to address the learning goals aligned with Vision II (see the review by Zeidler, 2014). SSI instruction provides students with meaningful contexts to reflect on how science may relate to their own lives and society. It also prompts students to consider the moral and ethical implications of complex societal issues that are central in personal judgement and decision-making in everyday life. In addition to achieving the goals for Vision II, an issue-based approach has been linked with improvement in learning outcomes associated with Vision I. Learning science in the context of SSI can promote students' understandings of scientific content knowledge and ideas about the nature of science (Khishfe, 2014; Sadler, Romine & Topcu, 2016; Zohar & Nemet, 2002).

Despite the potential for addressing both visions of SL, the promise of SSI demonstrated in the literature is rarely translated into science classrooms. Many teachers find it challenging to enact SSI in their classrooms for various reasons such as their discomfort with discussions about global problems without clear solutions (Ekborg, Ottander, Silfver, & Simon, 2013; Hancock, Friedrichsen, Kinslow & Sadler, 2019; Lee, Abd-El-Khalick, & Choi, 2006). In particular, teachers often hold a content-centered view of SSI and background the social-political aspects of the SSI approach (Lazarowitz & Bloch, 2005; Tidemand & Nielsen, 2017). For some teachers, a lack of instructional focus on the social dimensions of the issue is considered unproblematic because their goal for implementing an SSI approach is to motivate student learning in the given content. For other teachers who do aspire to embrace a Vision II learning goal, they view classroom discussions around the social implications of the issue not "scientific" enough to have a space in science classrooms. This limited view of SSI suggests that teachers consider SSI instruction and discipline-based instruction as mutually exclusive approaches to science teaching. In other words, for many teachers, SSI can only be taught effectively in sacrifice of students' gaining scientific knowledge and practices.

The dilemma teachers are facing with regard to balancing the goals of Vision I and Vision II signals an imperative need to reframe SSI instruction so that it can accommodate both visions. This is important because, in practice, most science educators must attend to a set of instructional goals that embrace a combination of SL Vision I and II (i.e., Vision I-II, Aikenhead, 2007). On one hand, science teachers would claim, with varying degrees, that they hope their students can apply what they learned in class to their daily lives. On the other hand, they are held accountable to help students achieve the academic levels defined by standards and/or curricula that predominantly promote Vision I SL. Hence, a framework for SSI teaching and learning with an explicit connection to disciplinary knowledge and practices can better prepare teachers to enact SSI instruction from a pragmatic perspective.

In this chapter, we present our framework for SSI instruction, *Socio-scientific Issue and Model Based Learning* (SIMBL), that emphasizes both disciplinary knowledge and norms and its social implications as an effort to address the aforementioned dilemma (Sadler, Friedrichsen, & Zangori, 2019). In particular, we argue that scientific modeling and socio-scientific reasoning (SSR) can work synergistically with each other, and the integration of scientific modeling and SSR can advance students' competencies in both areas, thus promoting students' scientific literacy for Vision I-II. We chose scientific modeling, among other scientific practices, as the focal disciplinary practices for the following reasons: First, modeling is a key epistemic practice that epitomizes how scientists work in the discipline and has been emphasized in recent reform efforts in science education around the world. Therefore, focusing on scientific modeling provides an entry point for us to tap into the learning needs that many teachers are trying to address in response to the call for SL Vision I. Second, there are important parallels between the practice of modeling and the negotiation of complex societal issues (i.e., socio-scientific reasoning); for example, both modeling and socio-scientific reasoning require systems thinking. As such, we presume that the commonalities shared by the two practices may make it easier for students to transfer their scientific understanding and practices to their reasoning for SSI (which addresses the call for SL Vision II). Finally, despite the central role of scientific modeling in our teaching framework of SSI, we recognize that modeling cannot be taught in isolation without connecting to other scientific practices such as constructing explanations and argumentation. In fact, we posit that students should engage in other practices as they see fit in combination with scientific modeling (e.g., analyzing and interpreting external data to decide if their models need to be revised or not) so that these scientific practices are meaningful to students. In other words, we see modeling as the anchoring practice, as opposed to the only epistemic practice, that students should engage in as they explore the scientific dimension of a complex issue.

In the subsequent sections, we first provide a brief background on research on scientific modeling and SSR and a rationale for integrating the two practices. We then present the SIMBL framework and an associated instructional sequence to show how scientific modeling and SSR can be integrated in instruction. We next present an illustrative example from our work with elementary students to demonstrate the connection between students' modeling practice and their SSR observed in student work. We conclude the chapter by introducing the epistemic tools we developed to support students' modeling practice and SSR as well as implications for classroom enactments.

BACKGROUND

Scientific Modeling

Scientific models are simplified representations that can illustrate, explain, and predict phenomena (Lehrer & Schauble, 2006; Schwarz et al., 2009). Scientific modeling, the practice of developing, revising, evaluating, and testing scientific models, is an epistemic practice that scientists constantly engage in to develop an understanding of the natural world. It is considered a cornerstone of science as the embodiment of theory lies at the heart of scientific endeavor (Clement, 2008; Nersession, 2002). Learning to develop models can also be pedagogically useful for students' knowledge building in science classrooms and is an essential goal of modern science education (Gilbert, 2005). In the U.S., the *Next Generation Science Standards* (NGSS, NGSS lead States, 2013) call for students' appropriation of scientific practices that resemble the work of scientists, and scientific modeling is one of the eight key practices that all students are expected to engage in, together with disciplinary core ideas and cross-cutting concepts, to make sense of the world.

In the area of model-based instruction, there is ample evidence in the literature suggesting that students across grade level are capable of engaging in modeling practice in productive ways (see the review by Louca & Zacharia, 2012). In particular, previous studies have found that, with appropriate instructional and curricular supports, students' model-based reasoning as well as understanding of content knowledge improved (Louca, Zacharia, & Constantinou, 2011; Manz, 2012; Passmore, Gouvea, & Giere, 2014; Zangori & Forbes, 2014). While scientific models can take many forms, such as a diagram, a computer simulation, or mathematical equation, model-based instruction is typified by a modeling process that encompasses multiple elements of modeling practice. Students might start with developing models to test a hypothesis or answer a question about a phenomenon and then revise, evaluate, and use models as they gain additional experiences with the phenomenon and associated evidence. In addition to the processes of scientific modeling, knowledge about the purposes and goals of modeling, or metamodeling knowledge (Schwarz & White, 2005), is also a critical part of model-based instruction. In order for model-based instruction to be meaningful in classroom communities, students need to understand the nature of models and modeling (e.g., What is a model? How are models used? What are models for? What criteria are used to evaluate model?) as they participate in the modeling processes. In essence, scientific modeling is an epistemic practice, and students need to understand what models are and how they can be used in order for the epistemic potential of model-based instruction to be fully realized. Otherwise, modeling could run the danger of becoming a procedural exercise where students put the "correct" things into the models without using models as tools to make sense of the phenomena or the underlying systems (Ke & Schwarz, 2020).

Socio-scientific Reasoning

To maximize the potential value of SSI teaching, students require opportunities to do more than learn science ideas and practices (Sadler, 2009); SSI-based teaching should help learners become better skilled at negotiating complex issues. While science content may help with this negotiation process to some extent, a focus on content alone underrepresents the knowledge and competencies required. However, SSI advocates have struggled to operationalize what the other knowledge and competencies are and how to assess them. In response to this challenge, the construct of socio-scientific reasoning was developed as a means of capturing the knowledge and competencies necessary for the consideration and negotiation of complex issues (Sadler, Barab & Scott, 2007). It is proposed as a set of interrelated reasoning practices that are epistemic in nature and incorporate the following five dimensions (Kinslow, Sadler & Nguyen, 2019):

- Recognizing the inherent *complexity* of SSI and how this complexity necessarily impacts the manner in which these issues can be addressed
- Appreciating the fact that we do not know everything that could be known about SSI and therefore SSIs remain open to *inquiry*

- Considering SSI from the *multiple perspectives* that stakeholders with different interests likely possess
- Exhibiting *skepticism* in the face of potentially biased information
- Recognizing the potential *affordances and limitations* of science as a way of understanding and resolving the issue

Empirical studies of learners developing more sophisticated socio-scientific reasoning (SSR) in association with SSI-based learning experiences have been documented with secondary students (Kinslow, 2018; Kinslow, Sadler & Nguyen, 2019) as well as undergraduates (Romine, Sadler & Dauer, in review). For example, in Kinslow's (2018) work, students participating in a high school course that integrated science and social studies through SSI as a pedagogical approach demonstrated significantly improved SSR after a semester of issues-based learning experiences. Other researchers have expanded the notion of SSR and focused more on the idea of "socioscientific decision-making" (Eggert, Ostermeyer, Hasselhorn, & Bogeholz, 2013; Simonneaux & Simonneaux, 2009) and found similar results. Overall, research in this area is still emerging, but findings suggest that SSR can develop in response to SSI teaching; however, learning SSR takes time and high-quality instruction and curriculum materials (Kinslow et al., 2019).

RATIONALE FOR THE INTEGRATION OF MODELING AND SOCIO-SCIENTIFIC REASONING

While we know much about how each of the practices of scientific modeling and socio-scientific reasoning can be effective in addressing goals aligned with Vision I and II SL respectively, the combination of the two for promoting Vision I-II has yet to be explored extensively. In this section, we present our theoretical argument that scientific modeling and SSR can be mutually supportive and a productive way to achieve goals associated with Vision I and II SL simultaneously when enacted in a synergistic way.

First, SSI provides students with a meaningful context to engage in scientific modeling for the purpose of explaining and predicting scientific phenomena in ways that are relevant to students' lives. SSI supports modeling practice as the relevancy of the issue affords students opportunities to leverage their personal goals (e.g., making decisions about an issue that matters) to align with the discipline-goals of science (e.g., figuring out the mechanisms of the phenomenon underlying the issue). For example, a student whose goal is to convince his parents to quit e-cigarette use, may use that goal as the driving force to study the mechanism of how nicotine impacts human body through modeling. Also, for students who are not interested in science or science-related professions, SSI may also serve as an interesting context that stimulates students, who would otherwise become disengaged in disciplinary-based activities such as scientific modeling, to explore the science aspects of the issue.

Second, we hypothesize that using the practices of modeling to understand the scientific knowledge relevant to the issue can in turn support students' SSR practice. Findings from previous studies support the claim that content knowledge underlying SSI is important for negotiation of complex issues (see the review by Sadler & Zeidler, 2005). We contend that students need to develop basic understandings of the relevant science content before they can start to engage in SSR. This point is demonstrated in Sadler and Donnelly's (2006) study of how high school students' content knowledge contributed to the quality of their SSI argumentation. Therefore, engaging students in scientific modeling to explore the underlying phenomena associated with the issue should advance their SSI reasoning skills given that model-based instruction promotes both students' content knowledge and their understandings of how science works.

Third, as previously mentioned, we see parallels between the underlying ideas constitutive of the subdimensions of each practice. We argue that the connections between those ideas play a key role in students' transferring the disciplinary knowledge as well as the ways of knowing to their decision-making or solutions to complex societal issues. For example, one key aspect of scientific modeling is related to explaining phenomena through the processes of identifying key components, articulating causal relationships among those components, and describing and predicting behaviors of the systems based on the identified causal relationships. Similarly, socio-scientific reasoning involves considering varied perspectives and analyzing the causal links and potential outcomes as perceived by different stakeholders. As such, a robust understanding of the underlying ideas about causal reasoning from a systems perspective shared by both practices (modeling and SSR) may support students in applying the complex reasoning from one context to the other. Another parallel we see between modeling and SSR is the notion that neither scientific models or SSR is static; they are open to inquiry and may need further revisions in light of additional evidence. It is also important to recognize that both scientific models and SSR have limitations concerning their utilities in explaining the phenomena or resolving the issues. Indeed, we hypothesize that the congruence between the two practices with respect to the underlying ideas that constitute the two practices may be the key to supporting students in making the transition from developing understanding of science to using that science in their lives.

SIMBL Framework

The SIMBL framework has been iteratively designed and informed by our prior research efforts in defining what teaching SSI should include (Friedrichsen, Sadler, Graham & Brown, 2016; Sadler, 2011). It highlights both scientific modeling and socio-scientific reasoning—the framework calls for learners to engage in modeling to make sense of the science underlying critical societal issues and develop their positions or solutions related to the issues. The framework includes a set of essential features for SSI teaching along with a suggested teaching sequence. The teaching sequence is a modified learning cycle, which has been an important instructional strategy for science teaching for decades (Bybee, 2015). The SSI teaching sequence comprises three successive phases (see Figure 1). It begins with students' exploration of the SSI that will serve as the focus of the unit. The point of this phase is for learners to become familiar with the issue and ideally develop interest in the issue which can support motivation to learn. The next phase of the sequence calls for students' engagement in scientific modeling relevant to the issue, creating opportunities for students to develop deep understandings of target science ideas while developing a key scientific practice (Zangori, Peel, Kinslow, Friedrichsen & Sadler, 2017). The second phase also involves students' socio-scientific reasoning as a means of developing an appreciation for the complex connections between the issue, the underlying science phenomena, and social challenges and implications associated with the issue. In the final phase, students generate and defend a position or solution to the issue by synthesizing the science ideas, modeling practices, and SSR competencies that they have been developing throughout the unit as well as their own personal commitments and values.

An Illustrative Example

In this section, we present an illustrative example from our prior work with elementary students to demonstrate the ways in which the integration of scientific modeling and socio-scientific reasoning may promote students' proficiencies in both areas. In particular, we trace one student's learning trajectory with regard to his models and SSI writings, in which he was asked to demonstrate SSR, to show how the young learner built upon the type of reasoning in his modeling work and applied it to his negotiation of complex issues. It is important to note that, we do not intend to examine the effectiveness of our teaching framework in the elementary context (see Zangori, Foulk, Sadler & Peel, 2018; Zangori, Ke, Sadler, & Peel, accepted – pending revisions) here in this chapter. Rather, the provided case of a single focus student serves as an example to illustrate how the practice of modeling and socio-scientific reasoning may interact in an SSI learning context.

Figure 1. SIMBL Framework (Figure from Sadler, Friedrichsen, & Zangori, 2019)

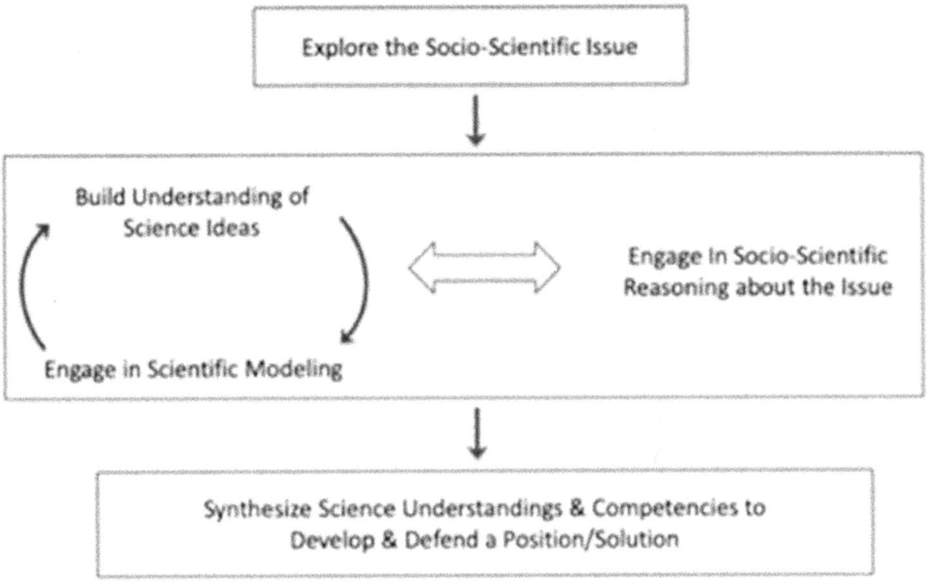

Context

The study took place in four 3rd grade classrooms with students aged 8 – 9 years old (n = 54) in a small Midwestern city in the U.S. The focal issue of the unit related to reductions in the Monarch butterfly population with a content focus on ecological relationships within an ecosystem. The issue was chosen because a decrease in Monarch population was identified as a local concern by the State Department of Conservation. The unit was co-designed, based on the SIMBL framework, by the four participating teachers and our research team in a 6-day professional development (PD) workshop with follow-up interactions among the teams throughout the school year. During the PD, the teachers were introduced to curricular design, background knowledge about SSI and modeling, and how the two can be integrated through instruction.

The unit began with a question: "Should our school turn one of our soccer fields into a garden to attract butterflies?" This was the first opportunity students had to write about the focal issue (SSI Writing 1). Next, students explored different types of models and how scientists might use models in the discipline of science in order for students to develop a better understanding of the nature of scientific models. Students then created diagrammatic models (Model 1) using paper and pencil to answer the question "How do all kinds of animals, insects, people, and plants interact with their environment?" Students were also provided with three reflective prompts, a) What does your model show? b) Why do you think this is how animals, insects, people, and plants interact with their environment? c) What observation have you made to help you understand how animals, insects, people, and plants interact with their environment? The prompts were designed to scaffold students towards using evidence to construct models that represent and explain underlying phenomena, a key epistemic aspect of the modeling practice.

Once students had developed initial understandings of the interrelationships within an ecosystem through scientific modeling, they explored the issue of Monarch butterfly declines as well as the social dimension of the issue through a series of instructional activities (e.g., watching a section of a movie showing butterfly migration, and a whole class activity of constructing a pictograph of the migration using external data). Following the SSI activities, students were asked to respond to the same question from the beginning of the unit: "Should our school turn one soccer field into a garden to attract butterflies?" (SSI Writing 2). It is important to note that the curriculum did not advocate a particular stance towards the issue; rather, students were encouraged to use ideas and evidence that they had learned to form their own positions on the issue.

After the SSI-focused lesson on butterfly migration and population decline, students learned about the life cycle of a butterfly, the interdependent relationship between different organisms (e.g., butterflies and plants) within a habitat before revising their initial models (Model 2). They were expected to revise their models based on the additional evidence they collected. The unit concluded with another SSI-focused lesson in which students were asked to write a letter to a principal of a different elementary school about whether or not the prairie in their schoolyard should be turned into a soccer field (SSI Writing 3). We next present our analysis of one focus student's (Ross) scientific models and SSI writings over the course of the unit to illustrate the link we theorized between students' modeling practice and their socio-scientific reasoning skills.

The Case of Ross

In his SSI Writing 1 at the beginning of the unit, Ross favored turning the soccer field into a garden because he "liked butterflies and could give them more milkweed." His response to the issue focused on his concrete reasoning about butterflies, but also included a human connection of caring for butterflies through feeding them. This is not uncommon among third-graders as young learners tend to make decisions about complex issues based upon personal emotions and dispositions. With reasoning patterns similar to that offered in Ross's response, we observed a large number of students articulating opposition to the proposal because they liked to play soccer.

In his initial model of ecological interactions (See Figure 2), Ross drew a bird in a tree with its eggs and nest while a leopard nearby was trying to catch the bird. Next to the scene of the bird and leopard was a pond composed of some fish, a crab, and algae. Within the initial model, Ross represented different species and indicated there were relationships among them. Through a semi-structured interview in which Ross talked about his initial model, we collected data that suggested that Ross was able to identify the relationship between the tree and the bird (e.g., "shelter relationship") as well as the one between the bird and the leopard (e.g., "eat relationship"). However, Ross was unsure about the other portions of his diagrammatical model, saying "I'm not really sure how they [organisms such as the fish and algae] interact with each other" and he "just want to draw something else beside the bird and the tree." This indicated that, for the pond system, Ross was capable of *representing* the system by identifying the components within it, but not *explaining* it by specifying the relationships among the components.

Figure 2. Ross's initial model of ecological interactions (Model 1)

In his second SSI writing, Ross maintained his position on the issue, but offered different reasoning in support of changing the soccer field into a garden. He wrote that "My opinion has not changed because we need more milkweed for butterfly to live and lay their eggs." In particular, Ross seemed to apply the two relationships he identified in his initial model, the "shelter relationship" (e.g., monarch lays eggs on milkweed) and the "eat relationship" (monarch eats milkweed) respectively, to his reasoning about the issue.

In his revised model of ecological interactions (See Figure 3), Ross again included the "eat relationship" and "shelter" relationships as he did in his initial model, but this time focused on butterflies instead of a bird and leopard. He drew a caterpillar eating milkweed on the right side and butterfly eggs hatching on the left side. He said in the interview afterwards that "I decided to draw butterflies because I learned more [about butterflies]." This implies that by exploring the issue of butterfly population decline through the SSI ecological unit, Ross gained more knowledge about butterflies, and used his new understandings as resources for revising his model.

Figure 3. Ross's revised model of ecological interactions (Model 2)

At the end of the unit, the issue was changed slightly to a conservation issue to further explore the students' SSR. We asked students to write a letter to a principal who was considering turning a prairie next to his school into a soccer field. Ross was asked to offer his position. Below is the interview excerpt in which Ross talked about his SSI Writing 3.

Interviewer: We are going to look at your writing about this issue. What changed about what you wrote?

Ross: I think I learned they [butterflies] only eat milkweed and they only lay their eggs on milkweed and I kinda learned more that they take pollen from flowers and they pollinate fruits and stuff and plants.

Interviewer: So it's a different school and they have one soccer field and one natural habitat and they are thinking about turning that habitat into a second soccer field. What do you think about that? What would you say?

Ross: I'd say I would not want him to turn the butterfly garden into a soccer field because then the butterflies won't have a home. Well, they might but they still wouldn't have enough milkweed and they would die. That's what I would say. I would say no.

Evident in the transcript, Ross used his butterfly knowledge, that they "only eat milkweed and they only lay their eggs on milkweed," in his reasoning about the issue. This reasoning was also consistent with what he included in his SSI Writing 2 and Model 2. However, this time he used both relationships (e.g., "eat relationship" and "shelter relationship") which resulted in a semi-quantitative causal relationship. He conjectures that the removal of a habitat or the decrease in the amount of milkweed would result in a decrease in butterfly population (e.g. "they would die"). In other words, his reasoning shifted from relational reasoning in SSI Writing 2 to causal reasoning in SSI Writing 3.

Summary

In Figure 4, we summarize Ross's learning trajectory regarding his engagement in modeling and socio-scientific reasoning. In terms of modeling competencies, Ross revised his model to focus more on explaining the interactions between butterflies and milkweed, as opposed to mainly representing the system as he did in part of his initial model (e.g., the pond system). With respect to responding to complex issues, Ross became increasingly sophisticated in his reasoning patterns over the course of the unit, shifting from personal reasoning in his SSI Writing 1, to relational reasoning in SSI Writing 2, and to causal reasoning in SSI Writing 3.

Figure 4. Ross's learning trajectory of modeling and socio-scientific reasoning

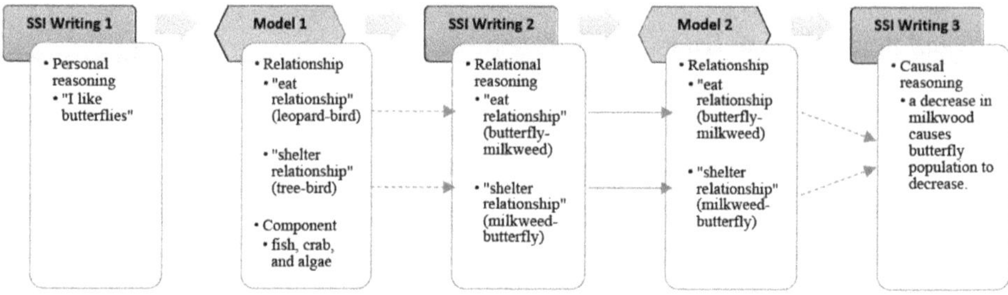

Moreover, we argue that the integration of scientific modeling and socio-scientific reasoning supported Ross's learning in both contexts. It was evident that the two relationships ("eat relationship" and "shelter relationship") Ross identified in his initial model were central ideas for both his modeling work and later SSI writings. We hypothesize that the initial modeling activity enabled Ross to recognize the underlying relationships before he could apply them to the SSI context. Note that while the two relationships identified in Model 1 and SSI Writing 2 were the same, the components were different. This indicates that Ross was able to transfer the reasoning patterns despite the contextual differences. In Figure 4, we used the dotted red arrows to denote this carry-over pattern across the two contexts. Further, the consistency between Ross's SSI Writing 2 and Model 2 (denoted by the two solid red arrows in Figure 4) suggests that the SSI contexts also played a key role in Ross's modeling work. From the interview data, we know that the exploration of the issue in the previous SSI lesson prompted Ross to include butterflies in his revised model. Finally, we theorize that the consolidation of the two relationships during the model revision process may largely contribute to Ross's shift from relational reasoning to semi-quantitative causal reasoning when negotiating about the issue: the "eat relationship" and "shelter relationship" were the building blocks of the causal claim that a decrease in milkweed would result in Monarch's population decline (denoted by the two dotted red arrows between Model 2 and SSI Writing 3 in Figure 4). To summarize, Ross's learning trajectory was consistent with our hypothesis regarding how scientific modeling and socio-scientific reasoning can be mutually supportive of each other as represented in the SIMBL teaching sequence (see Figure 1).

Some may argue that Ross just learned about the content knowledge and then put it into his models and SSI writings. While this alternative explanation is plausible, we believe this was not the case for Ross, and probably most of his peers for the following two reasons. First, we view both scientific modeling and socio-scientific reasoning as epistemic practices in which students engage in the practice for certain goals or purposes. Even though students were told that "butterflies eat milkweed and lay eggs on it," they may not have chosen to include it in their models or socio-scientific reasoning unless it fit with the goals of the activities. In our study, scientific models were framed as sense-making tools to figure out how things worked. Similarly, the task of SSI writing was aimed at helping students develop their own positions on complex issues with justifiable reasoning and evidence. Therefore, students needed to decide what to include in their knowledge products so that it was aligned with the purposes of the activities, as opposed to randomly throwing what they had learned into their models or SSI writings. Second, from the interview data, we know that Ross also learned about butterfly pollination. However, Ross did not include that piece of knowledge into his models or SSI writings. This serves as a counterevidence to the alternative explanation that it was primarily the gain of the content knowledge that promoted students' proficiencies in modeling and socio-scientific reasoning. In fact, the idea of butterflies interacting with other plants and flowers is a fruitful one that Ross could include in his models, which might deepen his understanding about the interrelationships within an ecosystem. It was also a potentially productive relationship that we hope Ross could build on so that he could see the impact of removing a habitat on species other than butterflies and milkweed as he engaged in the socio-scientific reasoning activities. However, the "pollination relationship" somehow was not taken up by Ross in his modeling work and socio-scientific reasoning, even though the idea was well circulated in the classroom community.

Epistemic Tools for Modeling and Socio-scientific Reasoning

Despite the great potential in integrating modeling and SSR, as shown in Ross's case, doing so remains challenging for teachers to enact in classrooms. This is largely due to the fact that most teachers are unfamiliar with how to support student engagement in modeling and/or SSR (Danusso et al., 2010; Justi & Gilbert, 2003), especially in regard to their epistemic aspects. In response to this challenge, our research team has designed epistemic tools to better support teachers in enacting modeling and SSR and students in shifting their reasoning between content and the issue. Epistemic tools are tools aimed at facilitating the construction of knowledge with a "symbolic pertinence that it becomes an intrinsic system of knowledge and thinking" (Magnusson, 2009, p.168). The tools are also flexible because teacher input into the tools is critical for their successful use in the classroom.

For scientific modeling, we have developed a modeling lesson and modeling packet, as described previously for the ecosystem unit, each of which are vital parts of our curriculum design. The modeling lesson starts by asking students to consider the definitions they know for the word model and to consider what a scientific model is and what it represents. Students explore different kinds of models within the classroom and develop a consensus model. They determine the purpose of different kinds of models and evaluate different kinds of models for how they support learning. These aspects of the modeling lesson focuses on metamodeling knowledge because they focus on the *nature of models* (what is a model? what do models represent? what is the relationship between models and reality?), the *process of modeling* (how are models constructed? are models ever changed?), the *evaluation of models* (how to determine if a model is useful for understanding something), and the *purpose of models* (what are models for? how are models useful to scientists and/or students?).

After the modeling lesson, students are introduced to the modeling packet which provides space for students to draw a beginning (pre), middle (mid), and end (post) model and answer questions about their models that guide students' modeling work in thinking about key epistemic considerations related to the practice (e.g., how and why do things work? What evidence do I have to support my model?). The modeling packet serves as a tool to help students engage in creating, revising, and evaluating models as they make sense of the underlying scientific phenomena. Students develop each model to the same question and reflective prompts. When students created their mid and post models, they are asked to first look at their previous model and write about how well it answers their modeling question and write about what they like and want to change about the previous model. The modeling packet always focuses on the underlying scientific phenomenon of the unit, so while individual lessons may focus on specific aspects of the phenomenon, when students return to their pre/mid/post models, they are asked to determine how the new pieces they have learned from the lessons fill in their knowledge about how and why the overarching phenomenon works. These links to the epistemic aspects of modeling practice, or metamodeling knowledge, within the modeling packet are essential for students to develop understandings of not only the science content about the phenomena, but also of how knowledge is constructed and justified with scientific models in scientific disciplines.

In terms of socio-scientific reasoning, our team has developed two epistemic tools, the *Star Chart* and the *Causal Map*. The *Star Chart* (see Figure 5) is a tool to help students appreciate the multi-dimensional nature of SSI. The intent is for students to use the *Star Chart* as they explore an issue by filling the points of the star, which represent the various dimensions (e.g., science, economics, politics, etc.) of the issue that students should take into consideration as they develop their positions. Teachers can choose what dimensions students should focus on in each point of the star, so that the chart is flexible and can be altered to align with any SSI. For example, in one of the high school SSI units our team developed around the issue of vaping (the use of e-cigarettes), students complete the *Star Chart* with information concerning the following five dimensions: science (e.g., chemical mixtures in e-cigarettes), politics (e.g., governmental regulation of the sale of e-cigarettes), economics (e.g., sales of vapes), history (e.g., history of smoking), and ethics (e.g., advertisement targeting at teenagers).

Figure 5. Star Chart

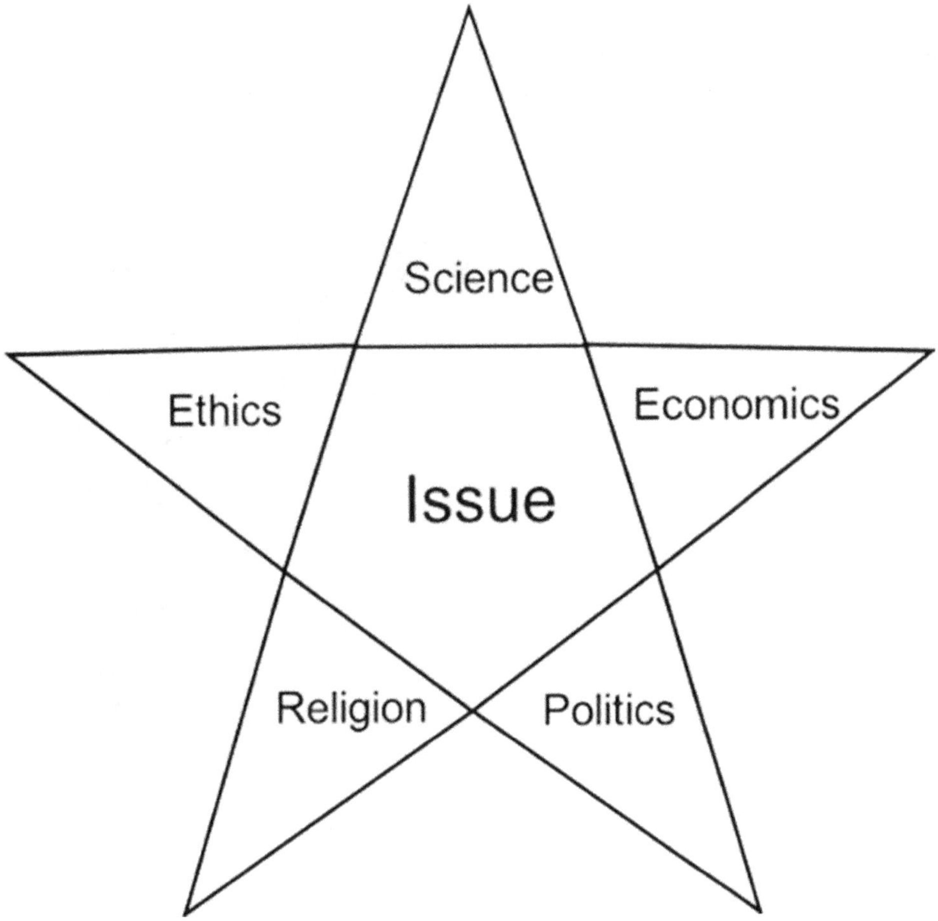

Students then use their reasoning about multiple dimensions to take a position on the issue and develop a *Causal Map*. The causal map is a conceptual representational tool that helps students recognize the complexity of an issue by identifying causal relationships among elements of systems in which the issue is embedded. Here systems refer to social systems and the elements are related to multiple social-political dimensions that are relevant to the focal issue. The intent is for students to use the *Causal Map* and consider how their position/solution might impact the overall system. For example, in the vaping unit, students are asked to propose their own regulation of vaping and work through the *Causal Map* to determine how their regulations might impact the whole system. Figure 6 is an example of *Causal Map* on regulation (e.g., no advertisement of vaping products) that one student came up with. Surrounding the center were different stakeholders the student identified: They were kids, teens, parents, adults, employees, vaping company, and advertisers. Further branching off those stakeholders were the impact the student thought his/her regulation had. For instance, the student identified a positive effect on parents, saying that they would "feel better and don't have to worry much about kids vaping."

Figure 6. An example of a Causal Map for the regulation of vaping

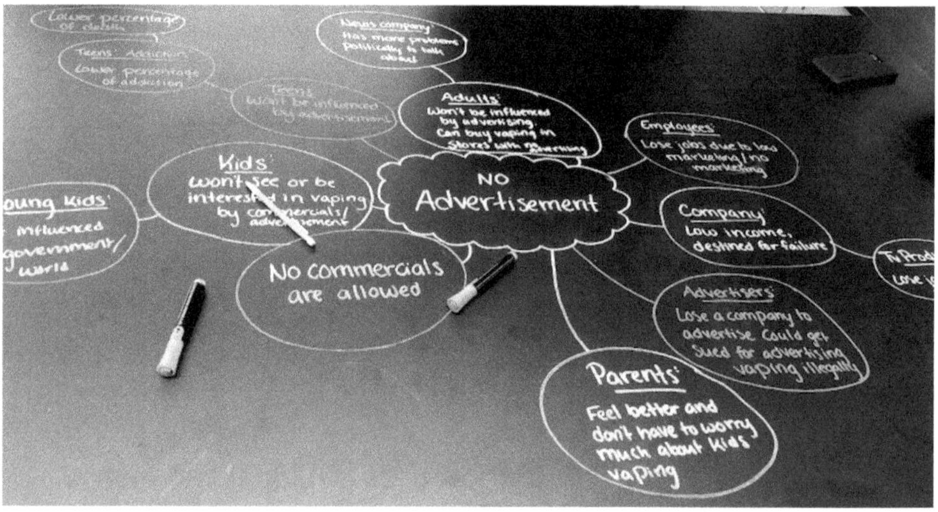

It is important to note that learners may appropriate tools in ways that are not reflective of the intended use of the tools. It depends on a number of factors including the learners' prior knowledge and learning goals and the social contexts of learning. For example, we found from our previous work that students' appropriation of the *Star Chart* and *Causal Map* can be categorized into three levels: 1) lack of appropriation, 2) appropriate surface features, and 3) appropriate epistemic purposes (Ke, Sadler, Zangori, & Friedrichsen, in press). While level 3 appropriation is consistent with the intended use of our tools, we observed that about half of the of the students were either not using the tools at all (level 1, a small portion of the student population) or using the tools as an alternative way to take notes (level 2). From interview data, it was apparent that the students mostly attended to school-based epistemologies (i.e., "doing school") when they used the tools. The findings suggest that it is critical for teachers to make explicit and guide students towards the epistemic goals of the practices as students use the tools in their learning.

CONCLUSION

Within science education, learning through an SSI requires students to build not only science understanding, but also to examine different global perspectives. While the science underlying the issue is fundamental for understanding the basis of the issue (e.g., species decline, antibiotic resistance, climate change), students must also explore political, economic, and/or global impacts. With this in mind, the intention of our work is to explore how the epistemic practices of science, modeling in particular, support students in building robust knowledge of the content and how they are able to transfer this knowledge to reason about and justify their positions on socio-scientific issues. Furthermore, we have developed tools that are integral to the SIMBL approach to support students' modeling practice to develop scientific knowledge about the issue, and how to incorporate their science knowledge to the multiple dimensions of the issue.

For this book chapter, we specifically chose a 3rd-grade student example to demonstrate how elementary students may productively engage in scientific modeling and socio-scientific reasoning simultaneously, as there are few such examples in the literature. It is important to note that elementary students are not blank slates and, through siblings, family, TV, and media are aware of and worry about global issues. Therefore, it is important that both modeling and SSI based instruction be introduced in elementary classrooms, in grade appropriate ways, so students' have opportunities to investigate these issues, ask questions, and reason about solutions. To build scientific literacy, students need opportunities to see how science and their lives are intertwined, rather than view them as two different silos, one that focuses on content while another that focuses on global issues. Our approach to scientific literacy is that both epistemic science practices and SSI instruction must be intertwined in science lessons if we are to realize both Vision I and Vision II scientific literacy.

REFERENCES

Aikenhead, G. S. (2007). Expanding the research agenda for scientific literacy. In *Promoting scientific literacy: Science education research in transaction, Proceedings of the Linnaeus Tercentenary Symposium held at Uppsala University*, (pp. 64–71). Uppsala: Geotryckeriet.

Bybee, R. W. (2015). *The BSCS 5E Instructional Model: Creating Teachable Moments*. doi:10.2505/9781941316009

Clement, J. (2008). The role of explanatory models in teaching for conceptual change. In S. Vosniadou (Ed.), *International handbook of research on conceptual change* (pp. 417–452). Routledge.

Danusso, L., Testa, I., & Vicentini, M. (2010). Improving Prospective Teachers' Knowledge about Scientific Models and Modelling: Design and evaluation of a teacher education intervention. *International Journal of Science Education, 32*(7), 871–905. doi:10.1080/09500690902833221

DeBoer, G. E. (2000). Scientific literacy: Another look at its historical and contemporary meanings and its relationship to science education reform. *Journal of Research in Science Teaching, 37*(6), 582–601. doi:10.1002/1098-2736(200008)37:6<582::AID-TEA5>3.0.CO;2-L

Eggert, S., & Bögeholz, S. (2010). Students' use of decision-making strategies with regard to socioscientific issues: An application of the Rasch partial credit model. *Science Education, 94*(2), 230–258. doi:10.1002ce.20358

Ekborg, M., Ottander, C., Silfver, E., & Simon, S. (2013). Teachers' Experience of Working with Socio-scientific Issues: A Large Scale and in Depth Study. *Research in Science Education, 43*(2), 599–617. doi:10.100711165-011-9279-5

Friedrichsen, P. J., Sadler, T. D., Graham, K., & Brown, P. (2016). Design of a Socio-scientific Issue Curriculum Unit: Antibiotic Resistance, Natural Selection, and Modeling. *International Journal of Designs for Learning, 7*(1). Advance online publication. doi:10.14434/ijdl.v7i1.19325

Gilbert, J. (2005). Visualization: A metacognitive skill in science and science education. In J. Gilbert (Ed.), *Visualization in science education. Models and modeling in science education* (pp. 9–27). Springer. doi:10.1007/1-4020-3613-2_2

Hancock, T. S., Friedrichsen, P. J., Kinslow, A. T., & Sadler, T. D. (2019). Selecting socio-scientific issues for teaching. *Science & Education, 28*(6), 639–667. doi:10.100711191-019-00065-x

Justi, R., & Gilbert, J. (2003). Teachers' views on the nature of models. *International Journal of Science Education, 25*(11), 1369–1386. doi:10.1080/0950069032000070324

Ke, L., Sadler, T., Zangori, L., & Friedrichsen, P. (2020). Students' perceptions of socio-scientific issue-based learning and their appropriation of epistemic tools for systems thinking. *International Journal of Science Education*, 1–23. Advance online publication. doi:10.1080/09500693.2020.1759843

Ke, L., & Schwarz, C. (2020). Using epistemic considerations in teaching: Fostering students' meaningful engagement in scientific modeling. In A. Upmeier zu Belzen, D. Krüger, & J. Van Driel (Eds.), *Towards a Competence-based View on Models and Modeling in Science Education* (pp. 181–199). Springer International Publishing.

Khishfe, R. (2014). Explicit Nature of Science and Argumentation Instruction in the Context of Socioscientific Issues: An effect on student learning and transfer. *International Journal of Science Education*, *36*(6), 974–1016. doi:10.1080/09500693.2013.832004

Kinslow, A. T. (2018). *The development and implementation of a heuristic for teaching reflective scientific skepticism within a socio-scientific issue instructional framework* (Unpublished dissertation). University of Missouri.

Kinslow, A. T., Sadler, T. D., & Nguyen, H. T. (2019). Socio-scientific reasoning and environmental literacy in a field-based ecology class. *Environmental Education Research*, *25*(3), 388–410. doi:10.1080/13504622.2018.1442418

Lazarowitz, R., & Bloch, I. (2005). Awareness of societal issues among high school biology teachers teaching. *Journal of Science Education and Technology*, *14*(5), 437–457. doi:10.100710956-005-0220-4

Lee, H., Abd-El-Khalick, F., & Choi, K. (2006). Korean science teachers' perceptions of the introduction of socio-scientific issues into the science curriculum. *Canadian Journal of Science. Mathematics and Technology Education*, *6*(2), 97–117. doi:10.1080/14926150609556691

Lehrer, R., & Schauble, L. (2006). Scientific thinking and science literacy: Supporting development learning in contexts. In W. Damon, R. M. Learner, K. A. Renninger, & I. E. Sigel (Eds.), Handbook of child psychology (6th ed., Vol. 4). John Wiley and Sons.

Louca, L. T., & Zacharia, Z. C. (2012). Modeling-based learning in science education: Cognitive, metacognitive, social, material and epistemological contributions. *Educational Review*, *64*(4), 471–492. doi:10.1080/00131911.2011.628748

Louca, L. T., Zacharia, Z. C., & Constantinou, C. P. (2011). In Quest of productive modeling-based learning discourse in elementary school science. *Journal of Research in Science Teaching*, *48*(8), 919–951. doi:10.1002/tea.20435

Magnusson, T. (2009). Of Epistemic Tools: Musical instruments as cognitive extensions. *Organised Sound, 14*(2), 168–176. doi:10.1017/S1355771809000272

Manz, E. (2012). Understanding the codevelopment of modeling practice and ecological knowledge. *Science Education, 96*(6), 1071–1105. doi:10.1002ce.21030

Nersessian, N. J. (2002). The cognitive basis of model-based reasoning in science. In P. Carruthers, S. Stich, & M. Siegal (Eds.), *The Cognitive Basis of Science* (pp. 133–153). Cambridge University Press. doi:10.1017/CBO9780511613517.008

NGSS Lead States. (2013). *Next Generation Science Standards: For states, by states*. National Academies Press.

Passmore, C., Gouvea, J. S., & Giere, R. (2014). Models in Science and in Learning Science: Focusing Scientific Practice on Sense-making. In M. R. Matthews (Ed.), *International Handbook of Research in History, Philosophy and Science Teaching* (pp. 1171–1202). Springer Netherlands. doi:10.1007/978-94-007-7654-8_36

Roberts, D. A. (2007). Scientific Literacy/Science Literacy. In S. K. Abell & N. G. Lederman (Eds.), *Handbook of Research on Science Education*. Lawrence Erlbaum Associates., doi:10.4324/9780203824696-32

Roberts, D. A., & Bybee, R. W. (2014). Scientific Literacy, Science Literacy, and Science Education. In N. G. Lederman & S. K. Abell (Eds.), *Handbook of Research on Science Education* (Vol. 2, pp. 545–558). Routledge. doi:10.4324/9780203097267-38

Romine, W., Sadler, T. D., & Dauer, J. (in review). *Measurement of socio-scientific reasoning (SSR) and exploration of how students' SSR competencies improve over time*.

Sadler, T. D. (2009). Situated learning in science education: Socio-scientific issues as contexts for practice. *Studies in Science Education, 45*(1), 1–42. doi:10.1080/03057260802681839

Sadler, T. D. (2011). Socio-scientific Issues-Based Education: What We Know About Science Education in the Context of SSI. In T. D. Sadler (Ed.), *Socio-scientific Issues in the Classroom: Teaching, Learning and Research* (pp. 355–369). Springer Netherlands. doi:10.1007/978-94-007-1159-4_20

Sadler, T. D., Barab, S. A., & Scott, B. (2007). What do students gain by engaging in socioscientific inquiry? *Research in Science Education, 37*(4), 371–391. doi:10.100711165-006-9030-9

Sadler, T. D., & Donnelly, L. A. (2006). Socioscientific Argumentation: The effects of content knowledge and morality. *International Journal of Science Education, 28*(12), 1463–1488. doi:10.1080/09500690600708717

Sadler, T. D., Friedrichsen, P., & Zangori, L. (2019). A framework for teaching for socio-scientific issue and model based learning (SIMBL). *Educação e Fronteiras/ Education and Borders, 9*(25), 8-26.

Sadler, T. D., Romine, W. L., & Topcu, M. S. (2016). Learning science content through socio-scientific issues based instruction: A multi-level assessment study. *International Journal of Science Education, 38*(10), 1622–1635. doi:10.1080/09500693.2016.1204481

Sadler, T. D., & Zeidler, D. L. (2005). The significance of content knowledge for informal reasoning regarding socioscientific issues: Applying genetics knowledge to genetic engineering issues. *Science Education, 89*(1), 71–93. doi:10.1002ce.20023

Schwarz, C. V., Reiser, B. J., Davis, E. A., Kenyon, L., Achér, A., Fortus, D., Shwartz, Y., Hug, B., & Krajcik, J. (2009). Developing a learning progression for scientific modeling: Making scientific modeling accessible and meaningful for learners. *Journal of Research in Science Teaching, 46*(6), 632–654. doi:10.1002/tea.20311

Schwarz, C. V., & White, B. Y. (2005). Metamodeling Knowledge: Developing Students' Understanding of Scientific Modeling. *Cognition and Instruction, 23*(2), 165–205. doi:10.12071532690xci2302_1

Simonneaux, L., & Simonneaux, J. (2009). Students' socio-scientific reasoning on controversies from the viewpoint of education for sustainable development. *Cultural Studies of Science Education, 4*(3), 657–687. doi:10.100711422-008-9141-x

Tidemand, S., & Nielsen, J. A. (2017). The role of socioscientific issues in biology teaching: From the perspective of teachers. *International Journal of Science Education, 39*(1), 44–61. doi:10.1080/09500693.2016.1264644

Zangori, L., & Forbes, C. T. (2014). Scientific Practices in Elementary Classrooms: Third-Grade Students' Scientific Explanations for Seed Structure and Function. *Science Education, 98*(4), 614–639. doi:10.1002ce.21121

Zangori, L., Foulk, J., Sadler, T. D., & Peel, A. (2018). Exploring Elementary Teachers' Perceptions and Characterizations of Model-Oriented Issue-Based Teaching. *Journal of Science Teacher Education, 29*(7), 555–577. doi:10.1080/1046560X.2018.1482173

Zangori, L., Ke, L., Sadler, T. D., & Peel, A. (2020). Exploring primary students causal reasoning about ecosystems. *International Journal of Science Education*. DOI: 10.1080/09500693.2020.1783718

Zangori, L., Peel, A., Kinslow, A., Friedrichsen, P., & Sadler, T. D. (2017). Student development of model-based reasoning about carbon cycling and climate change in a socio-scientific issues unit. *Journal of Research in Science Teaching*, *54*(10), 1249–1273. doi:10.1002/tea.21404

Zeidler, D. L. (2014). Socioscientific Issues as a Curriculum Emphasis: Theory, Research, and Practice. In N. G. Lederman & S. K. Abell (Eds.), *Handbook of Research on Science Education* (Vol. 2, pp. 711–740)., doi:10.4324/9780203097267-45

Zohar, A., & Nemet, F. (2002). Fostering students' knowledge and argumentation skills through dilemmas in human genetics. *Journal of Research in Science Teaching*, *39*(1), 35–62. doi:10.1002/tea.10008

Section 2

Socioscientific Issues for Scientific Literacy in the Middle Grades

Chapter 3
Impact of Socioscientific Issues on Middle School Students' Character and Values for Global Citizenship

Wardell A. Powell
https://orcid.org/0000-0002-3486-2077
Framingham State University, USA

Mark H. Newton
https://orcid.org/0000-0002-3616-319X
East Carolina University, USA

Dana L. Zeidler
University of South Florida, USA

ABSTRACT

This chapter demonstrates the impact of an animal cloning socioscientific issues instructional unit on a group of middle school students' abilities to use their ecological worldview, social and moral compassion, and sense of socioscientific accountability to determine the permissibility of animal cloning. Seventy-seven 7th grade students at a public middle school in the Southeastern region of the United States participated in this investigation. Results from a non-parametric two-tailed Wilcoxon test indicated the students' social and moral compassion ($Z = -2.505$, $p = .012$) and socioscientific accountability scores ($Z = -2.381$, $p = .017$). In contrast, the results did not demonstrate a statistically significant difference between students' pre and post ecological worldview ($Z = -1.185$, $p = .236$). Qualitative analyses of the data revealed several interesting trends and themes discussed in the chapter. The findings from this investigation support the use of SSI as key pedagogical strategies in promoting character and values for global citizens among middle school students.
DOI: 10.4018/978-1-7998-4558-4.ch003

INTRODUCTION

A perennial goal of science education is to develop students with the capacity to contribute to the resolution of contentious issues in socially and environmentally sustainable manner. This aim is evident in the global recommendations for science education (Lead States, 2013: UNESCO, 2019). This mission is paramount in helping students to act responsibly when confronted with opportunities to support social justice movements aimed at preserving the environment and supporting others in need. Citizens are expected to make decisions on the environment, our health, and the safety of others, whether through engagement in the democratic process or through direct decisions based upon their employment in today's society, which has been deemed functional scientific literacy (Zeidler, 2003). Making informed decisions requires careful examination of data in order to evaluate evidence. Studies that examined the types of evidence that are essential in the evaluation of controversial issues have reported that participants made fewer requests about the research findings, who conducted the research, and where the research was conducted (Korpan, Bisanz, & Henderson, 1997). This finding is troubling in light of calls from researchers for students to exhibit skepticism to carefully evaluate information that is potentially biased (Sadler, 2007; Sadler, et al., 2011). Equally troubling are reports that students tend to accept information without evaluation of evidence (Owens, Sadler, & Zeidler, 2017; Ratcliffe, 1999). At a time when policy-makers can ignore a city's water contamination in the United States as in the case of Flint, Michigan, when lawmakers deny climate change (Fisher, Waggle, & Leifeld, 2012; McCright & Dunlap, 2011), or when lives are endangered due to where natural gas pipelines are erected (Brito, de Almedia, & Mota, 2009; Sklavounos & Rigas, 2006), it is critical for students to learn to think for themselves, evaluate evidence, and make appropriate and informed decisions. The acceptance of claims without evaluation only adds to the difficulty of making informed decisions.

The spread of COVID-19 around the world is a case in point of why students in general, and the public in particular need to be able to make informed decisions on scientific phenomena. During the midst of the pandemic, we have seen regular citizens, faith-based leaders, and politicians make egregious claims based on feelings and ideology, rather than on known scientific facts. Some who heeded to these false claims paid the ultimate price with their life. For example, many pastors and politicians in the United States down played the threats COVID-19 pose. They encouraged their followers to ignore the social distancing guidelines that were initiated by members of the scientific community. Some who followed this ill-fated advice made the ultimate sacrifice by losing their life to the virus. Such a blatant lack of respect for science and what it has to offer must not become commonplace in our society, because science matters, and scientific knowledge is imperative.

Therefore, teachers must provide their students with opportunities to develop their skills in making informed scientific decisions on SSI.

Scholars have suggested a socioscientific issues (SSI) framework as a potential approach to develop the skills requisite of a functionally scientific literate citizenry capable of negotiating complex issues (Zeidler, Berkowitz, & Bennett, 2013; Zeidler & Sadler, 2008; Zeidler, Applebaum, & Sadler, 2011; Zeidler, Sadler, Applebaum, & Callahan, 2009). SSI are typically contentious in nature, can be considered from a variety of perspectives, do not possess simple conclusions, and frequently involve morality and ethics (Zeidler, Walker, Ackett, & Simmons, 2002). SSI call upon individuals to use moral considerations, empathetic concerns, perspective taking, and feelings of responsibility as they evaluate evidence and make informed decisions on controversial issues connected with scientific phenomena (Kahn & Zeidler, 2014). SSI are well positioned to allow students to better understand their own socioscientific responsibility to the environment as well as their moral responsibility to others (Sadler, Klosterman, & Topcu, 2011). Students' abilities to freely discuss, evaluate evidence, and determine a path of action to complicated questions represents the hallmark of a democratic system (Aikenhead, Orpwood, & Fensham, 2011; Kolstø, 2001; Miller, 1983; Miller, 1998; Sadler, 2004; Shamos, 1995).

In an effort to develop functional scientific literacy, attention must be paid to individual's character and values to promote responsible and altruistic decisions while resolving SSI (Choi et al., Lee et al. 2012; 2013, Herman, Zeidler, Newton, 2018). Three subcategories have emerged within the construct of character and values that provide insight into development of character and values: ecological worldview, socioscientific accountability, and social and moral compassion. These are discussed below:

Ecological Worldview

It is crucial for the public in general, and students in particular, to understand that no event in nature take place in isolation, but rather through a series of interdependent relationships. Such understanding is paramount to developing an ecological worldview that results in actions that are aimed at sustainability. The term ecology relates to the interactions among organisms and their physical environment. Thus, an ecological worldview is a belief that humans and nature are interconnected in a global social-ecological system and as co-creative partners in the process of development and evolution (Du Plessis, 2009, Du Plessis, 2015). Our existence is dependent on the sustainable future of the numerous ecological systems. For example, global air and water pollution is detrimental to the health and well-being of human population

(World Health Organization [WHO], 2010b).

A high ecological worldview helps students to make informed decisions on such issues as deciding on the optimal location to build landfills, in whose communities should natural gas pipelines be constructed, should scientists engage in embryonic stem cell research, and if scientists should be allowed to clone animals to alleviate food shortages. To make sound decisions, students need to understand the interconnectedness of different systems in our environments. They must also become aware of actions required for sustainable development of different environmental systems. The debate over climate change, for example, provides an illustration of the importance of an extensive ecological worldview. While the data suggest that carbon dioxide levels have increased in the atmosphere over the last century, many policymakers throughout the world, and especially in the United States, have denied the notion of global climate change. This denial represents a failure of these individuals to understand the interconnectedness of various systems on our planet. This failure contributes to the current lack of sustainable developmental actions required to preserve many of Earth's systems threatened by climate change.

Interconnectedness. Interconnectedness refers to the understanding that human beings are embedded in and inseparably linked with nature (Dunlap et al., 2000; Zeidler, Berkowitz, & Bennett, 2013; Smith & Williams, 1999). For example, communities that have challenges with air pollution generally have more people with related illness. Worldwide epidemiological studies have shown a strong correlation between particulate matter and cardiac and respiratory morbidity and mortality (Du. Et al, 2016; Lee & Lee, 2014). In essence, the more we pollute the air around us, the more our health deteriorates. The reciprocal relationship between humans and nature is the defining characteristic of interconnectedness. Individuals who demonstrate high levels of interconnectedness consider the impact of environmental resource decisions on both humans and the natural environment.

Sustainable Development. Multiple definitions exist for the term sustainable development (Redclift, 2005). The concept is commonly regarded as an attempt to combine an expanding range of environmental concerns with socio-economic issues (Hopwood, Mellor, & Geoff, 2005). Individuals demonstrating a high ecological worldview advocate for solutions that meet the needs of current and future generations of humans, while also respecting and preserving nature (Dillion, 2014). Therefore, solutions to environmental problems should be beneficial to humans and the environment (Colucci-Gray, Giuseppe Barbiero, & Gray, 2006; Mueller, 2008; 2009).

Social and Moral Compassion

The ill-structured nature of SSI makes it ideal for students to consider various perspectives and be empathic toward the others as they engage in moral reasoning on matters that lack clear cut solutions. Therefore, school science curriculum must be a vehicle to prepare students to make informed decisions on the scientific issues they will confront on a daily basis. To provide students with skills to make sound decisions on these issues, values, moral education, and character development must be included in all instruction. A review of research from the mid-1980s to the mid-2000s showed that students often use personal experiences, emotive considerations, social considerations, and morality to drive their decision-making on SSI (Zeidler, Sadler, Simmons, & Howes, 2005b). In highlighting the moral purpose of school, Dewey (1909) stressed that "moral" does not designate a special region or portion of life. He believed it was imperative that we translate the moral into the conditions and forces of our community life and into the impulses and habits of the individual. Moral education is necessary if students are to develop social and moral compassion towards each other and to the environment (Herman, Zeidler, & Newton, 2018).

Moral and Ethical Sensitivity

Science education researchers have long called for the inclusion of real-world issues with conceptual ties to science to become part of all science curricula (Driver et al., 1996; Kolsto, 2001; Siebert & McIntosh, 2001; Zeidler & Keefer, 2003). These researchers believe curriculum can build students' capacity to use morality when making decisions on SSI. According to Sadler (2004), moral sensitivity describes the tendency for an individual to recognize that some aspects of an issue possess moral implications. Asking students to make decisions on issues that challenge their entrenched beliefs often stirs students' morality.

Perspective Taking

It is quite common for one to mistakenly asked another about their position on an issue while mistakenly thinking they are asking about perspective (Kahn & Zeidler, 2016). When requesting that a person share their perspective, the answer given must explain the belief behind the stated position.

The explanation given is regarded as one's perspective since it requires a more in-depth understanding of the point of view of the one whose perspective is being considered. Kahn and Zeidler (2016) further stated that to "take" another's perspective requires a psychological shift between one's outside, or etic viewpoint, to another's inside, or emic view (p. 9). In this sense, perspective is viewed as a being more personal or private, whereas, one's position is described as more general and public (Kahn & Zeilder, 2016). For example, asking students whether animals should be clone requires them to take a position on animal cloning. The thoughts behind any response to this question are more general or public. However, asking students to provide their perspectives on animal cloning will create opportunities for the students to think about their position on issues of cloning as well as provide explanations of why they took such a stance in the first place. In this case, the response would be more personal, since it requires the student to search their attitudes, feelings, values, moral beliefs, and emotions.

Such formation of perspective is rooted in the features of the SSI framework (Kahn & Zidler, 2019). The ability to engage in this type of perspective-taking is critical to argumentation, which others have described as a process of using the practice of science to engage scientific discourse (Jiménez-Aleixandre and Erduran 2007; Latour and Woolgar 1986). When we asked students to make decisions on whether animals should be cloned, our question should require them to consider the various perspectives of cloning. This will enable students to weigh the pros and cons before arriving at a decision.

Empathetic Concern

Empathy is a vicarious emotional response that is identical or very similar to that of the other person (Eisenberg et al. 1994). The ability to put oneself in another's position, especially when they are in a time of need, often evokes emotions. The notion of separating an individual from their emotions is a tricky proposition since one's emotion has a bearing on their interest or disengagement. Therefore, educators need to understand that educational exercises that enable students to evoke positive or negative emotions can impact their students' empathetic concerns for the well-being of others. In so doing, this can have a significant influence on students' engagement. The SSI framework seeks to understand as well as promote perspective-taking, awareness of emotional states, concerns for the well-being of others, and prosocial behaviors in response to the plights of others (Herman et al. 2018; Kahn & Zeidler, 2016). So, the notion of people dying due to circumstances like starvation or disease outbreaks may raise students' empathetic concerns to the point where they might support causes that goes against their long-held traditional views. These types of experiences may determine a course of action that may impact those with or in similar circumstances.

SOCIOSCIENTIFIC ACCOUNTABILITY

Our public education system must produce graduates capable of taking a social justice stance as they advocate for better living conditions around the globe. For too long, we have seen citizens in impoverished communities around the globe not having a voice in important environmental decisions that directly impact their lives. The public education system has the responsibility to enhance students' abilities to become effective advocates for themselves, others, and the environment. Students must have the ability to advocate for safer conditions in their environment when policymakers determine where to build nuclear power plants, landfills, natural gas pipelines, and to decide whether to engage in fracking. Not taking up the social justice mantle will lead to the demise of the poor.

The push for making America "Great Again" may come at the cost of rolling back safeguards created to protect the environment. If and when this happens, we should not be surprised if we see lead-contaminated drinking water and more related health effects due to increased air pollution. Improving students' abilities to effectively make decisions on these important environmental issues represents the hallmark of democratic systems (Aikenhead, Orpwood, & Fensham, 2011; Kolstø, 2001; Miller, 1998; Owens, Sadler, & Zeidler, 2017; Sadler, 2004; Shamos, 1995). In the current study, students were asked if they believed it was their responsibility to put aside personal views, such as religious beliefs and family values, to support the process of cloning for foods purposes.

Feelings of Responsibility

Our conscience can create feelings of responsibility due to our actions or inactions when confronted with the task of making decisions on moral and social issues (Green, 1985). As a result, it is quite common for people to blame themselves for events over which they have no control. Thomas Green (1985) highlighted the importance of conscience and its relationship to moral education below:

...there is something in favor of speaking about the formation of conscience rather than moral education. It is a simple fact that each of us has the capacity to judge our own conduct and even to stand in judgment on what we discern to be the composition of our own affections. Our conscience can influence judgment that each of us makes in his or her own case. In short, it is reflexive judgment. (p. 3)

Making decisions on SSI sometimes requires a sense of responsibility. Research has found that students can moderate their concern, particularly in environmental contexts, by abdicating their responsibility of environmental SSI resolution (Herman, Zeidler, & Newton, 2018). Students should be aware that simple actions, such as voting might influence the creation of laws to prevent scientists from engaging in embryonic stem cell research. This may delay the breakthroughs that researchers need to effectively treat genetic disorders for millions of patients. Students need to feel responsible for their actions or inactions when it comes to making decisions on these issues. Our inability to generate feelings of responsibility to act and determine that all human beings are entitled to drink lead-free water is simply bringing out the brute in us (Green, 1985; Zeidler, 2007).

The majority of studies conducted on SSI have examined various attributes of high school students, college students, or university professors who engage in SSI instruction, there has been a dearth of research on the extent to which middle school students can utilize constructs of morality and empathy to drive their social justice intuitions as they make decisions on SSI. The aim of this study is to shed light on how students' empathetic reasoning may influence their actions/inactions towards the use of biotechnology to meet the needs of individuals, society, and the environment.

Willingness to Act

Researchers have long reported that the primary goal of science education is to cultivate students with abilities to act at the personal, societal, and global levels (Lee et al., 2013). To preserve the environment for future generations, students need to become advocates for the environment. They need to champion causes that will protect drinking water resources, biotechnology research that detect, and in some cases treat human diseases, ways to reduce pollutants in the oceans, and how to prevent toxic gas emissions that are damaging to the ozone layer. However, to be successful at championing the above causes will require students to become literate about the environment. Being environmentally literate could potentially impact the students' willingness to engage in actions that are aimed at environmental sustainability. While this is the case, research evidence is scarce on middle school environmental literacy (McBeth & Volk, 2010). This lack of research makes it more difficult for education policymakers and educators to determine what essential tenets should constitute middle-grade ecological education programs to enhancing students' willingness to act on preserving the environment.

However, several studies that investigated secondary school students' willingness to act to reduce climate change have reported that students usually identify actions that they could take such a turning off lights and unplugging electrical appliances when not in use (Chhokar et al., 2010; Chu et al., 2007; Hermans & Korhonen, 2017; Skamp, Boyes, & Stanisstreet, 2009). Some of these same studies also reported that students are less willing to sell or purchase things at Thrift stores or even to walk moderate distances (Hermans & Korhonen, 2017). These results are an indication that more robust environmental education programs are needed in the K-12 system.

PURPOSE OF STUDY AND RESEARCH QUESTIONS

The purpose of this investigation was to determine the extent to which SSI instruction on animal cloning impacted students' ecological worldview, social and moral compassion, and sense of socioscientific accountability regarding animal cloning. Specifically, this study sought to address the following research questions:

1. To what extent might socioscientific issues impact middle school students' ecological worldview on the permissibility of animal cloning?

Rationale: While biotechnology has the potential to increase agricultural productivity, studies have reported that most Americans are unaware of the extent to which biotechnology has become part of their food supply (Haban, 2004; Pew, 2001). SSI instruction provides opportunities for students to critically examine differing perspectives on animal cloning, including the potential consequences cloning may have on the natural environment. Existing SSI scholarship has demonstrated that postsecondary students have considered their ecological worldview and have developed more sophisticated ecological worldviews, as a result of effective SSI instruction (Herman, Zeidler, & Newton, 2018; Lee et al. 2012;2013). What is not known is whether middle school students (13-14 years old) experience similar changes to their ecological worldviews after engaging in SSI instruction.

2. To what extent might socioscientific issues impact middle school students' social and moral compassion on the permissibility of animal cloning?

Rationale: Protein-energy malnutrition (PEM) accounts for almost half of all premature deaths from nutrition-related disease (World Health Organization, 2000). However, the use of agricultural biotechnology applications to assist in food production has received little support from the general public (Knight, 2006). To help instill in the next generation the social and moral compassion necessary to guide their actions/inactions towards others in need will require values, moral education, and character education to become part of all K-12 science education curriculum. This is of paramount importance in helping researchers to understand how students' empathetic reasoning influence their actions/inactions towards people whose needs are not met due to their health, zip code, or financial situations. The use of SSI in K-12 science education curriculum will provide opportunities for students to engage in discourse on controversial scientific issues that will promote students' development of values, moral education, and character education. This may also provide opportunities for science education researchers to understand how students' empathetic reasoning influences their actions towards others in need.

3. To what extent might socioscientific issues impact middle school students' sense of socioscientific accountability regarding the cloning of animals?

Rationale: As future global citizens, students will be confronted with responsibilities to vote on issues that may have a long-term impact on the fate of biotechnologies in the United States and other regions of the world. For students to make informed decisions on whether genetically modified plants and animals should be consumed, it is imperative that they understand socioscientific accountability towards others. Socioscientific accountability is defined as the feelings of accountability and personal responsibility manifested within stakeholders as they engage with global SSI and demonstrate a willingness to take sociopolitical action when necessary (Zeidler, Berkowitz, & Bennett, 2013). To develop students' understanding of their socioscientific accountability, K-12 policymakers should consider including SSI into the general science education curriculum. This change in curriculum will provide opportunities for students to engage in classroom discourse that may enhance their sense of socioscientific accountability.

In seeking solutions to the many formidable scientific challenges of the 21st century world and beyond, the public educational system has been tasked with the responsibility to produce global citizens with the skills necessary to find solutions to these problems. Empirical evidence has demonstrated that effective SSI instruction promotes the development of the skill necessary to resolve complex problems in secondary and postsecondary students. The current study seeks to determine the extent to which SSI instruction can develop these similar skills in grade 6-8 students.

METHODOLOGY

Participants

The participants of this study were 77 seventh-grade students (13-14 years old; 43 females, 34 males) enrolled in five comprehensive science courses that were taught by the same teacher, at a large public middle school in the southeastern United States. The school enrolls students in grades 6-8 with a total enrollment of over 1100 students. Minority students make up 46% of the student body. Overall, the school is rated in the bottom 50% of schools in the state on reading and math proficiency.

Instrumentation

Quantitative data was collected using the *Character and Values as Global Citizens Assessment* (CVGCA) survey. The CVGCA survey developed by Zeidler, et al., (2013) determines students' abilities to use character and values as they engaged in SSI instruction. Terminology referencing cloning was inserted in the CVGCA to better align the instrument with the SSI unit investigated in this study. The survey measured students' ecological worldview through the dimensions of interconnectedness and sustainable development; social and moral compassion through moral and ethical sensitivity, perspective-taking, and empathetic concerns; and socioscientific accountability through the feeling of responsibility and willingness to act.

Qualitative data was also collected from students' responses to open-ended questions, answers on KWL charts, comments made during class discussions, and persuasive essays.

Coding of Data

We employed the use of *a priori* coding (Stemler, 2001) and applied the various codes to the written text. We developed these codes from the CVGCA survey. Two graduate students with extensive experience in analyzing qualitative data along with the first author coded and analyzed a subset of the students' data (10 students). All codes obtained were revised to reveal connections between them so that they could be categorized and classified. The students' data sets were then split and shared among the two analysts. Each graduate student then carried out their analysis separately, and then presented the common themes and trends they generated to the first author for evaluation. The themes were categorized via consensus into the following groups: 1) interconnectedness; 2) sustainable development; 3) moral and ethical sensitivity; 4) perspective-taking; 5) empathetic concerns; 6) feeling of responsibility; 7) willingness to act.

Design and Procedures

A concurrent parallel mixed method design was implemented to collect data (Greene, 2007). This instructional unit used in this study took place over 10 class periods. Each class period lasted 55 minutes. Table I below illustrates the events during each period and the data collection process.

Table 1. Class activities

Class Periods	Data Collection
1	• Pretest CVGCA survey administered • Students completed the "I Know" and "I Want to Know" portions of a KWL chart about cloning • Whole group discussion on "I Know" and "I Want to Know" responses
2	• The students completed a *See-Think-Wonder* exercise after watching cloning 101 (https://www.youtube.com/watch?v=q0B9Bn1WW_4) • Whole-class discussion on the cost and benefits associated with cloning
3	• Responses to open-response questions • Students grouped according to their support or opposition to animal cloning • Whole group discussion on the cost and benefits of animal cloning.
4	• All students read the *Benefit, and Problems with Cloning Animals* (Smith, Bordignon, Babkline, Fecteau, & Keefer, 2000). • Students who were supported cloning animals read the potential applications of cloning • Students who opposed cloning read about the health hazards associated with cloning • Both groups of students read the conclusion • Students summarized the main points of assigned readings
5	• Whole-class discussion on the pros and cons of animal cloning • The students used the information from their assigned readings to strengthen their stance on whether animals should be cloned.
6	• Preparations of arguments on Westminster Champion Show Dog Cloning Dilemma
7 & 8	• Students engaged in argumentation on Westminster Champion show dog cloning.
9	• Completion of the "What I Learned" portion of KWL chart • Responses to open-ended questions about animal cloning
10	• Post-test CVGCA survey administered

In this chapter, the quantitative findings of the CVGCA survey will be reported first. These will be followed by more detailed explanations of the qualitative results. A more robust discussion will follow the overall findings of the interventions used in this chapter.

QUANTITATIVE FINDINGS

Ecological Worldview

The CVGCA survey measured students' ecological worldview through their responses to question related to the interconnectedness of nature, and their beliefs in sustainable efforts to protect Earth' systems. A Wilcoxon Signed-ranks test was conducted to determine if differences existed between the students' pre and post-tests scores. Table 2 below shows the comparison of the pre-CVGCA and post-CVGCA conceptual components.

Table 2.

Conceptual Components	Pre-test Median (pre-Md)	Post-test Median (post-Md)	*Z*	*p*	*r*
Ecological Worldview	3.171	3.33	-1.185	.236	.320

The results from the Wilcoxon Signed-Ranks Test indicated no statistically significant change from pre-test to post-test (Z = - 1.185, p = .236).

Interconnectedness

Results from the Wilcoxon Signed Ranks Test on students' pre/post interconnectedness scores measured through CVGCA items 1-3 were not statistically significant (pre-Md 3.00, post- Md 3.33, Z -1.503, p .133, r = .264). As reported by other researchers (Lee, et al., 2013), these non-significant results on students' interconnectedness may be explained by the fact that students already had strong ecological worldviews prior to this investigation. The frequency with which students expressed interconnectedness as part of their ecological worldview in determining relationships between research in biotechnology and its impact on the balance of nature, health, food production, and wildlife management are illustrated in Figure 1 below.

Figure 1.

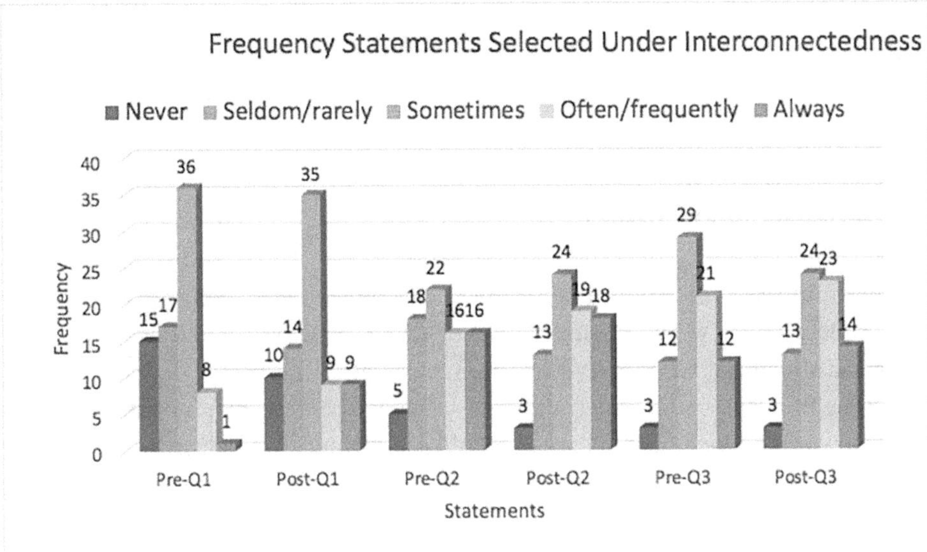

Figure 2.

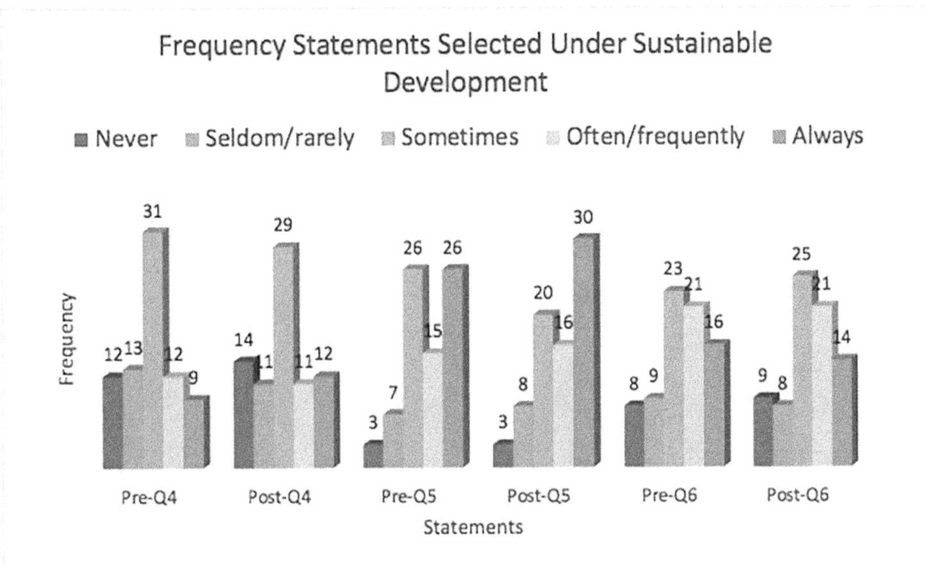

Table 3.

Conceptual Components	Pre-test Median (pre-Md)	Post-test Median (post-Md)	*Z*	*p*	*r*
Social and Moral Compassion	3.14	3.29	-2.505	.012	- .067

Sustainable Development

Results from the Wilcoxon Signed Ranks Test on students' pre/post sustainable development scores were measured through CVGCA items 4-6 and were found to be not statistically significant (pre-Md 3.33, post- Md 3.33, $Z = -.179$, $p = .858$, $r = .289$). The frequency with which students selected questions that inquire about the different ways in which human beings should use research to produce things needed to sustain life and the environment are illustrated in Figure 2 below.

Social and Moral Compassion

The CVGCA survey measured the students' social and moral compassion through their responses to question related to moral and ethical sensitivity, perspective taking, and empathetic concerns. A Wilcoxon Signed-ranks test was conducted to determine

Figure 3.

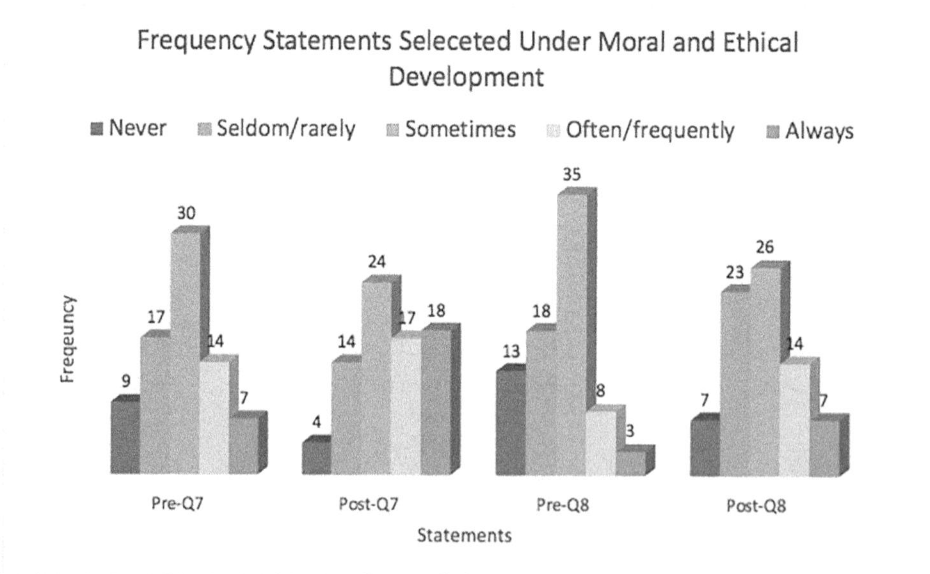

the difference between the students' pre and post-tests scores. Table 3 below shows the comparison of the pre-CVGCA and post-CVGCA conceptual components.

Figure 4.

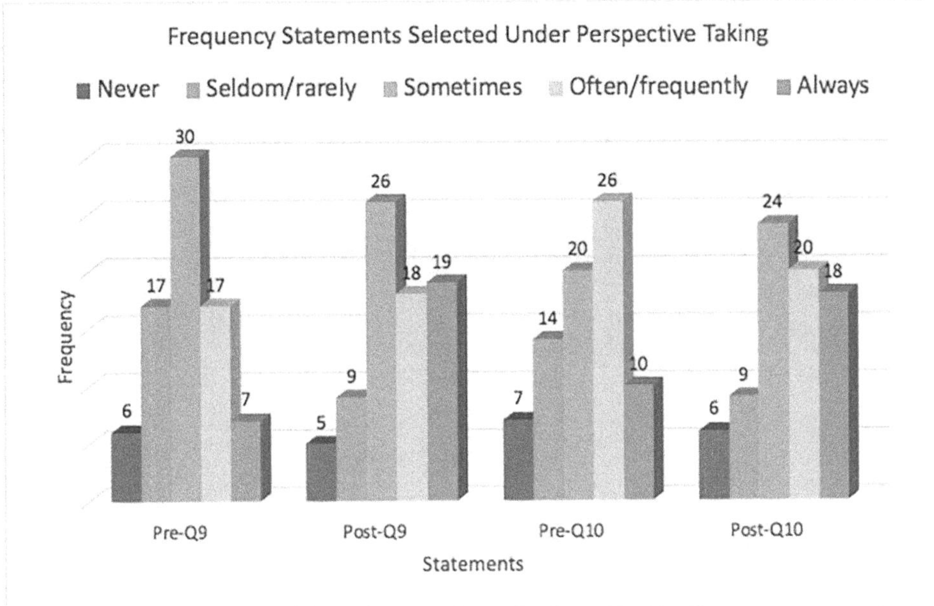

The results from the Wilcoxon Signed-Ranks Test indicated that the median post-test scores were statistically significantly higher than the median pre-test scores Z = -2.505, p = .012.

Moral and Ethical Sensitivity

Results from the Wilcoxon Signed Ranks Test on students' pre/post scores on their moral and ethical sensitivity were measured through CVGCA items 7-13 and were found to be statistically significant (pre-Md 3.00, post-Md 3.00, $Z = -2.448$, $p = .014$, $r = -.27$). The frequency with which students selected these items that inquire about ways research in biotechnology can raise ethical concerns and conflicts are illustrated in Figure 3 below.

Figure 5.

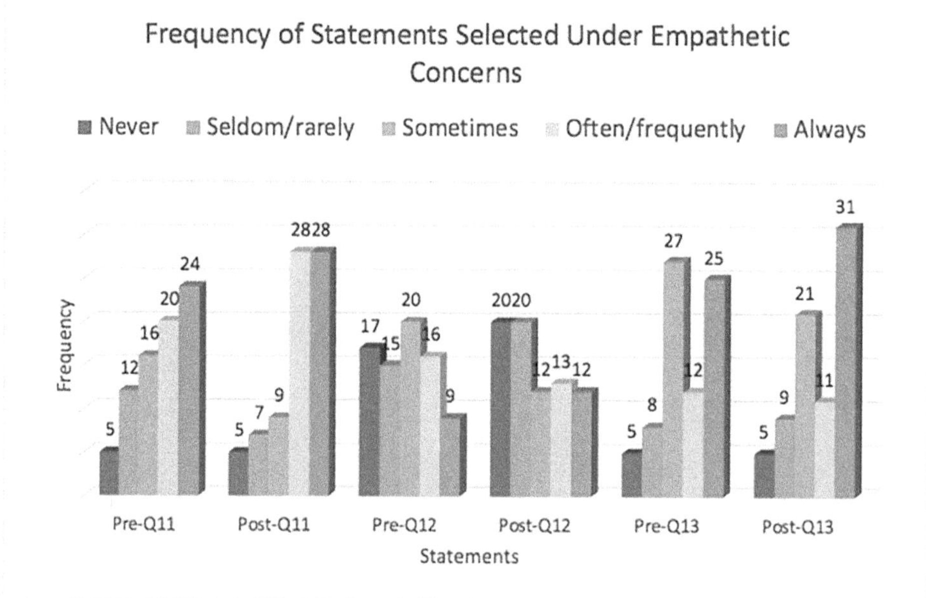

Perspective-Taking

Results from the Wilcoxon Signed Ranks Test on students' perspective taking pre/post scores were measured through CVGCA items 9-10 and were found to be statistically significant (pre-Md 3.00, post-Md 3.50, $Z = -2.471, p = .013, r = .09$). The frequency with which students selected items about considering the diverse opinions and perspectives of those involved when deciding which sides to take on controversial research in biotechnology are illustrated in Figure 4.

Empathetic Concerns

Results from the Wilcoxon Signed Ranks Test on students' empathetic concern pre/post scores were measured through CVGCA items 11-13 and were found not to be

Table 4.

Conceptual Components	Pre-test Median (pre-Md)	Post-test Median (post-Md)	*Z*	*p*	*r*
Socioscientific Accountability	2.86	3.14	-2.381	.017	.264

statistically significant (pre-Md 3.33, post-Md 3.33, $Z = -.494, p = .622, r = .005$). The frequency with which students selected items that inquire about how they feel for others who are suffering from the loss of their pets, agricultural resources, and finances due to the impacts of research in biotechnology (CVGCA items 11-13) are illustrated in Figure 5.

Figure 6.

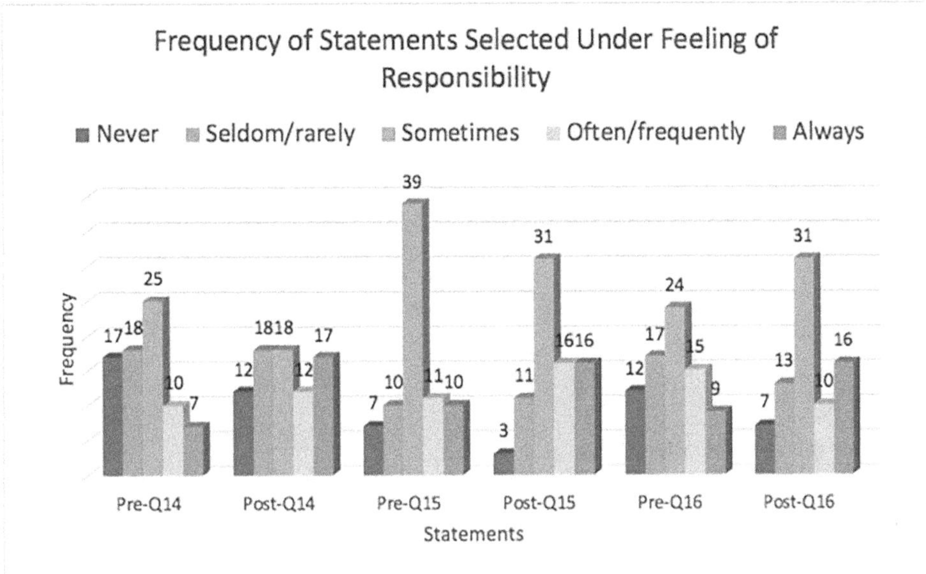

Socioscientific Accountability

The CVGCA survey measured the students' socioscientific accountability through their responses to question related to feelings of responsibility and willingness to act are discussed on issues relating to biotechnology. A Wilcoxon Signed-ranks test was conducted to determine the difference between the students' pre- and post-tests scores. Table 4 below shows the comparison of the pre-CVGCA and post-CVGCA conceptual components.

The results from the Wilcoxon Signed-Ranks Test indicated that the median post-test scores were statistically significantly higher than the median pre-test scores $Z = -2.381, p = .017$

Figure 7.

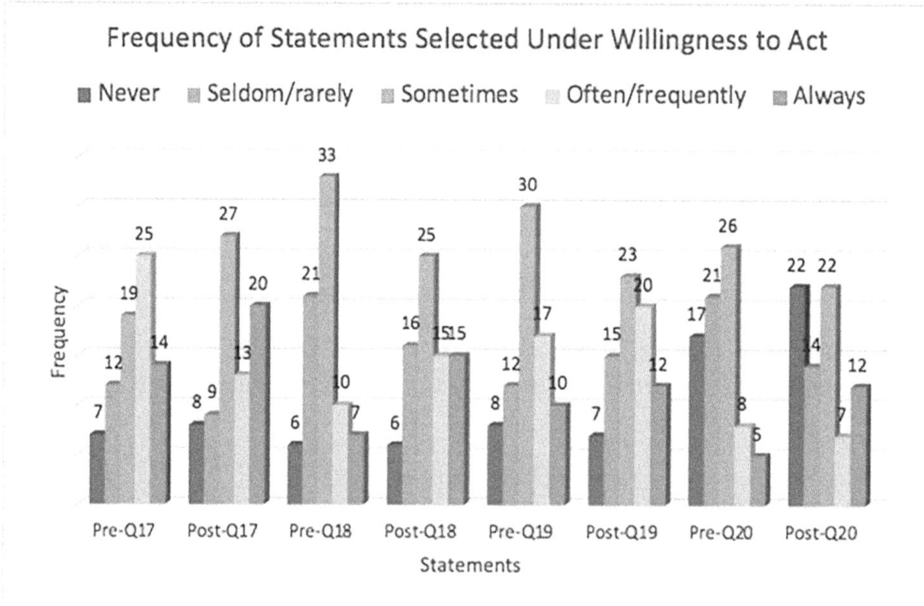

Feeling of Responsibility

Results from the Wilcoxon Signed Ranks Test on students' pre/post scores regarding their feelings of responsibility were measured through CVGCA items 14-16 and were found to be statistically significant (pre-Md 3.00, post-Md 3.00, $Z = -2.952, p = .003, .202$). The frequency with which students selected items that inquire about feeling responsible for causing social and moral issues based on their views regarding research in biotechnology, and if they believe their actions can contribute to solving some of these social and moral issues are illustrated in Figure 6.

Willingness to Act

Results from the Wilcoxon Signed Ranks Test on students' pre/post scores regarding their feelings of responsibility were measured through CVGCA items 17-20 and were found not to be statistically significant (pre-Md 3.00, post-Md 3.00, $Z = -.949, p = .343, r = .177$). The frequency with which students selected items that inquired about their willingness to lead and support efforts to use biotechnology to solve social and health related issues (CVGCA items 14-16) are illustrated in Figure 7.

QUALITATIVE FINDINGS

Ecological Worldview

There were no apparent changes in students' qualitative statements regarding Ecological Worldview. This might be due to students' core beliefs, prior exposures to media reports on biotechnology, and class instructions on the relationships between research in biotechnology and its impact on the balance of nature, health, food production, and wildlife management. For example, research findings on public perceptions of genetically modified foods have found that Americans pay little attention to agricultural biotechnology, do not have much knowledge about agricultural biotechnology, and that their opinion on the acceptability of genetically modified foods is split (Hallman, et al., 2003). There are other studies that found that people in the United States believe the genetically modified foods are bad for the health and well-being of human (Funk, 2020; Kennedy, Hefferon, & Funk, 2018; Anderson, 2015).

Interconnectedness

Qualitative responses indicated a lack of interconnectedness between humans and non-humans. For example, *Scientists should be allowed to cloning [sic] animals because it can result in improvement in our technology to save lives* and *Scientists should be allowed to clone animals because cloning can increase the population of endangered animals.* The first student exemplar demonstrates a belief that science and technology should be used for the betterment of humans and certain species without any consideration for the impact cloning may have on other organisms. Both statements demonstrate a linear thought process where students failed to consider the secondary or tertiary consequences of animal cloning.

Many students opposed to cloning used moral and ethical sensitivity to decide if scientists should be allowed to clone special animals for transplantation purposes. When asked if scientists should clone animals to prevent malnutrition and deaths among infants in low-and-middle income countries, these same students expressed support for cloning. For example:

Cloning should be done to provide more animals that will provide more food for the children who are dying due to malnutrition in low to middle-income countries. Everyone deserves to eat and if they don't have food they should be allowed to clone animals for them to eat since that will help.

This argument underscores the notion of human beings using biotechnology to produce beneficial services for other humans without considering the consequences cloning animals may have on the natural environment. The notion of using clones to provide nutritional supports shows an understanding of the potential use of biotechnology to impact on food production and ultimately the mitigation of human death due to starvation. Repeatedly, the students failed to demonstrate the "man as part of nature" attitude demonstrated by those with high feelings of interconnectedness. As seen in the exemplar above, the student places a greater priority on the survival of humans without considering the impact on the non-human species. These results support the lack of statistical significance in the quantitative data related to interconnectedness.

Sustainable Development

Written responses revealed intriguing results related to sustainable development. For example, during day 4, the students who supported cloning read about the potential applications of cloning, while those who opposed cloning read about the health hazards that are associated with cloning. Students' responses showed a lack of consideration for sustainable development as they considered the appropriateness of cloning animals to increase the food supply. For example, students who favored cloning before and after exposure to the lesson provided comments such as:

- *We are all human beings living on the same planet so we should clone animals until they have enough food to support and prevent deaths. Cloning is not harming anything.*
- *Cloning animals is fine because we kill and eat meat daily for survival, which means we eat the organs of animals.*

These arguments suggested that human beings should use cloning to produce what is needed to support human lives. Students who opposed cloning at the beginning of the lesson but became in favor of the practice gave a similar response.

- *Cloning will produce more animals, which means we will be able to provide more food for these people.*
- *Cloning will provide opportunities for people to feed their children.*

Again, these students considered cloning as a means-to-an-end to alleviate the issue of death of humans due to malnutrition. The students' arguments were void of the possibility of using research in biotechnology to seek to development mutually beneficial solutions for human beings and nature. These results support the lack of statistically significant differences between the students' pre-test and post-test scores that were obtained for Sustainable Development and the students' broader ecological worldview on the CVGCA survey.

Social and Moral Compassion

The majority of changes between the students' pre- and post- statements were associated with the elements of social and moral compassion. The students seem to realize that while there are potential risks that are associated with biotechnology, they also consider many rewards of the applications of biotechnology in providing essential things such as foods for human consumption. As a result, the students' moral and ethical beliefs on whether scientists should be allowed to clone animals created cognitive dissonance. For example, students who stated that their religion teaches against human interfering with God's creation was able to set aside such beliefs and support cloning for food production. These students believed that cloning would provide enough food to people, and thus prevent deaths due to malnutrition.

Moral and Ethical Sensitivity

Inductive analyses on students' responses to the question of whether scientists should clone animals to prevent malnutrition and deaths among infants revealed that students used moral and ethical sensitivity to determine a course of action. The argument below was from a student who was undecided about cloning prior to the lesson, but who supported the practice after the lesson.

- *We should stop the inequality among the rich and poor and clone animals to help these people get enough food. This will prevent unnecessary deaths.*

This student believed that income inequality was the driving force behind the malnutrition deaths of children. This argument is an expression of the student's use of moral and ethical sensitivity to explain why scientists should be allowed to clone animals for food. The student's argument shows how research in biotechnology can raise ethical concerns and conflicts.

Perspective-Taking

Analyses of students' responses to the question of whether scientists should be allowed to clone animals to prevent malnutrition and deaths among infants revealed that even students staunchly opposed to cloning on religious grounds used varying perspective taking when considering this issue. The responses below illustrate this varying perspective taking:

- *My religion teaches that God is the creator of life. Although cloning goes against my religion I imagine how they feel. As a Christian, I believe it is my job to serve and help people when they are in need. Others has helped [sic] us so we should help now in whatever ways we can.*

This student believed that although cloning was against their Christian faith that their duty was to help others in need. The student states they imagined how they would feel if they were dying of malnutrition and so were willing to apply aspects of their religion to support cloning. The statement provides evidence of a moral context and the beginnings of an etic/emic shift that (Author, 2016; 2019) identify as necessary for socioscientific perspective taking to occur.

Empathetic Concerns

Emotions generated as a result of empathy often influence people to become advocates for social justice (Cohen, 2006). For example, students in this investigation were able to put themselves in the shoes of the children dying from malnutrition. These students rejected entrenched religious beliefs and their parent's values to help the children affected by malnutrition. This reaction was common among students who had varying opinions on cloning. For example, a response from students who supported cloning included:

- *If we were in their position, we would want others to help. Clone other things to send as well.*

This student went further and suggested that cloning extend beyond animals to eliminate the issue of death due to malnutrition. Clearly, this student believed we need to use research in biotechnology such as cloning to help those in need.

Other comments from a student who expressed empathy:

- *Because of my love for people, I am able to put myself in their shoes and put aside my religious beliefs.*

The love this student felt for people in general allowed her to identify with those suffering from malnutrition. The student was able to empathize with those who are suffering and suggested that she was willing to put aside personal beliefs and religious views to help those suffering.

Socioscientific Accountability

The public, in general, and school-age children, in particular, must have the basic scientific knowledge to advocate for and engage in actions that are aimed at environmental sustainability. While this is the case, the literature of the public understanding of science over the last seven decades has reported severe deficits in basic scientific knowledge among adults living across many continents (Bauer, 2010). The Committee on the Conceptual Framework for New K-12 Science Education Standards got the responsibility of developing a framework that articulates a broad set of expectations for students in science (NRC, 2012) to reverse this lack of scientific knowledge. The committee's statement of the overarching goal is described below:

...framework for K-12 science education is to ensure that by the end of 12th grade, all students have some appreciation of the beauty and wonder of science; possess sufficient knowledge of science and engineering to engage in public discussions on related issues; are careful consumers of scientific and technological information related to their everyday lives; are able to continue to learn about science outside school; and have the skills to enter careers of their choice, including (but not limited to) careers in science, engineering, and technology (NRC, 2012, p 1).

To fulfill the committee's overarching goal will require students to use the knowledge gained in school science to enhance their sense of responsibility and willingness to engage in actions that are necessary to protect the beauty and wonders of the natural world. The students in this investigation mostly put aside their personal views and advocated for animal cloning to provide foods for people who are at risk of dying from malnutrition.

Feelings of Responsibility

Qualitative analyses were conducted on students' written responses to the question. *Do you believe it is your responsibility to put aside personal views, such as religious beliefs and family values, to support the process of cloning for foods?* Responses, like the one provided below, showed the extent to which students felt responsible for resolving the issue:

- *If your family is against cloning, it is your responsibility to convince them to support cloning. This process will help people from dying of starvation.*

This student believed he was responsible for convincing his parents to support cloning to assist in food production. This argument suggested that this student believed that taking small actions could contribute to resolving social and moral issues such as cloning. The qualitative findings associated with Feelings of Responsibility were consistent with the qualitative responses, both of which indicated a change in students' feelings of responsibility.

Willingness to Act

Researchers have long reported that the main goal of science education is to cultivate students with abilities to act at the personal, societal, and global levels (Lee et al., 2013). To preserve the environment for future generations, students need to become advocates for the environment. They need to champion causes that will preserve drinking water resources, reduce pollutants in the oceans, and prevent toxic gas emissions that are damaging the ozone layer.

While students' CVGCA pre- and post- scores did not yield statistically significant differences, students' qualitative responses identified various actions they would take to help children dying due to malnutrition. The students in this investigation were asked to identify what they would specifically do to support or not to support cloning. The majority of students were able to identify varying ways they will support cloning. For example:

- *I would use social media to advertise and persuade people to help by donating things.*

This student expressed a willingness to act by using social media to inform others about children dying due to malnutrition in low to middle-income countries.

Other examples included:

- *Protest, make commercials, and ask news media to bring awareness to the issue.*

Another student suggested that he would engage in civil disobedience in addition to making television commercials to bring awareness to the cause.

- *Raise money and get people to help transport animals to them.*

This student expressed a willingness to act by engaging in fundraising activities to provide money for transporting animals to be used as food in countries affected by malnutrition.

DISCUSSION

This study provides answers regarding the extent to which middle school students' ecological worldview, social and moral compassion and sense of socioscientific accountability were impacted when considering the role of animal cloning for various purposes as a function of being immersed in an intensive SSI curriculum. The results of this study showed that students' ecological worldview did not change as a result of the SSI intervention. The quantitative data revealed no statistically significant changes in either *Interconnectedness* or *Sustainable Development.* The qualitative data indicated that students consistently provided responses that supported the notion that humans are above nature and neglected to consider the consequences that cloning may have on the environment. Likewise, qualitative responses demonstrated that students failed to discuss resolutions to animal cloning that examined the long-term health of the environment. Some students tacitly mentioned that animal cloning could improve biodiversity and prevent extinction, which we assume is an indication that these students believed that enhancement in species diversity is essential in sustaining ecological systems in the environment. The data overwhelmingly indicated that students prioritized the needs of humans instead of thinking about sustainable practices that are beneficial to both human beings and the environment. For example, the students who support cloning at the beginning of the intervention, as well as those who opposed it, conclude that human beings are above nature and that we should do what it takes to provide enough food to prevent deaths.

The findings related to *Social and Moral* Compassion are more less straight-forward than *Ecological* Worldview. The quantitative data indicated statistically significant changes in students' *Moral and Ethical Sensitivity* and *Perspective Taking*, while there was no statistical change associated with their *Empathetic Concern.* However, qualitative responses indicated students demonstrated empathy when considering those impacted by famine, often while considering the religious impacts related to their decisions. Some of these students indicated increased empathy towards the suffering children and made statements about believing that as a Christian, it was their duty to serve and help those in need. The students stated that others had helped them in their time of need, for example, in disasters such as hurricanes, so they can emphasize and support the practice of cloning to produce food for those who are suffering. Clearly, these statements would suggest that the SSI lessons these students were exposed to create a profound effect on their moral considerations, empathetic concerns, and perspective taking as students evaluated the evidence and made informed decisions on controversial scientific phenomena. These results lend support to the theoretical model of Socioscientific Perspective Taking (SSPT), which suggests that individuals must be engaged with the issue within a moral context and experience an etic/emic shift (Zeidler, Herman, & Sadler, 2019). In the current study, students were clearly engaged in an authentic learning experience (i.e., genetic modification). The data from this study indicated that the students were able to identify the moral context within which animal cloning should be considered and the etic/emic shift that students demonstrated when they considered animal cloning from the perspective of someone who was impacted by famine. This study is among the first to provide empirical evidence to support SSPT.

Similar to *Social and Moral Compassion*, the results related to *Socioscientific Accountability* provided conflicting results. The quantitative data indicated statistically significant changes in students' *Feelings of Responsibility* while no significant changes to their *Willingness to Act* were reported. However, the qualitative data revealed that many of the students in this study who were opposed to cloning animals were able to put aside personal beliefs and expressed support for the practice of cloning to prevent deaths due to starvation. Several of these students also stated that they felt it was their responsibility to support the practice since it would provide food. When these students were asked what specific actions they would take to help those in need, they gave examples such as donating food, spreading the word via social media about those who are dying, petitioning their government for help, or even helping with the cloning process themselves.

The idea of middle school students expressing a willingness to raise money to help with the transportation of animals to those affected by malnutrition, engaging in protests to bring awareness to the cause, making informative commercials, and asking news media to bring awareness to the issue of animal cloning highlight the potential impact that SSI may have on students' feelings of responsibility. However, we should be cautious with these students' responses, since other research related to pro-environmental behaviors has shown that students, in general, are often reluctant to engage in meaningful actions to sustain the environment.

Resolving contentious issues that include, but are not limited to, climate change, environmental sustainability, disease pandemic, and where to build natural gas pipelines will require citizens to make informed decisions that are based on scientific evidence. For example, the closure of the United States economy due to the COVID-19 pandemic and the subsequent push by many politicians to re-open the economy against the advice of the scientific community is a case in point of why students need to be exposed to SSI in the classroom settings. Such exposure to SSI will develop students' abilities to engage in evidence-based reasoning and informed decision making. The practice of exposing students to SSI will also enable the enhancement of their moral and ethical decision-making. This paramount in developing character and values which are needed for global citizens.

The interventions used and the findings presented in this study could serve as a blueprint for teachers in the K-12 system who are interested in using SSI instructions in their classroom settings. We hope that teachers in the K-12 system use the SSI framework presented in this chapter to develop their students' character and values as global citizens.

REFERENCES

Aikenhead, G., Orpwood, G., & Fensham, P. (2011). Scientific literacy for a knowledge society. In Exploring the landscape of scientific literacy. New York, NY: Routledge / Taylor & Francis Group.

Batson, C. D., Early, S., & Salvarani, G. (1997). Perspective taking: Imagining how another feels versus imagining how you would feel. *Personality and Social Psychology Bulletin*, *68*, 619–631. doi:10.1037/0022-3514.68.4.619

Bauer, M. W. (2009). The evolution of public understanding of science - discourse and comparative evidence. *Science, Technology & Society*, *14*(2), 221–240. doi:10.1177/097172180901400202

Brito, A. J., de Almedia, A. T., & Mota, M. M. C. (2009). A multicriteria model for risk sorting of natural gas pipelines based on ELECTRE TRI integrating Utility Theory. *European Journal of Operational Research*, *200*(3), 812–821. doi:10.1016/j.ejor.2009.01.016

Chokker, K., Dua, S., Taylor, N., Boyes, E., & Stanisstreet, M. (2011). Indian secondary students' views about global warming: Beliefs about the usefulness of actions and willingness to act. *International Journal of Science and Mathematics Education*, *9*(5), 1167–1188. doi:10.100710763-010-9254-z

Cohen, J. (2006). Social, emotional, ethical, and academic education: Creating a climate for learning, participation in democracy, and well-being. *Harvard Educational Review*, *76*(2), 201–237. doi:10.17763/haer.76.2.j44854x1524644vn

Dewey, J. (1975). Moral principles in education. Carbondale: Southern Illinois University Press. (Originally published 1909)

Driver, R., Leach, J., Millar, R., & Scott, P. (1996). *Young people's images of science*. Open University Press.

Du Plessis, C. (2009). *An approach to studying urban sustainability from within an ecological worldview* (Unpublished doctoral thesis). University of Salford.

Du Plessis, C., & Brandon, P. (2015). An ecological worldview as basis for a regenerative sustainability paradigm for the built environment. *Journal of Cleaner Production, 109*, 53–61. doi:10.1016/j.jclepro.2014.09.098

Dunlap, R. E., Van Liere, K. D., Mertig, A. G., & Jones, R. E. (2000). Measuring endorsement of the new ecological paradigm: A revised NEP scale. *The Journal of Social Issues, 56*(3), 425–442. doi:10.1111/0022-4537.00176

Eisenberg, N., Fabes, R., Murphy, B., Karbon, M., Maszk, P., Smith, M., O'Boyle, C., & Suh, K. (1994). The relations of emotionality and regulation to dispositional and situational empathy-related responding. *Journal of Personality and Social Psychology, 66*(4), 776–797. doi:10.1037/0022-3514.66.4.776 PMID:8189352

Feldman, M. P., Morris, M. L., & Hoisington, D. (2000). Why All the Controversy? *Choice (Chicago, Ill.), 15*(1), 8–12.

Fisher, D. R., Waggle, J., & Leifeld, P. (2012). Where Does Political Polarization Come From? Locating Polarization Within the U.S. Climate Change Debate. *The American Behavioral Scientist*, 1–23.

Green, T. F. (1985). The formation of conscience in an age of technology. *American Journal of Education, 94*(1), 1–32. doi:10.1086/443829

Haban, T. (2004). *Public Attitude towards Agricultural Biotechnology*. ESA Working Paper.

Herman, B. C., Sadler, T. D., Zeidler, D. L., & Newton, M. (2018). A socioscientific issues approach to environmental education. In G. Reis & J. Scott (Eds.), *International perspectives on the theory and practice of environmental education: A reader* (pp. 145–161). Springer International Publishers. doi:10.1007/978-3-319-67732-3_11

Hermans, M., & Korhonen, J. (2017). Ninth graders and climate change: Attitudes towards consequences, views on mitigation, and predictors of willingness to act. *International Research in Geographical and Environmental Education, 26*(3), 223–239. doi:10.1080/10382046.2017.1330035

Hopwood, B., Mellor, M., & O'Brien, G. (2005). Sustainable Development: Mapping different approaches. *Sustainable Development, 13*(1), 38–52. doi:10.1002d.244

Knight, A. J. (2006). Does application matter? An examination of public perception of agricultural biotechnology applications. *AgBioForum, 9*(2), 121–128.

Kolstø, S. D. (2001a). Scientific literacy for citizenship: Tools for dealing with the science dimension of controversial socioscientific issues. *Science Education, 85*(3), 291–310. doi:10.1002ce.1011

Korpan, C., Bisanz, G., Bisanz, J., & Henderson, J. (1997). Assessing literacy in science: Evaluation of scientific news briefs. *Science Education, 81*(5), 515–532. doi:10.1002/(SICI)1098-237X(199709)81:5<515::AID-SCE2>3.0.CO;2-D

Lee, H., Yoo, J., Choi, K., Kim, S., Krajcik, J., Herman, B., & Zeidler, D. L. (2013). Socioscientific Issues as a Vehicle for Promoting Character and Values for Global Citizens. *International Journal of Science Education, 35*(12), 2079–2113. doi:10.1080/09500693.2012.749546

Loureiro, L. M., & Bugbee, M. (2005). Enhanced GM foods: Are consumers ready to pay for the potential benefits of biotechnology? *The Journal of Consumer Affairs, 39*(1), 52–70. doi:10.1111/j.1745-6606.2005.00003.x

McBeth, W., & Volk, T. L. (2009). The National Environmental Literacy Project: A baseline study of middle grade students in the United States. *The Journal of Environmental Education, 41*(1), 55–67. doi:10.1080/00958960903210031

McCright, A. M., & Dunlap, R. E. (2011). Cool dudes: The denial of climate change among conservative white males in the United States. *Global Environmental Change, 21*(4), 1163–1172. doi:10.1016/j.gloenvcha.2011.06.003

Miller, J. (1983). Scientific Literacy: A Conceptual and Empirical Review. *Daedalus, 112*, 29–48.

Miller, J. (1998). The measurement of civic scientific literacy. *Public Understanding of Science (Bristol, England), 7*(3), 203–223. doi:10.1088/0963-6625/7/3/001

Miller. (1987). Scientific Literacy in the United States. In *Communicating Science to the Public*. London: Wiley.

Miller. (1995). Scientific Literacy for Effective Citizenship. In *Science/Technology/Society as Reform in Science Education*. New York: State University Press of New York.

National Research Council (NRC). (2012). *A framework for k-12 science education: Practices, crosscutting concepts, and core ideas*. The National Academy Press.

NGSS Lead States. (2013). *Next generation science standards: For states, by states*. The National Academy Press.

Pew Initiative on Food and Biotechnology. (2000). *Public Sentiment about Genetically Modified Food.* Available at: http://www.pewagbiotech.org

Pew Research Center. (2015). *Amid debate over labeling GM foods, most Americans believe they're unsafe*. Retrieved from http://pewrsr.ch/1MhCFSN

Pew Research Center. (2018). *Americans are narrowly divided over health effects of genetically modified foods*. Retrieved from https://pewrsr.ch/2FDG8RB

Pew Research Center. (2020). *About half of U.S. adults are wary of health effects of genetically modified foods, but many also see advantages*. Retrieved from https://pewrsr.ch/2w5UmGU

Ratcliffe, M. (1999). Evaluation of abilities in interpreting media reports of scientific Research. *International Journal of Science Education, 21*(10), 1085–1099. doi:10.1080/095006999290200

Redclift, M. (2005). Sustainable Development (1987–2005): An Oxymoron Comes of Age. *Sustainable Development, 13*(4), 212–227. doi:10.1002d.281

Sadler, T. (2004). Moral sensitivity and its contribution to the resolution of socio-scientific issues. *Journal of Moral Education, 33*(3), 339–358. doi:10.1080/0305724042000733091 PMID:16317839

Sadler, T. (2004). Informal reasoning regarding socioscientific issues: A critical review of the research. *Journal of Research in Science Teaching, 41*(5), 513–536. doi:10.1002/tea.20009

Sadler, T. (2007). *The aims of science education: Unifying the fundamental and derived senses of scientific literacy*. Paper presented at the Linnaeus Tercentenary 2007 Symposium "Promoting Scientific Literacy," Uppsala University, Uppsala, Sweden.

Sadler, T. (2011). Situating socio-scientific issues in classrooms as a means of achieving the goals of science education. In T. D. Sadler (Eds.), Socio-scientific issues in the classroom (pp. 1-9). New York, NY: Springer. doi:10.1007/978-94-007-1159-4_1

Sadler, T., Klosterman, M., & Topcu, M. (2011). Learning science content and socio-scientific reasoning through classroom explorations of global climate change. In T. D. Sadler (Ed.), *Socio-scientific issues in the classroom: Teaching, learning and research* (pp. 45–77). Springer. doi:10.1007/978-94-007-1159-4_4

Shamos, M. (1995). *The myth of scientific literacy*. Rutgers University Press.

Siebert, E. D., & McIntosh, W. J. (Eds.). (2001). *College pathways to the science education standards*. NSTA Press.

Skamp, K., Boyes, E., & Stannistreet, M. (2009). Global warming responses at the primary secondary interface 1. students' beliefs and willingness to act. *Australian Journal of Environmental Education, 25*, 15–30. doi:10.1017/S0814062600000379

Sklavounos, S., & Rigas, F. (2006). Estimation of safety distances in the vicinity of fuel gas pipeline. *Journal of Loss Prevention in the Process Industries*, *19*(1), 24–31. doi:10.1016/j.jlp.2005.05.002

Smith, G. A., & Williams, D. R. (1999). *Ecological education in action: On weaving education, culture, and the environment*. State University of New York Press.

UNESCO. (2019). *Science for a sustainable future*. Retrieved from https://en.unesco.org/themes/science-sustainable-future

World Health Organization. (2000). *Turning the tide of malnutrition: responding to the challenge of the 21st century*. WHO.

World Health Organization (WHO). (2010b). *The World Health Report - Health Systems Financing: The Path to Universal Coverage*. Retrieved from http://www.who.int/entity/whr/2010/whr10_en.pdf

Zeidler, D., Sadler, T., Applebaum, S., & Callahan, B. (2009). Advancing reflective judgment through socioscientific issues. *Journal of Research in Science Teaching*, *46*(1), 74–101. doi:10.1002/tea.20281

Zeidler, D., Walker, K., Ackett, W., & Simmons, M. (2002). Tangled up in views: Beliefs in the nature of science and responses to socioscientific dilemmas. *Science Education*, *3*(86), 343–367. doi:10.1002ce.10025

Zeidler, D. L. (2007). *An inclusive view of scientific literacy: Core issues and future directions*. Paper presented at "Promoting Scientific Literacy: Science Education Research and Practice in Transaction," LSL Symposium, Uppsala, Sweden.

Zeidler, D. L., Applebaum, S. M., & Sadler, T. D. (2011). Enacting a socioscientific issues classroom: Transformative transformations. In T. D. Sadler (Ed.), *Socio-scientific issues in science classrooms: Teaching, learning and research* (pp. 277–306). Springer. doi:10.1007/978-94-007-1159-4_16

Zeidler, D. L., Berkowitz, M. W., & Bennett, K. (2013). Thinking (scientifically) responsibly: the cultivation of character in a global science education community. In *Assessing schools for generation R (responsibility), contemporary trends and issues in science education 41* (pp. 83–99). Springer.

Zeidler, D. L., Herman, B. C., & Sadler, T. D. (2019). New directions in socioscientific issues research. *Disciplinary and Interdisciplinary Science Education Research*, *1*(11), 1–9. doi:10.118643031-019-0008-7

Zeidler, D. L., & Kahn, S. (2014). *It's debatable: using socioscientific issues to develop scientific literacy, K-12*. NSTA Press.

Zeidler, D. L., & Keefer, M. (2003). The role of moral reasoning and the status of socio-scientific issues in science education. In D. L. Zeidler (Ed.), *The role of moral reasoning on socio- scientific issues and discourse in science education* (pp. 7–38). Kluwer Academic Publishers.

Zeidler, D. L., & Sadler, T. D. (2008). The role of moral reasoning in argumentation: conscience, character and care. In S. Erduran & M. Pilar Jimenez-Aleixandre (Eds.), *Argumentation in science education: perspectives from classroom-based research* (pp. 201–216). Springer Press.

Zeidler, D. L., Sadler, T. D., Simmons, M. L., & Howes, E. V. (2005). Beyond STS: A research-based framework for socioscientific issues education. *Science education, 89*(3), 357–377. doi:10.1002ce.20048

APPENDIX

Supplementary Material. Character and Values as Global Citizens Assessment (CVGCA)

Please indicate the extent that you agree with the following statements by circling the number associated with the following descriptions.

1= never, 2 = seldom/rarely, 3 = sometimes, 4 = often/frequently, 5 = always

Ecological Worldview

Interconnectedness (I)

I believe scientific research in biotechnology probably will not disrupt the balance in nature. 1 2 3 4 5

I believe research in biotechnology will have long-term impact on health, food production and wildlife management. 1 2 3 4 5

If human beings use research in biotechnology to manipulate and change nature for their benefit, it might cause devastating results. 1 2 3 4 5

Sustainable Development (S)

I believe human beings should use research in biotechnology to produce what they want to sustain life and the environment. 1 2 3 4 5

I believe we have to manage natural resources within the scope of not tempering with natural ecosystems. 1 2 3 4 5

I believe it is possible to use research in biotechnology to seek development mutually beneficial for human beings and nature. 1 2 3 4 5

Social and Moral Compassion

Moral and Ethical Sensitivity (M)

I believe research in biotechnology (ex: cloning, stem cell research) can raise ethical concerns and conflicts. 1 2 3 4 5

I can predict the social, ethical and moral impacts that research in biotechnology might cause. 1 2 3 4 5

Perspective-taking (P)

I try to consider the diverse opinions and perspectives of those involved, when deciding which sides to take on controversial research in biotechnology. 1 2 3 4 5

I try to think in others' perspectives to imagine, "What if I were in their situation?" before criticizing different opinions on research in biotechnology. 1 2 3 4 5

Empathetic Concerns (E)

I genuinely feel sorry for those who suffer (ex: from loss of pets, agricultural resources and finances) from the results of research in biotechnology. 1 2 3 4 5

I am unable to feel like it is my own experience when I see those who are suffering (ex: from loss of pets, agricultural resources and finances) from the impacts of research in biotechnology. 1 2 3 4 5

I believe we have to take care of (e.g. provide financial compensation, cultural preservation) those who suffer (e.g. from loss of pets, agriculture and finances) because of results from research in biotechnology. 1 2 3 4 5

Socioscientific Accountability

Feeling of Responsibility (R)

1. I do not feel responsible for causing social and moral issues based on my views regarding research in biotechnology.

I believe small actions I take can contribute to resolving social and moral issues regarding research in biotechnology. 1 2 3 4 5

1 2 3 4 5

I am willing to endure personal inconveniences in order to resolve social and moral issues caused by research in biotechnology because I feel responsible for those issues. 1 2 3 4 5

Willingness to Act (W)

I believe cooperation and supports from members of community I live to use research in biotechnology are required to solve social and health related issues (ex: degenerative diseases, cancers, infectious diseases). 1 2 3 4 5

I will participate in supporting the resolution of social issues related to research in biotechnology at the national and global levels (e.g. through inter-country cooperation and international convention). 1 2 3 4 5

I will make efforts to initiate community movements and communicate with community members to resolve social issues related to research in biotechnology. 1 2 3 4 5

I do not believe global cooperation and support from various nations are required to solve social issues related to research in biotechnology. 1 2 3 4 5

Chapter 4
A Socioscientific Issue Approach to Understanding Middle School Students' Beliefs and Intentions Toward Climate Change

Wardell A. Powell
https://orcid.org/0000-0002-3486-2077
Framingham State University, USA

ABSTRACT

This chapter is a demonstration of how to use socioscientific issues to impact middle school students' beliefs and intentions towards climate change. Fifty-one middle school students from a summer enrichment program in the Northeastern United States participated in this study. The duration of this curricular unit took place over six consecutive one-hour class periods. The researcher utilized quantitative and qualitative procedures to analyze the students' abilities to think critically and to argue persuasively about their beliefs and intentions about climate change. The results indicate that the students concluded that human actions are a significant factor in climate change. The students' intentions to act, as well as their desire to encourage others to take actions necessary to mitigate climate change, were compelling. Additionally, the knowledge the students gained from the interventions used enhanced their abilities to write persuasively to the chief executive officers from power plants and waste treatment facilities to a round table discussion on ways to mitigate climate change.

DOI: 10.4018/978-1-7998-4558-4.ch004

INTRODUCTION

Finding solutions to societal problems, such as climate change, will require citizens to understand two fundamental principles. One, they must first believe that the climate is indeed changing, and two, they must have the intention to act upon the known causes of climate change. Scientific literacy is central to helping such citizenry first to believe, and then to adequately act upon climate change. In its quest to determine what is essential for citizens to know and be able to do, the Programme for International Student Assessment (PISA) assesses the extent to which 15-year-old students have acquired key knowledge and skills that are essential for full participation in modern societies (Organization for Economic Co-operation and Development (OECD), 2015). One would agree that the P-12 system cannot wait until students become 15 years of age before providing them with opportunities in the classroom settings to develop scientific literacy. To solve scientific problems of the 21st-century world and beyond will require students to be exposed to activities that are aimed at improving their scientific literacy from day one. Interestingly, the term scientific literacy has gone through several iterations in its definition since it was first introduced in the late 1950s (American Association for the Advancement of Science [AAAS], 1989; Bybee, 1997; Holbrook & Rannikmae, 1997; Hurd, 1958; Laugksch, 2000; National Science Education Standards [NSES], 1996; Organisation for Economic Cooperation and Development [OECD], 2003; 2006; 2007; 2009, & 2012). However, one thing that seems evident is that a scientifically literate population matters more and more as humanity faces significant fallout from challenges such as climate change (UNEP, 2012). For example, the recent climate science special report (USGCRP, 2017), states that global annual averaged temperature measured over both land and oceans has increased by about 1.8°F (1.0°C) from 1901 to 2016, and by 1.2°F (0.65°C) for the period 1986–2015 as compared to 1901–1960. The report further states that the last few years have also seen record-breaking, climate-related weather extremes. These events from global climate change will create problems for local communities, and as a result, individuals will be faced with decisions about actions to take to combat these problems. Such actions will require individuals to have particular scientific awareness (European Commission, 1995:28).

The concept of scientific literacy is critical in developing a populace with the ability to make informed decisions on scientific and technological issues. According to the recent PISA report, individuals who are scientifically literate must have a knowledge of the major concepts and ideas that form the foundation of scientific and technological thought; how such knowledge has been derived; and the degree to which such knowledge is proved by evidence or theoretical explanations (OECD, 2015). Expectedly, the challenges created by climate change will require innovative solutions that have a basis in scientific thinking and discovery (OECD, 2015). While on the one hand, many would agree that an abundance of well-educated scientists are needed to find solutions to climate change and the crises it creates, on the other hand, many would also agree that school science must prepare current and future students to become some of these well-educated scientists. For these reasons, the notion of scientific literacy should be of paramount importance for the P-12 system.

This study focuses on a socioscientific issues curricular unit of instruction that uses scaffolds to enhance middle students' beliefs in climate change, and their intentions to enact positive environmental change aimed at combating climate change. The primary goal of this research is to demonstrate for science education researchers, policymakers, and practitioners how to use socioscientific issues to promote middle school students' ability to become the change agent for their communities.

THEORETICAL FRAMEWORK

What is Scientific Literacy?

In his review of the literature on scientific literacy that has developed since the 1950s, Roberts (2007) proposed two broad overall visions of scientific literacy, which he termed vision 1, and vision 2. Vision 1 is aimed at enabling students to build knowledge and skills that will empower them to think about scientific situations as a professional scientist (Roberts, 2007, 2014). Professional scientists worked towards solving real-world scientific problems, and so the aim of school science with this vision is to prepare future scientists who would study more advance science (Roberts, 2007, 2014). According to Roberts, Vison 1 looks inward at science, to build the curriculum from its rich and well-established array of techniques and methods, habits of mind, and well-tested explanations for the events and objects of the natural world (Roberts, 2104, p. 546). Roberts further states that vision 2 looks outside of science to build the curriculum that illuminates how science permeates and interacts with many areas of human endeavor and life situations (Roberts, 2104, p. 546). Societal problems such as climate change, where to build landfills, how to protect drinking water resources, in whose communities should natural gas pipelines be constructed, deforestation, genetically modified crops, and ocean acidification, etc., are some of the examples of how science permeates and interact with our life. As such, these are some of the topics that the school science curriculum should be built around. This approach to building the school science curriculum is paramount in developing the next generation of scientists who can help in solving these environmental issues. Socioscientific issues as part of the school science curriculum are pivotal in helping us to achieve the type of scientific literacy advocated in vision 2.

The type of scientific literacy that has been advocated by PISA's 2015 Assessment and Analytical Framework aligns with Roberts (2007, 2014) vision 2 of scientific literacy. In the report, PISA (2015) defines scientific literacy is the ability to engage with science-related issues, and with the ideas of science, as a reflective citizen (PISA, 2015, p. 22). PISA definition further states that:

A scientifically literate person is willing to engage in reasoned discourse about science and technology, which requires the competencies to:

- Explain phenomena scientifically – recognize, offer, and evaluate explanations for a range of natural and technological phenomena.
- Evaluate and design scientific inquiry – describe and appraise scientific investigations and propose ways of addressing questions scientifically.
- Interpret data and evidence scientifically – analyze and evaluate data, claims, and arguments in a variety of representations and draw appropriate scientific conclusions (PISA, 2015, p. 22).

The above definition of scientific literacy does require teachers to provide students with opportunities to engage in the scientific process, and curriculum developers to build curriculum that illuminate how science permeates and interacts with many areas of human endeavor and life situations (Roberts, 2104, p. 546). Finding solutions to everyday social issues with moral or ethical implications embedded within scientific contexts (Author, 2014; Yap, 2014; Zeidler, 2007) does require individuals to have competencies in explaining phenomena scientifically. Students' ability to evaluate and analyze data, claims, explanations, design scientific inquiry, and interpret data and evidence scientifically are the types of competencies that are required for them to question claims from elected officials who believe that scientific issues such as climate change is a hoax.

What are Socioscientific Issues?

I have argued elsewhere (Author, 2020, Author, 2019, Author, 2014) that socioscientific issue seeks to involve students in discourse opportunities on everyday scientific problems that are undergirded with moral and ethical implications. The denial of the anthropogenic effects of climate change from many of our elected officials, particularly in the United States, has created many challenges to teaching about climate change in the K-12 system. Frequently, the discussion of climate change is seen by many as a discourse in politics, although students are presented with historical data of the increase in greenhouse gases in the atmosphere, the increase in global temperature, rise in sea level, etc. The current push by elected officials to rolling back of environmental regulations, pulling the United States out of the

Paris climate accord, and the practices of turning the term "global warming" into a punch line rather than a prognosis (Davenport & Landler, 2019), has sowed many seeds of disbelief and inactions in finding solutions to climate change and the fallout it creates. To reverse this trajectory, school science must provide opportunities for students to become the next generation of policymakers, teachers, and scientists, etc., to learn how to analyze, synthesize, and evaluate information (Author, 2020). Socioscientific issues as part of the school science curriculum will help to enhance students' beliefs and intentions to act towards finding solutions to climate change and the crises it creates.

Adolescents Beliefs and Intentions about Climate Change

Reaching consensus on what influences adolescents' beliefs and intentions about climate change have not been reached among researchers in the science education community. Recent studies have reported that communication between adolescents and authority figures, such as scientists, teachers, and families, has impacted adolescents' engagement with climate change issues (Arnold, Cohen, & Warner 2009; Feldman et al. 2010; Vreede Warner, & Pitter 2017). In a study that investigated middle school students' climate change behavior and concerns in the family, Lawson et al., (2019) reported that family discussions about climate change impacted parent behaviors towards climate change and children's concerns about climate change. The study found that family discussions are strong are predictors of the degree to which children will engage in activities to mitigate climate change. Results from other studies have also reported that children whose families are concerned about the human-induced cause of climate change are generally concerned themselves (Mead et al. 2012; Stevenson et al. 2016). The results from other studies that have investigated adolescents' beliefs and concerns about climate change have reported greater consensus in students' climate change beliefs and concerns regardless of worldview (Bofferding and Kloser 2015; Flora et al. 2014; Reinfried, Aeschbacher, and Rottermann 2012; Stevenson et al. 2014). While these results are encouraging, the findings from other studies have reported that students, as well as the general public, often have alternative conceptions about climate change (Leiserowitz & Smith, 2010; Shepardson, Choi, Niyogi, & Charusobat, 2011). Some of these alternative conceptions include, but are not limited to, increases in solar irradiance (Shepardson et al., 2011), hole in the ozone (Osterlind, 2005), and pollution other than greenhouse gas emission (Papadimitriou, 2004). These alternative beliefs about what is the cause behind climate change are indicators of the complex nature of the science behind climate change. To address the alternative conception that increasing amounts of solar irradiation is the cause of climate change, Lombardi, Sinatra, and Nussbaum (2013) conducted a study with one hundred and sixty-nine

(169) middle school students in the Southwestern United States. In this study, the students in the treatment group were exposed to an instructional activity that promotes critical evaluation of two competing climate change models, while the students in the comparison group were exposed to the regular science curriculum materials that discuss climate change. Lombardi et al., (2013), then followed up with a delayed post instruction phase six months after the students were first exposed to the treatment. Lombardi et al. (2013) reported that the students in the treatment group enhanced their ability to critically evaluate evidence, which resulted in a higher scientifically accurate conception of climate change than alternative conception. The study further finds that the students in the treatment group maintained these changes even six months after instruction. The results from this investigation show that carefully crafted instruction can result in changes to students' alternative conceptions of the cause of behind climate change.

While it is essential to see that the science curriculum can impact students' alternative conception of climate change, it's equally important that school science also affects students' intentions towards acting to mitigate climate change. Studies have reported that while students might enhance their understanding of the causes and consequences of climate, there is no clear link between students' knowledge about the environment, and their willingness to engage in actions that are aimed at mitigating climate change (Kollmus & Agyeman, 2002; Ozdem et al., 2014). For example, Skamp, Boyes, and Stanissstreet (2004) conducted a large scale (N=1001) study of middle and high school students' ideas about the composition of unpolluted air, the nature of air pollution, the biological consequences of air pollution, acid rain, and greenhouse effects. These authors reported that while the students agreed to environmental statements regarding taxation and legislation of companies and others who pollute, the students disagreed with these statements when they were directed at them. The students believe that reducing air pollution is the responsibility of others, rather than themselves. This earlier finding is similar to the reports of recent studies (Patterson, 2014; Ozdem et al., 2014). While other studies have reported that students believe they cannot personally influence changes to the environment (Connell, 1999; Jenkins & Pell, 2006), the results from a recent study reported that students were more willing to engage in actions necessary to mitigate climate change to a certain degree (Hermans and Korhonsen (2017). The students in this study reported that they were willing to turn off lights and electrical appliances when not in use. The study finds that while about 50% of the students were willing to sell things at secondhand and to walk or ride bikes moderate distances, they were less willing to purchase items second and to refrain from driving in motor vehicles. This study shows the dilemma of engaging in actions to reduce greenhouse gas emissions.

CURRENT STUDY

The study aims to examine the effects of a socioscientific issues unit of instruction on middle school students' beliefs and intentions to act toward climate change. The following research questions and corresponding rationale are addressed:

1. To what degree might the use of socioscientific issues instruction impact middle school students' beliefs about climate change?

The research question examined students' beliefs about climate change. This question is important because if students don't believe in human-induced climate change, then they are generally less likely to engage in environmental actions to mitigate climate change. When educators know what their students believe in, such knowledge can become useful in providing learning opportunities that are aimed at addressing misconceptions that might be influenced by students' beliefs. A recent study conducted by Leiserowitz et al. (2017) reported that a mere 12% of the American public is aware that over 90% of scientists believe in human-induced climate change. This misconception on the part of the public is referred to as a consensus gap (Cook, 2019).

Interestingly, this consensus gap exists between science teachers (Plutzer et al., 2016) as well as journalists (Wilson, 2000). In order to reduce this gap, school science must provide more opportunities for students to identify the misconceptions in the beliefs about climate change. This is imperative in helping students to engage in environmental actions that are aimed at mitigating climate change.

2. What roles might socioscientific issues instruction play in middle school students' intentions to mitigate climate change?

Research has discussed the term action in several different ways (Vesterinen, Tolppanen, & Aksela, 2016). For example, Jensen and Schnack (1997) define action as a student's ability to act and address unforeseen situations. Mogensen and Schnack (2010) see action as that which is influenced by intentions, motives, and reasons. Vesterinen et al. (2016) define action as intentional, conscious, and purposive behavior and that such definition of the notion of action links it to humanistic concepts of personhood, responsibility, democracy, and education (Vesterinen, 2016, p. 34). For students to have the intention to engage in environmental actions that are aimed at mitigating climate change will require students to understand the purpose of stewardship to our environment. The use of socioscientific issues as a critical pedagogical strategy might play an essential role in motivating students to engage in actions to curb known behaviors that contribute to the climate change crisis.

METHODS

Participants

The study took place with 51 middle school students (11–12-years-old; 25 males, 26 females) in 2018 at a summer enrichment program operated by a non-profit organization in the Northeastern United States. The purpose of this program is to provide enrichment opportunities in science, mathematics, reading, and writing for students in underserved communities. The program prepared students for acceptance to the top independent, Catholic, and public exam schools in the region. The students are generally admitted to the program in the spring before their fifth or sixth-grade year. These students were from 40 distinct feeder schools in the region. Seventy-seven percent of the students attended public schools, nineteen percent attended charter schools, and four percent attended parochial schools. The racial breakdown of the students is as follows: 21% Asian, 32% Black, 32% Hispanic, 5% Multiracial, 4% White, and 6% other. The average family income was $43,586, and fifty-five percent of the students qualify for free and reduced meals.

Study Design

This study employed the use of a mixed methods quasi-experimental design approach by using students enrolled in three science classes in a summer enrichment program operated by a non-profit organization in the northeastern United States. Pre- and post-test quantitative data were collected with the use of a 5-point Likert scale climate change attitude survey (Christensen & Knezek, 2015). Qualitative data were collected from students' responses on open-ended questionnaires, students' interpretation of changing volume in greenhouse gases over the different time frames, carbon footprints calculations, video discussions, and persuasive writing activities over five consecutive days. Both quantitative and qualitative data were gathered to better understand the effects of a socioscientific issues curricular unit of instruction on the students' beliefs and intentions towards climate change.

Data Collection and Analysis

The students' pre and post-tests responses to the climate change attitude survey (Christensen & Knezek, 2015) were analyzed using a paired t-test quantitative method. Each participant's response to this survey was categorized using a ranking method where 0 meant strongly disagree; one meant disagree, two undecided, three agree, and four strongly agree. The paired t-test was employed because it is ideal for comparing before-and-after observations on the same subject (Wilkerson, 2008). The overall alpha was set at 0.05.

Qualitative data for this study was collected using written texts in which the students expressed their thoughts regarding their beliefs in climate change, and their intentions to actions toward climate change. The students were asked to express their opinions in written format on designated topics before and after exposure to various instructional activities on greenhouse gases. Two graduate students with experience in data analysis were involved with the coding and analysis of the students' data.

During the qualitative data analysis process, the students' written responses were read line by line by each graduate student to determine any significant dimensions of the raw data. The inductive approach to analyzing qualitative data was used to generate code lists of data (Strauss and Corbin, 1998). All codes obtained were revised to reveal connections between them so that they could be categorized and classified. After both graduate students carried out this process separately, they presented the categories they generated to the author for evaluation. After discussion and consensus, the codes were then finalized.

Findings

Day 1

The Climate Change Attitude Survey (Christensen & Knezek, 2015) was administered before and after the socioscientific issues climate change unit of instruction. This survey is composed of 15 Likert-type attitudinal items selected to measure students' beliefs and intentions toward the environment with a focus on climate change. A paired-samples t-test was conducted to compare the students' beliefs and intentions to enact positive environmental change before exposure to the socioscientific issues unit on climate change and after exposure to the unit of instruction. Table 1 below shows the difference between the pre and posttest scores.

Table 1.

Variable	Mean	Std. Dev.	t-value	df	p-value
1. I believe our climate is changing.	-.333	.589	-4.043	50	.000*
2. I am concerned about global climate change.	-.039	1.216	-.230	50	.819
3. I believe there is evidence of global climate change.	-.471	.946	-3.554	50	.001*
4. Global climate change will impact our environment in the next 10 years.	-.471	.987	-3.405	50	.001*
5. Global climate change will impact future generations.	-.176	.953	-1.322	50	.192
6. The actions of individuals can make a positive difference in global climate change.	-.235	1.176	-1.429	50	.159
7. I can do my part to make the world a better place for future generations.	-.333	.909	-2.618	50	.012*
8. Human activities cause global climate change.	-.412	1.004	-2.930	50	.005*
9. Climate change has a negative effect on our lives.	-.353	1.671	-1.508	50	.138
10. There is not much I can do that will help solve environmental problems.	.412	.726	4.050	50	.000*
11. It is a waste of time to work to solve environmental problems.	.176	.684	1.842	50	.071
12. Things I do have no effect on the quality of the environment.	.314	1.068	2.099	50	.041*
13. I think most of the concerns about environmental problems have been exaggerated.	.176	.888	1.419	50	.162
14. Knowing about environmental problems and issues are important to me.	-.294	1.045	-2.010	50	.050*
15. We cannot do anything to stop global climate change.	.392	1.115	2.512	50	.015*

* Indicates statistically significant results at $p < 0.05$

Christensen and Knezek (2015) reported that initial trials with middle school students revealed that the instrument is reliable and valid. The current study's set of data showed acceptable reliability as measured by Cronbach's Alpha (.716 for Beliefs and .654 for Intentions). Beliefs consisted of 9 items and related to statements such as "I believe there is evidence of global climate change." Intentions consisted of 6 items and related to statements such as "We cannot do anything to stop global climate change."

Table 2.

What do I know about climate change	What do I want to learn about climate change
1. Usage of fossil fuels are contributing to climate change	1. How does carbon-dioxide traps heat in the atmosphere?
2. Climate change will have an adverse effects on life and the environment	2. Can carbon-dioxide be stopped?
	3. How much time do we have before we cannot do anything about climate change
3. Climate change is contributing to extreme weather	4. Why people don't care about climate enough about climate change?

Table 3.

What did you see?	What do you think?	What do you wonder?
• Cars and trucks driving and producing carbon-dioxide • Charts showing increasing temperature and carbon dioxide • Heard that diseases and other health issues are increasing because of climate change • Humans are absolutely the reason for climate change	• More actions should be taken to stop climate change, like stop using so much gas • We need to switch to renewable energy like wind • Find different ways to power our cars and trucks • Everyone should help • Have a day when we all don't use gas • Countries should try to reduce the amount of carbon dioxide	• If we know the dangers of climate change, why are we not doing enough to stop it? • When are people going to care about climate change? • Is it too late to stop climate change? • What can kids do? • Why can't we just try to fix the problem? • How to reduce the amount of gas we use?

To identify students' beliefs in climate change, and to build their intentions to act upon known causes of climate change, the students were asked to complete the "What I Know" and "What I Want to Know" portions of a KWL chart. Analyses of the students' responses are shown in table 2.

The responses in table 2 indicated that the students believe that our use of fossil fuel is contributing to climate change. They also believe that climate change is contributing to extreme weather which will have deleterious effects on life and the environment. Interestingly, the students wanted to learn about the science behind how carbo-dioxide gets trapped in the environment and what are some of the actions that could potentially be taken to reduce the amount of carbon-dioxide in the atmosphere. These are important questions to ask in order to develop the understanding necessary to influence our actions in mitigating climate change. The students were interested in the amount of time we have left before it is too late to take the necessary steps of reducing climate change and its effects. The students were also alarmed about why aren't people doing more to stop climate change. They asked questions such as, what is the president doing about climate change, does he know about climate change, and how can we stop climate change. These are important questions to ask, as the students think about potential actions that they can take to help in solving the climate change crisis.

To provide the students with a brief overview of the key takeaways from the 2018 climate assessment, they were asked to watch the brief CBSN news broadcast (https://www.youtube.com/watch?v=xiU6y12kZ80). After watching this broadcast, the *see-think-wonder* technique was used to engage the students in inquiry-based thinking through close observation. This technique was also used to enable the students to generate a list of their beliefs and intentions about climate change. Table 3 below shows students beliefs and intentions on what they think and wonder after watching the CBSN news broadcast.

After viewing the news report, the students were convinced that human actions are directly related to climate change. They reported that they saw automobiles producing carbon-dioxide, increasing Earth's temperature and carbon-dioxide levels. The students also reported that they believe diseases and other health issues are directly associated with climate change. While the students comment about what they saw were more belief based, their comments under what they think and wonder were more aligned with their intent to act. For example, the students suggested that more actions should be taken and inquired about what children their age can do to combat climate change.

In order to provide the students with a deeper understanding of greenhouse gases, how these gases are produced, and how long these gases remain in the atmosphere they were asked to use the Environmental Protection Agency's link provided (https://www.epa.gov/ghgemissions/overview-greenhouse-gases) to determine what are greenhouse gases, identify the four types, how these gases are produced, how long they remain in the atmosphere, and how strongly do they believe these gases negatively impact the atmosphere. The latter question was for the students to think about the length of time these gases remain in the atmosphere and the potential impact they have on climate change. The students' responses to these questions fell under three main categories:

1. It depends on which gas it is. Some has more impact than others
2. They make the planet warmer
3. These gases thicken the Earth's blanket which warms up the earth more

The students attributed these responses to what they learned about the length of time carbon-dioxide, methane, nitrous oxide, and fluorinated gases remained in the atmosphere. For example, methane remains in the atmosphere for four years, while carbon dioxide remains for 100 year (https://www.epa.gov/ghgemissions/overview-greenhouse-gases#CO2-lifetime). The students used this knowledge to suggested that different greenhouse gas will have different impact on climate change. However, they believe that all four types of greenhouse gases are contributing to increasing global temperature.

In the 2015 report, PISA identified four aspects of the scientific literacy assessment framework. These include context which involves personal, local/national, global issues which requires individuals to display competencies. According to PISA, such competencies involve the individual ability to explain phenomena scientifically, evaluate and design scientific inquiry, and interpret data and evidence scientifically. The PISA (2015) report further state that individuals will show competences by their attitudes, that include interest in science, their value for scientific approaches to enquiry, and environmental awareness. Additionally, PISA (2015) report further states that the individual understanding of the major facts, concepts and explanatory theories form the basis of scientific knowledge.

In order to use the 2015 scientific literacy framework outlined by PISA (2015), the students were asked to respond to the four questions below:

1. How might climate change affect the commonwealth of Massachusetts?
2. What activities do you engage in that might be contributing to climate change?
3. What actions, if any, can you take to reduce climate change?
4. Pose one or two questions that you would like to get answered about climate change.

The first question was used to make the issue of climate change personal and relevant in the lives of the students. If the students can identify personal connections between their lives and climate change, then they are generally more interested to learn about the subject. The students overwhelmingly (100%) stated that climate change could cause flooding, increase precipitation, and an increase in temperature in Massachusetts.

The second question was used to get the students to consider ways they might be contributing to climate change. This question was used as a springboard to get the students to start thinking about potential actions that they and others could take to mitigate climate change. For this question, the students identify the following actions behaviors that they engaged in, which could potentially contribute to climate change. Their behaviors include leaving lights on when no one is using them, leaving the television on when no one is watching, excessive cooking, and unnecessary use of cars. The students overwhelming stated that the actions that they can take to mitigate climate change include walking or biking to places that are near, support renewable sources of energy, and protest against climate change. It's interesting to hear middle school students taking on a social justice plan, to the point where they overwhelming wanted to protest against climate change. It is this type of advocacy that others believe is necessary to sustain the environment. Finally, when the students were asked to pose on a question they would like to get answered about climate change, the students overwhelming asked: Is it too late to stop climate change?

Day 2

The activities on the first day introduced the students to the concept of greenhouse gases. To help the students make further connections between human actions and greenhouse gas emission, and to build their scientific literacy competencies (PISA, 2015), the students are arranged into small groups to investigate greenhouse gas emission in Massachusetts. The question below was used to help the students to start thinking about total emissions.

1. What are the two major sources of greenhouse gases? Why?

The students stated that transportation is one major source because we need to travel. They further stated that electricity is the other because we need lights, heating, and other things in our homes. The students were then asked to review the total U.S. greenhouse gas emissions by economic sector in 2017 (https://www.epa.gov/ghgemissions/sources-greenhouse-gas-emissions), and explain how it relates to their responses to the above question. The students were excited to see that their responses match that of the United States Environmental Protection Agency (EPA).

To make the assignment in day 2 more personally relevant to the students, they were teamed into working pairs to gather data on greenhouse gas emissions from powerplants and waste treatment facilities in their communities, or neighboring communities in Massachusetts. The students used the 2017 EPA Greenhouse Gas Reporting Program data (https://www.epa.gov/ghgreporting/ghg-reporting-program-data-sets) to investigate the amount of greenhouse gas emissions from these facilities. Table 4 below outlines the greenhouse gas emission data the students gathered from ANP Blackstone Energy Company in New England.

Table 4.

Emission from Power Plant in Massachusetts (Metric Tons)	Greenhouse Gas Reporting Program for 2010			Greenhouse Gas Reporting Program for 2011			Greenhouse Gas Reporting Program for 2012			Greenhouse Gas Reporting Program for 2013		
	Carbon Dioxide	Nitrous oxide	Fluorinated Gases	Carbon Dioxide	Nitrous oxide	Fluorinated Gases	Carbon Dioxide	Nitrous oxide	Fluorinated Gases	Carbon Dioxide	Nitrous oxide	Fluorinated Gases
ANP Blackstone Energy Company, 204 Elm St, Blackstone, MA 01504	699,127	324	387	815,186	378	451	802,154	372	444	739,250	343	409
Emission from Power Plant in Massachusetts (Metric Tons)	Greenhouse Gas Reporting Program for 2014			Greenhouse Gas Reporting Program for 2015			Greenhouse Gas Reporting Program for 2016			Greenhouse Gas Reporting Program for 2017		
	Carbon Dioxide	Nitrous oxide	Fluorinated Gases	Carbon Dioxide	Nitrous oxide	Fluorinated Gases	Carbon Dioxide	Nitrous oxide	Fluorinated Gases	Carbon Dioxide	Nitrous oxide	Fluorinated Gases
ANP Blackstone Energy Company, 204 Elm St, Blackstone, MA 01504	836,723	388	463	815,627	378	451	597,722	277	330	785,363	365	434

The students were asked to use the data they collected on their assigned power plants and waste treatment facilities to describe what the data indicate about greenhouse gas emissions.

The students who investigated greenhouse gas emissions between 2010 – 2017 at ANP Blackstone Energy Company, LLC Power Plant in Massachusetts, reported that emission levels of carbon-dioxide increased over this timeframe. When asked what they believe is responsible for this increase, it indicated that population growth is causing us to use more energy for electricity. While not directly attributed to the power company, the students restated that the increasing population growth is also causing people to use more transportation, which also contributes to climate change. When the students were asked to state some questions that would like to ask the management of ANP Blackstone Energy Company, they overwhelmingly said they would ask why is carbon-dioxide increasing, and how can we reduce emission? The question of how we can reduce carbon-dioxide emissions is not new to these students. On the first day of this instructional unit, the students asked several times what actions can be taken to reduce carbon-dioxide emissions, what can children do to mitigate climate change, and why people seem to not care too much about carbon-dioxide emission and climate change.

However, when asked to explain the actions they believe individuals should take to reduce the amounts of greenhouse gas emitted from your assigned facilities such as ANP Blackstone Energy Company, they stated we must try to save electricity. The students believed that unnecessary energy usage results in power companies producing more electricity from burning fossil fuels. They think if we use less, then these companies will also burn fewer amounts of fossil fuels to create electricity. The

Table 5.

Question	Students' Responses	Researchers' Interpretation
1. What did you learn about your carbon-foot print?	• Home energy is the most • If I reduce my usage I can reduce my carbon footprint	The students identified that home energy usage consumed and identified that if steps are taken, the carbon footprint will be lowered.
2. What questions do you have based on what you learned from this process?	• How to get my carbon footprint to the lowest it can be? • Will small actions prevent climate change? • How can I encourage others to try to reduce their carbon dioxide emissions? • How can I help the environment and stop climate change? • Do people know about their carbon footprint?	The students' inquired about actions they can take to become good stewards to the environment. Also, they wanted to know how to encourage others to become more responsible in sustaining the environment.
3. What, if any, actions that should be taken to reduce global climate change?	• Don't drive everywhere • Drive electric vehicles • Turn off lights when not in use • Don't use AC that high • Use cold water to wash clothing • Recycle more often • Stop using so much energy • Give food to people who want it instead of sending it to landfills • Produce less greenhouse gases • Not use too much plastic • Not overuse things	The students suggested changes in driving habits, types of vehicles to purchase, less consumptions of resources, recycle, and amounts of materials to landfills.

students believe that society, on the whole, should carpool more, seek out alternative energy sources, and sign petitions to reduce greenhouse gases.

Day 3

If the students can measure how their actions directly relate to climate, then their attitudes can be influenced. Therefore, the activities assigned during the third day of the instructional was used to influenced the students' attitudes towards climate change. The students were asked to calculate their carbon footprint (https://www3.epa.gov/carbon-footprint-calculator/), and respond to four questions. Table 5 below are the common responses the students gave.

Calculation

The calculation of carbon footprint activities allowed the students to examined how their habits at home contribute to climate change. This allowed them to questioned how they can become better stewards to the environment, in addition to encouraging others to do the same. Furthermore, the students also suggested steps that they and others can do to mitigate climate change. They advocated for donating excess foods to others so that unnecessary food items are not thrown out in landfills. This argument shows the transfer of knowledge from the previous day's activity, where students learned about the volume of greenhouse gases that are emitted from landfills.

Day 4

Table 6.

What did you see?	What do you think?	What do you wonder?
• Glaciers in Antarctica are melting six times faster than it was 40 years ago because of climate change • Glaciers melting at to fill three Olympic size swimming pool every second • Average temperature in the last 70 years has increased by 3.5 degrees	• We should produce a lot less greenhouse gases • Human are part of the reason • Climate change will cause flooding • Boston will be flooded	• How are we going to relocate people who lives close to the ocean when sea level rise? • What are we doing to find a way to solve the problem of climate change? • If we know we are causing global warming, why are we not stepping up • Will the ice in Antarctica disappear entirely? • How do we stop it?

The students watched the short PBS News Hour report on Glacial melting in Antarctica (https://www.youtube.com/watch?v=YRe1ymYR45k). The *see-think-*

wonder technique was used for the students to close observation. Table 6 below shows the common responses form the students on the melting Antarctica Glaciers.

Although this video was over nine minutes long, all the students reported that they saw that glaciers in Antarctica are melting six times faster than they were 40 years ago because of climate change. The students also reported that the rate at which these glaciers are melting is enough to fill three Olympic-size swimming pools every second. Finally, the students stated that the average temperature in the last 70 years in Antarctica has increased by 3.5 degrees. The figures and imagery in this report were presented in a manner that the students could understand. As a result, they were all alarmed by these data. Therefore, when asked to share what they think, the students repeated several statements that they have made in days 1 and 2. For example, we should produce less greenhouse gases and human beings are contributing to climate change. Additionally, the students relate the effects of Antarctica Glaciers melting to their own lives. They stated that they think climate change will cause flooding due to sea level rise and that Boston will be one of the areas that are flooded due to climate change. The threat of flooding due to sea level rise caused the students to wonder how will the authorities relocate people who lives close to the ocean when sea level rise, why aren't we doing enough to mitigate climate change, and will Antarctica Glaciers disappear entirely?

Table 7.

Northeast	
a. Summarize how climate change could impact your assigned region of the country.	• Shorter winter and longer and hotter summers • Decrease in the fishing industry • More deaths and health risk from contaminated floods • Sea level rising • Heat waves
b. Explain what concerns you might have based on your summary.	• It will affect us because we live in the northeast • We are causing it our self • The flooding and deaths • Concerned over longer summer and shorter winter
c. What actions, if any, should be taken to reduce the risk described in your summary?	• Produce less greenhouse gas • Reduce our carbon footprints • Reduce trash in landfills
d. Identify who, if anyone, is responsible for carrying out the action(s) identified.	• We are responsible because of the greenhouse gases we produced daily • Recycle more

Table 8.

	Communities	Economy	Interconnected Impacts
Main Idea			
Detail			

To allow more opportunities for the students to think about how climate change could potentially affect various regions across the United States, they were asked to watch a brief CNN report (https://www.cnn.com/2018/11/27/health/climate-change-impact-by-region/index.html). The students then worked in pairs to respond to the flowing:

1. Summarize how climate change will impact your assigned region of the country.
2. Explain what concerns you might have based on your summary.
3. What actions, if any, should be taken to reduce the risk described in your summary?
4. Identify who, if anyone, is responsible for carrying out the action(s) identified.

Table 7 below is a summary reports from the students who were assigned the Northeast region.

The students believe that human beings are contributing to climate change. Again, the students advocated for the production of less greenhouse gas, reduction in our carbon footprints, and recycling as measures to reduce the amounts of trash in we send to landfills. The activities the students did in day 2 where they investigate the amount of greenhouse gases from power plants and waste treatment facilities influenced these responses from the students. This is a clear indication that the students were able to use the knowledge they have gained earlier in this unit of study to drive their decision making as they think about ways to mitigate climate change.

Table 9.

Question: The recent report from the Intergovernmental Panel on Climate Change (IPCC) recently reported on the impacts of global warming of 1.5 degree Celsius above pre-industrial levels. While the panel report indicates that human activities are the main result of the increase in global temperature, many policy makers at the federal level in the United States dismiss global warming as a hoax. Based on what you have learned about climate change, do you believe human activities contribute to climate change? Your job is to outline your ideas to this question in the graphic organizer below.			
FOR		**AGAINST**	
Reasons to be FOR the issue	**List at least two facts (evidence) for each reason**	**Reasons to be AGAINST the issue**	**List at least two facts (evidence) for each reason**
Reason 1	Fact	**Reason 1**	Fact
	Fact		Fact
Reason 2	Fact	**Reason 2**	Fact
	Fact		Fact
Reason 3	Fact	**Reason 3**	Fact

Day 5

The students were introduced to the Fourth National Climate Assessment Summary Findings (https://nca2018.globalchange.gov/). The students worked in small group to complete main ideas and supporting details on the impact of climate change on

Figure 1a.

Dear Ceo of MNP Blackstone energy Company, 204 christ, Bladstin NM 01504,

I would like to inform you that natural gas is killing our world. Our infrastructures are being destroyed So is our tourist rates. Lastly our tide levels are rising also due to climate change. Which as you probably know is due to us. Now with my understanding climate change is very harmful to the environment which is the reason I would like to invite you to a round table discussion. Please except this offer because I have many things to discuss in person. Thankyou for listening, now here are my reasons.

The first thing natural gas is destroying are our world's infrastructures. Our infrastructures are on very high risks of disappearing as we speak. These infrastructures disappearing are mainly caused by climate change but climate change is caused by us. A website on google states "Our nation's aging and deteriorating infrastructures". This article also states "Without adaption climate change will continue to degrade infrastructures". These reasons is part of why I would love to have a discussion very soon. All in all that is how natural gas and climate change are destroying our infrastructures.

Another thing climate change is doing is making our tourist rates go down. The touristh economy will go down because of heat and flooding. Natural disasters will also become a problem because it makes people not want to come here. The same article states "outdoor recreation tourist economy".

the twelve items in the summary findings. Table 8 below is an example of how the students engaged in this activity.

The students were asked to think about the flowing as they gather their main idea and detail:

Figure 1b.

Right under it, it states "that will be degraded by the impacts of climate charge" These tourist rates going down also means less money, which can trouble a lot of people. Basically that is another way climate charge is hurting our world.

Our harbors will also be hit by the impacts of climate charge because of tide levels rising. These tide levels are caused by ice melting into the ocean, adding more unneded water into our harbors. The last pieces of evidence from the website state "Threatened by the impacts of climate changing." The last reason is "A chronic high tide flooding." With all these reason I ask you once again to accept my invite to a table discussion. As this states climate charge is also affecting our harbors and causing floods.

In summary those are some of the ways climate charge is killing our world. How our infrastructures are getting destroyed by climate charge. As well as our tourist rates as they go down more problems are caused. Lastly our harbors are flooding with water, flooding cities along with it. I would like to ask you one final time to accept my invite to a round table discussion. In conclusion that is how natural gas and climate charge is killing our world.

1. You will use the main ideas and details identified from the risks posed by climate change to help you in preparing an essay or letter to the Chief Executive Officer of the power or waste treatment company you studied in day 2 of the unit.

Figure 2.

Dear Sir/Madam,

Did you know our world is dying? It is, the truth, and we need to help it. We emit greenhouse gases and over time the amount builds up and right now, we are emitting too much. It's causing too much climate change and we really need to stop it!

At school, I learned our world is breaking. Greenhouse gas emissions are to blame. No individual is to blame, but everyone. Everyone uses energy, but some more than others. Help save the earth! We are all responsible for climate change.

In my science class we all researched a energy company and their gas emicsions every year. My group reseached ANP Blackstone Energy Company and the emissions were high. We all need to work together to stop global warming and everyone needs to do their part.

We would like to invite you to a round table discussion on what we can all do to help the world. We will be discussing how do help and stop emitting greenhouse gases. We also need to conference about what we can do to help those in need. We hope you can come!

We need to save the earth. We need to stop emitting greenhouse gases and do our port to save the world!

Helping save the world,

2. Your letter to the Chief Executive Officer must explain the research you have done in this unit. You will then invite the Chief Executive Officer or their representatives to a round table discussion on what individuals, the community, and society can do to combat climate change.

Figure 3a.

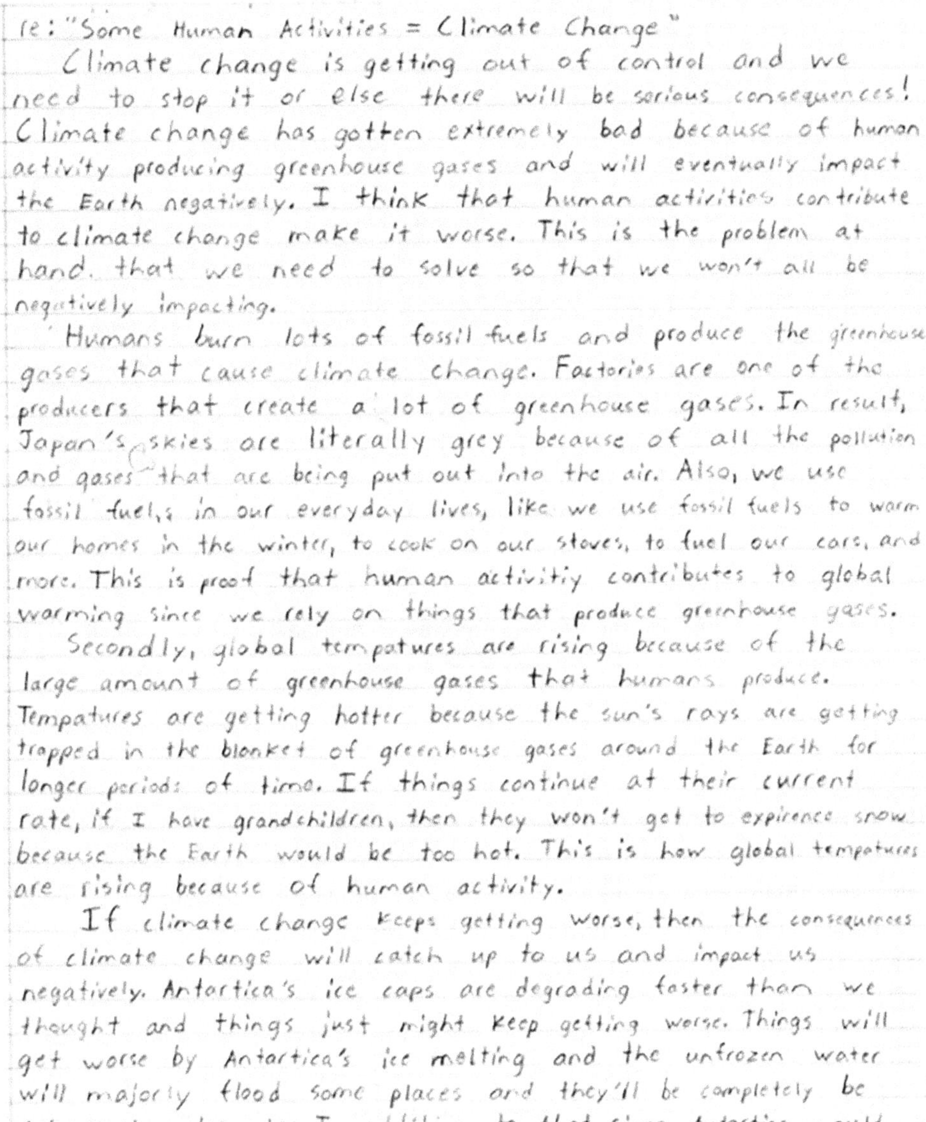

re: "Some Human Activities = Climate Change"

Climate change is getting out of control and we need to stop it or else there will be serious consequences! Climate change has gotten extremely bad because of human activity producing greenhouse gases and will eventually impact the Earth negatively. I think that human activities contribute to climate change make it worse. This is the problem at hand. that we need to solve so that we won't all be negatively impacting.

Humans burn lots of fossil fuels and produce the greenhouse gases that cause climate change. Factories are one of the producers that create a lot of greenhouse gases. In result, Japan's skies are literally grey because of all the pollution and gases that are being put out into the air. Also, we use fossil fuels in our everyday lives, like we use fossil fuels to warm our homes in the winter, to cook on our stoves, to fuel our cars, and more. This is proof that human activitiy contributes to global warming since we rely on things that produce greenhouse gases.

Secondly, global tempatures are rising because of the large amount of greenhouse gases that humans produce. Tempatures are getting hotter because the sun's rays are getting trapped in the blanket of greenhouse gases around the Earth for longer periods of time. If things continue at their current rate, if I have grandchildren, then they won't get to expirence snow because the Earth would be too hot. This is how global tempatures are rising because of human activity.

If climate change keeps getting worse, then the consequences of climate change will catch up to us and impact us negatively. Antartica's ice caps are degrading faster than we thought and things just might keep getting worse. Things will get worse by Antartica's ice melting and the unfrozen water will majorly flood some places and they'll be completely be submerged underwater. In addition to that, since Antartica would

3. Before you write your letter, use the graphic organizer below to help in organizing your thoughts. This will help them to write a more compelling letter. Table 9 below is a graphic organizer that the students used to outline their thoughts.

Figure 3b.

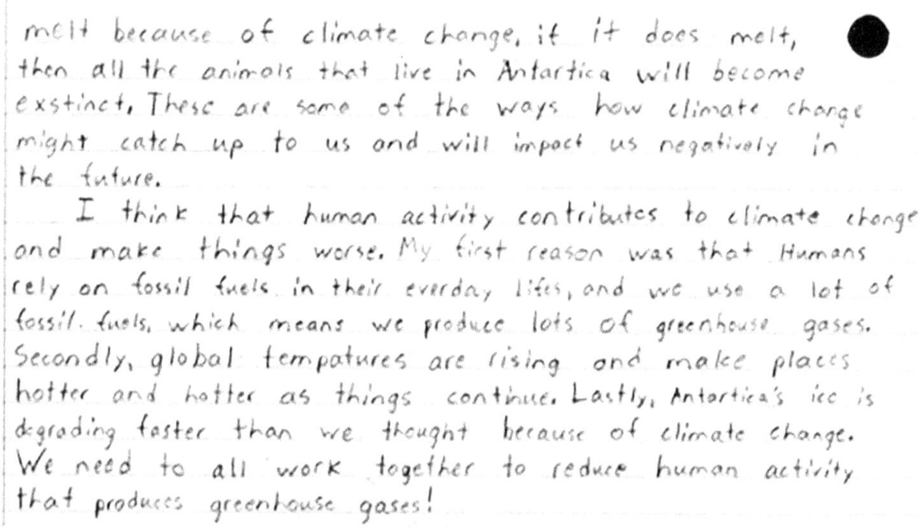

melt because of climate change, if it does melt, then all the animals that live in Antartica will become exstinct. These are some of the ways how climate change might catch up to us and will impact us negatively in the future.

I think that human activity contributes to climate change and make things worse. My first reason was that Humans rely on fossil fuels in their everday lifes, and we use a lot of fossil. fuels, which means we produce lots of greenhouse gases. Secondly, global tempatures are rising and make places hotter and hotter as things continue. Lastly, Antartica's ice is degrading faster than we thought because of climate change. We need to all work together to reduce human activity that produces greenhouse gases!

Interestingly, 100% of the students believed human activities contribute to climate change.

Day 6

The activities on day 6 were the culmination of this instructional unit. The students were asked to use all of the evidence that they have gathered about the Intergovernmental Panel on Climate Change (IPCC) report, to write an essay or letter on the impact of climate change on the environment, or to write a letter to the Chief Executive Officer of the power plant or waste treatment facility that they researched in day 2. The students were reminded to that their letter must explain the research they engaged in, what you have learned from the study, and their desire to invite him/her to a round table discussion on what individuals, communities, and society can do to help reduce their carbon footprint. The students were provided with a rubric and a template (see appendix) to use as a guide as they write their persuasive letters. Below are three exemplars letters of student letters.

DISCUSSION

To find solutions to the many scientific problems that threatened the environment and human existence, we need citizenry who are scientifically literate. This is evident in the United Nations Environment Programme (UNEP, 2012) report. According to the report, scientific literacy has become more critical at the local, national, and international levels as humanity faces significant challenges in providing enough clean water and food for the growing population. Additionally, the report further states controlling diseases, generating sufficient energy, and adapting to climate change are challenges that we cannot afford to ignore (UNEP, 2012). However, for this to be possible, the science education curriculum in the P-12 system needs to provide opportunities for the students to develop the type of scientific literacy advocate by PISA (2015). In this report (2015), PISA defines scientific literacy as the ability to engage with science-related issues, and with the ideas of science, as a reflective citizen. PISA's (2015) definition of scientific literacy advocated that individuals must able to explain phenomena scientifically, possesses the abilities to evaluate and design scientific inquiry, and interpret data and evidence scientifically. PISA's 2015 definition of scientific literacy emphasizes the importance of being able to apply scientific knowledge in the context of real-life situations. The instructional unit used in this study provided the opportunity for the students to evaluate and interpret data and evidence scientifically.

As a result, the students were able to develop their knowledge and understanding of climate change and proposed real-world solutions to mitigating climate change. For example, the students understood that individual actions might not be enough to reverse climate change, but collectively we can make a difference. The students advocated that we should take actions that included, but are not limited driving less, purchasing electric vehicles, turning off lights when not in use, recycle more, and finding alternative sources of energy. These are some of the actions that are advocated in the recent IPCC report on the responses to climate change Fourth (National Climate Assessment, 2018). The students in this investigation were able to use what they learned from their research on greenhouse gas emissions to write persuasively on the effects of climate change that are already being felt across many regions of the United States. These students discussed the current effects that greenhouse gases are having on increasing global temperature, melting glaciers, rising sea level tides, and flooding. As these students imagine the potential long-term impacts of climate change, they make finding solutions to mitigate climate change personal. For example, in exemplary 2 above, this student stated that "Global temperatures are rising because of the large amounts of greenhouse gases that humans produce. Temperature is getting hotter because the sun's rays are getting trapped in the blanket of greenhouse gases around the Earth for longer periods. If things continue at their current rate, if I have grandchildren, then they won't get to experience snow because the Earth would be too hot. This is how global temperatures are rising because of human activity."

The idea that students who ranged in age from 11–12-years-old, are making these profound statements after learning about climate change and its impact is perhaps what is required for policymakers in general, and the public, in particular, to engage in actions necessary to mitigate climate change. However, our science education curriculum at the P-12 level must provide opportunities for the students to participate in investigations that enhance the students' beliefs and intentions toward climate change. Throughout this instructional unit, students inquired about why aren't people engaging in actions to combat climate change, what they can do to help reduce the effects, and how they might include others to take steps. A science education curriculum that enables students to act upon these questions is paramount in solving the environmental challenges of the 21st century and beyond.

CONCLUSION

This instructional unit represent a modeled for how to engage middle school students in the process of researching data, interpreting the findings, and developing action steps for mitigating the disastrous impact of climate change on the environment. In this investigation, 100% of the students believe that human actions are contributing to climate change. While at the start of this instructional unit, the students thought that the use of fossil fuels played a significant role in climate change, they were unsure of the mechanism of this process. Throughout the various activities in this instructional unit, the students enhanced their understanding of how and why excessive burning of fossil fuels is associated with increasing global temperature. The students were able to use the knowledge they gained from the research they conducted to suggest ways to human beings can mitigate climate change. Additionally, the students were also able to combine the results from their study with the use of summary findings from the 2018 fourth national climate assessment reported to write persuasively about climate change and its impact.

The results indicate that this socioscientific issues curricular unit of instruction was very impactful in the students' beliefs and intentions towards climate change. The students enhanced their beliefs that human actions are contributing to changes in the climate, and they identify several steps that should mitigate climate change.

LIMITATIONS

This investigation was conducted with fifty-one (n=51) students who reside in the Northeastern United States. This region is not typically known for oil and gas production when compared to other parts of the country. Therefore, the finding of this investigation cannot be generalized. The small sample size of this study is also a major limiting factor. Such a small sample size makes it inappropriate to make any generalization. Therefore, science education researchers who want to determine the impact of this curricular unit of instruction on middle school students' beliefs and intentions toward climate change may want to replicate this study with a much larger population of students in different regions of the United States.

REFERENCES

American Association for the Advancement of Science. (1989). *Science for all Americans: a Project 2061 Report on Literacy Goals in Science, Mathematics and Technology*. AAS Publishing. www.project2061.org/publications/sfaa/online/sfaatoc.htm

Arnold, Cohen, & Warner. (2009). Youth and Environmental Action: Perspectives of Young Environmental Leaders on Their Formative Influences. *Journal of Environmental Education, 40*(3), 27–36. doi:.3.27-36. doi:10.3200/JOEE.40

Bofferding, L., & Kloser, M. (2015). Middle and High School Students Conceptions of Climate Change Mitigation and Adaptation Strategies. *Environmental Education Research, 21*(2), 275–294. doi:10.1080/13504622.2014.888401

Bybee, R. W. (1997). Towards an understanding of scientific literacy. In W. Gräber & C. Bolte (Eds.), Scientific literacy. An international symposium (pp. 37-68). Institut für die Pädagogikder Naturwissenschaften (IPN).

Choi, S., Niyogi, D., Shepardson, D. P., & Charusombat, U. (2010). Do earth and environmental science textbooks promote middle and high school students' conceptual development about climate change? Textbooks' consideration of students' misconceptions. *Bulletin of the American Meteorological Society, 91*, 889-898. . doi:10.1175/2009BAMS2625.1

Christensen, R., & Knezek, G. (2015). The climate change attitude survey: Measuring middle school student beliefs and intentions to enact positive environmental change. *International Journal of Environmental and Science Education, 10*(5), 773–788.

Cook, J. (2019). Understanding and countering misinformation about climate change. In I. Chiluwa & S. Samoilenko (Eds.), *Handbook of Research on Deception, Fake News, and Misinformation Online* (pp. 281–306). IGI-Global. doi:10.4018/978-1-5225-8535-0.ch016

Davenport, C., & Landler, M. (2019). Trump Administration Hardens its Attack on Climate Science. *New York Times*. Available from: https://www.nytimes.com/2019/05/27/us/politics/trump-climate-science.html

European Commission. (1995). *Teaching and learning: Towards the learning society*. White Paper on Education and Training, Office for Official Publications in European Countries. https://europa.eu/documents/comm/white_papers/pdf/com95_590_en.pdf

Feldman, L., Nisbet, M., Leiserowitz, A., & Maibach, E. W. (2010). *The Climate Change Generation? Survey Analysis of the Perceptions and Beliefs of Young Americans*. Yale Project on Climate Change Communication.

Flora, J. A., Saphir, M., Lappé, M., Roser-Renouf, C., Maibach, E. W., & Leiserowitz, A. A. (2014). Evaluation of a National High School Entertainment Education Program: The Alliance for Climate Education. *Climatic Change*, *127*(3–4), 419–434. doi:10.100710584-014-1274-1

Holbrook, J., & Rannikmae, M. (Eds.). (1997). Supplementary teaching materials promoting scientific and technological literacy. Tartu, Estonia: ICASE (International Council of Associations for Science Education).

Hurd, P. D. (1958). Science literacy: Its meaning for American schools. *Educational Leadership*, *16*(1), 13–16.

Jensen, B. B., & Schnack, K. (1997). The action competence approach in environmental education. *Environmental Education Research*, *3*(2), 163–178. doi:10.1080/1350462970030205

Kollmuss, A., & Agyeman, J. (2002). Mind the gap: Why do people act environmentally and what are the barriers to pro-environmental behaviour? *Environmental Education Research*, *8*(3), 239–260. doi:10.1080/13504620220145401

Laugksch, R. C. (2000). Scientific Literacy: A conceptual overview. *Science Education*, *84*(1), 71–94. doi:10.1002/(SICI)1098-237X(200001)84:1<71::AID-SCE6>3.0.CO;2-C

Lawson, D. F., Stevenson, K. T., Peterson, M. N., Carrier, S. J., Seekamp, E., & Strnad, R. (2019). Evaluating climate change behaviors and concern in the family context. *Environmental Education Research*, *25*(5), 1–13. doi:10.1080/13504622.2018.1564248

Leiserowitz, A., Maibach, E., Roser-Renouf, C., Rosenthal, S., & Cutler, M. (2017). *Climate change in the American mind: May 2017. Yale University and George Mason University*. Yale Program on Climate Change Communication.

Leiserowitz, A., & Smith, N. (2011). *Knowledge of climate change among visitors to Science & Technology Museums.* Retrieved from Yale Program on Climate Change Communication website: https://environment.yale.edu/climate/fles/MuseumReport.pdf

Lombardi, D., Sinatra, G. M., & Nussbaum, E. M. (2013). Plausibility reappraisals and shifts in middle school students' climate change conceptions. *Learning and Instruction*, *27*, 50–62. doi:10.1016/j.learninstruc.2013.03.001

Mead, E., Roser-Renouf, C., Flora, J. A., Maibach, E. W., Leiserowitz, A., & Rimal, R. N. (2012). Information Seeking about Global Climate Change among Adolescents: The Role of Risk Perceptions, Efficacy Beliefs and Parental Influences. *Atlantic Journal of Communication*, *20*(1), 31–52. doi:10.1080/15456870.2012.637027 PMID:22866024

Mogensen, F., & Schnack, K. (2010). The action competence approach and the 'new' discourses of education for sustainable development, competence and quality criteria. *Environmental Education Research*, *16*(1), 59–74. doi:10.1080/13504620903504032

National Science Education Standards. (1996). *National Academy of Sciences.* National Academy Press.

OECD. (2003). *The PISA 2003 assessment framework.* Retrieved November 2008 from: http://www.pisa.oecd.org/dataoecd/46/14/33694881.pdf

OECD. (2006). *Assessing Scientific, Reading and Mathematical Literacy: A Framework for PISA 2006, PISA*. OECD Publishing. doi:10.1787/9789264026407-

OECD. (2007). PISA 2006: Science Competencies for Tomorrow's World: Volume 1: Analysis, PISA. OECD Publishing. doi:10.1787/9789264040014-en

OECD. (2009). *PISA 2006 Technical Report, PISA*. OECD Publishing. doi:10.1787/9789264048096-

OECD. (2012). What kinds of careers do boys and girls expect for themselves? In *PISA in focus, No 14.* OECD Publishing. doi:10.1787/5k9d417g2933-

OECD. (2017). *PISA 2015 assessment and analytical framework: Science, reading, mathematics, financial literacy and collaborative problem solving* (Rev. ed.). Program for International Student Assessment, Organisation for Economic Co-operation and Development. doi:10.1787/9789264281820-

Österlind, K. (2005). Concept formation in environmental education: 14-year olds' work on the intensified greenhouse effect and the depletion of the ozone layer. *International Journal of Science Education, 27*, 891-908. . doi:10.1080/09500690500038264

Ozdem, Y., Dal, B., Öztürk, N., Sönmez, D., & Alper, U. (2014). What is that thing called climate change? An investigation into the understanding of climate change by seventh-grade students. *International Research in Geographical and Environmental Education, 23*(4), 294–313. doi:10.1080/10382046.2014.946323

Papadimitriou, V. (2004). Prospective primary teachers' understanding of climate change, greenhouse effect, and ozone layer depletion. *Journal of Science Education and Technology, 13*, 299-307. :JOST.0000031268.72848.6d doi:10.1023/B

Plutzer, E., McCaffrey, M., Hannah, A. L., Rosenau, J., Berbeco, M., & Reid, A. H. (2016). Climate confusion among U.S. teachers. *Science, 351*(6274), 664–665. doi:10.1126cience.aab3907 PMID:26912845

Reinfried, S., Aeschbacher, U., & Rottermann, B. (2012). Improving Students' conceptual Understanding of the Greenhouse Effect Using Theory-based Learning Materials That Promote Deep Learning. *International Research in Geographical and Environmental Education, 21*(2), 155–178. doi:10.1080/10382046.2012.672685

Roberts, D. A. (2007). Scientific literacy/Science literacy. In S. K. Abell & N. G. Lederman (Eds.), *Handbook of research on science education* (pp. 729–780). Lawrence Erlbaum Associates.

Roberts, D. A., & Bybee, R. (2014). Scientific literacy, science literacy, and science education. In N. G. Lederman & S. K. Abell (Eds.), *Handbook of research on science education* (Vol. 2, pp. 545–558). Routledge.

Shepardson, D. P., Choi, S., Niyogi, D., & Charusombat, U. (2011). Seventh grade students' mental models of the greenhouse effect. *Environmental Education Research, 17*, 1-17. . doi:10.1080/13504620903564549

Skamp, K., Boyes, E., & Stanisstreet, M. (2004). Students' ideas about air quality. *Research in Science Education, 34*(3), 313–342. doi:10.1023/B:RISE.0000044643.24770.5c

Stevenson, K. T., Peterson, M. N., & Bondell, H. D. (2016). The Influence of Personal Beliefs, friends, and Family in Building Climate Change Concern among Adolescents. *Environmental Education Research, 4622*, 1–14. doi:10.1080/13504622.2016.1177712

Stevenson, K. T., Peterson, M. N., Carrier, S. J., Strnad, R. L., Bondell, H. D., Kirby-Hathaway, T., & Moore, S. E. (2014). Role of Significant Life Experiences in Building Environmental Knowledge and Behavior Among Middle School Students. *The Journal of Environmental Education*, *45*(3), 163–177. doi:10.1080/00958964.2014.901935

UNEP. (2012). *21 Issues for the 21st Century: Result of the UNEP Foresight Process on Emerging Environmental Issues*. United Nations Environment Programme (UNEP). www.unep.org/pdf/Foresight_Report-21_Issues_for_the_21st_Century.pdf

USGCRP. (2017). Climate Science Special Report: Fourth National Climate Assessment, Volume I. U.S. Global Change Research Program. doi:10.7930/J0J964J6

Vesterinen, V.-M., Tolppanen, S., & Aksela, M. (2016). Towards Citizenship Science Education: What Students do to make the World a Better Place? *International Journal of Science Education*, *38*(1), 30–50. doi:10.1080/09500693.2015.1125035

Vreede, C. D., Warner, A., & Pitter, R. (2017). Facilitating Youth to Take Sustainability Actions: The Potential of Peer Education. *The Journal of Environmental Education*, *45*(1), 37–46. doi:10.1080/00958964.2013.805710

Wilkerson, S. (2008). Application of the Paired t-test. *Xavier University Of Louisiana's Undergraduate Research Journal, 5*(1).

Wilson, K. M. (2000). Drought, debate, and uncertainty: Measuring reporters' knowledge and ignorance about climate change. *Public Understanding of Science (Bristol, England)*, *9*(1), 1–13. doi:10.1088/0963-6625/9/1/301

APPENDIX

Day 1

Introduction to Climate Change

1. KWL on climate change [Students complete K&W portions of the graphic organizer. L portion will be completed at the end of the unit]

Table 10.

K **(What I know)**	**W** **(What I want to know)**	**L** **(What I have learned)**
What do you know about climate change?	What do you want to know about climate change?	

2. CBSN on the National Climate Assessment published on November, 24, 2018

https://www.youtube.com/watch?v=xiU6y12kZ80

Table 11.

What did you See?	*What do you Think?*	*What do you Wonder?*

3. Facts about Greenhouse Gases. Explore the site below, then create a fact sheet on greenhouse gases

https://www.epa.gov/ghgemissions/overview-greenhouse-gases

- What are greenhouse gases?
- What are the 4 types of greenhouse?
- How are greenhouse gases produced?

- How much of these gases are in the atmosphere?
- How long do they stay in the atmosphere?
- How strongly do they impact the atmosphere?

Homework

5. How might climate change affect the commonwealth of Massachusetts?
6. What activities do you engage in that might be contributing to climate change?
7. What actions, if any, can you take to reduce climate change?
8. Pose one or two questions that you would like to get answer about climate change.
9.

Day 2

Greenhouse Gas Emission in Massachusetts

1. Go to the website below and identify the sources of greenhouse emissions in the United States.

https://www.epa.gov/ghgemissions/sources-greenhouse-gas-emissions

2. What are the two major sources of greenhouse gases? Why?

Activity:

Working in pairs, you will be gathering data on greenhouse gas (carbon-dioxide) emissions from Power Plants and Waste Treatment facilities in your communities or neighboring communities in Massachusetts.

1. Go to the website below, and then click on the 2017 Data Summary Spreadsheets. Download the spreadsheets and complete their assigned tables. You will be collecting greenhouse gas levels under the "Direct Emitters) column.

https://www.epa.gov/ghgreporting/ghg-reporting-program-data-sets

2. After completing your assigned table, describe the patterns seen in the data collected, formulate questions, and generate hypotheses on what can be done to effect changes in carbon-dioxide emissions from your assigned facilities.

Groups 1-8
https://www.epa.gov/ghgreporting/ghg-reporting-program-data-sets

Table 12. Greenhouse gas emission from Berkshire Power Plant in Massachusetts

Emission from Power/ Waste Plant in Massachusetts (Metric Tons)	Greenhouse Gas Reporting Program for 2010			Greenhouse Gas Reporting Program for 2011			Greenhouse Gas Reporting Program for 2012			Greenhouse Gas Reporting Program for 2013		
	Carbon Dioxide	Nitrous oxide	Fluorinated Gases	Carbon Dioxide	Nitrous oxide	Fluorinated Gases	Carbon Dioxide	Nitrous oxide	Fluorinated Gases	Carbon Dioxide	Nitrous oxide	Fluorinated Gases

Emission from Power/ Waste Plant in Massachusetts (Metric Tons)	Greenhouse Gas Reporting Program for 2014			Greenhouse Gas Reporting Program for 2015			Greenhouse Gas Reporting Program for 2016			Greenhouse Gas Reporting Program for 2017		
	Carbon Dioxide	Nitrous oxide	Fluorinated Gases	Carbon Dioxide	Nitrous oxide	Fluorinated Gases	Carbon Dioxide	Nitrous oxide	Fluorinated Gases	Carbon Dioxide	Nitrous oxide	Fluorinated Gases
Berkshire Power 36 Moylan Lane Agawam, MA 01001												

1. What does the data tell you about greenhouse gas emissions at your assigned facility between 2010 - 2017?
2. What do you think is responsible for the changes in the data observed?
3. Based on the data, what questions would you like to ask the management of your assigned facility?
4. What, if any, action(s) can be taken by an individual to reduce the amounts of greenhouse gas emitted from your assigned facility?
5. What, if any, action(s) can be taken by society to reduce greenhouse gas emitted from your assigned facility?

Day 3

Calculating Carbon Footprints

Students will go to the website below to calculate their carbon footprint. The students will use the answers from their **Day 2** homework to assist them in completing the carbon footprint questionnaire. https://www3.epa.gov/carbon-footprint-calculator/

1. What did you learn about your carbon-foot print?
2. How does your carbon footprint compare to the national average?
3. What actions, if any, can you take to reduce your carbon footprint?
4. What impact, if any, might your actions have on global climate change?
5. What questions do you have based on what you learned from this process?
6. How might your actions impact the power companies?

7. How might your actions reduce the amount of trash that are sent to waste treatment facilities? What, if any, impact will this have on global climate change?
8. What is the overall impact that your actions might have on global climate change?
9. If we all participate in the reduction of our carbon footprint, what overall impact will this have on global climate change? Why?
10. What, if any, actions that should be taken to reduce global climate change?

Day 4

PBS News Hour

Published on April 10, 2019

Students will watch the video at the link below.

https://www.youtube.com/watch?v=YRe1ymYR45k

1. After watching the video, the students will respond to the following:

    a. What did you hear?
    b. What do you think?
    c. What do you wonder?

Activity

2. In pairs, read through the events in your assigned region then do the following:

https://www.cnn.com/2018/11/27/health/climate-change-impact-by-region/index.html

3. Events in your assigned region:

    a. Summarize how climate change will impact your assigned region of the country.
    b. Explain what concerns you might have based on your summary.
    c. What actions, if any, should be taken to reduce the risk described in your summary?
    d. Identify who, if anyone, is responsible for carrying out the action(s) identified.

Assigned Regions

Northeast
- Southeast
- Midwest
- California and Southwest
- Northwest
- Caribbean
- Alaska
- Hawaii and Pacific Islands

Day 5

Name: ____________________ Section: ____________________ Date: __________

Fourth National Climate Assessment Summary Findings

1. Student work in small groups.
2. Go to the Fourth National Climate Assessment Summary findings below https://nca2018.globalchange.gov/
3. Complete the information for their assigned group.

Group 1

Table 13.

	Communities	Economy	Interconnected Impacts
Main Idea			
Detail			

Group 2

Table 14.

	Actions to Reduce Risks	Water	Health
Main Idea			
Detail			

Group 3

Table 15.

	Indigenous Peoples	Ecosystems and Ecosystem Services	Agriculture and Food
Main Idea			
Detail			

Group 4

Table 16.

	Infrastructure	Oceans and Coasts	Tourism and Recreation
Main Idea			
Detail			

4. You will use the main ideas and details identified from the risks posed by climate change to help you in preparing a letter to the CEO of the power or waste treatment company you studied in day 2 of the unit.
5. Your letter to the CEOs must explain the research you have done in this unit. You will then invite the CEOs or their representatives to a round table discussion on what individuals, the community, and society can do to combat climate change.
6. Before you write your letter, use the graphic organizer below to help in organizing your thoughts. This will help them to write a more compelling letter.

Organize your thoughts in the table below

Table 17.

Question: The recent report from the Intergovernmental Panel on Climate Change (IPCC) recently reported on the impacts of global warming of 1.5 degree Celsius above pre-industrial levels. While the panel report indicates that human activities are the main result of the increase in global temperature, many policy makers at the federal level in the United States dismiss global warming as a hoax. Based on what you have learned about climate change, do you believe human activities contribute to climate change? Your job is to outline your ideas to this question in the graphic organizer below.			
FOR		**AGAINST**	
Reasons to be FOR the issue	**List at least two facts (evidence) for each reason**	**Reasons to be AGAINST the issue**	**List at least two facts (evidence) for each reason**
Reason 1	Fact	**Reason 1**	Fact
	Fact		Fact
Reason 2	Fact	**Reason 2**	Fact
	Fact		Fact
Reason 3	Fact	**Reason 3**	Fact

Day 6

Putting It All Together: A Persuasive Essay

Using all of the evidence that you have gathered about the Intergovernmental Panel on Climate Change (IPCC) report, write a letter to the Chief Executive Officer (CEO) of the power plant or waste treatment facility you researched in day 2. Your letter must explain the research you engaged in, what you have learned from the study, and your desire to invite him/her to a round table discussion on what individual, communities, and society can do to help reduce their carbon footprint. Use the template below to help out in organizing your thoughts. Please remember to refer to the reasons and facts you compiled in day 5 to help with your letter.

Final Persuasive Letter Rubric

Essay Rubric: Checklist

Table 18.

Description	Possible Points	Points Earned
A title that clearly defines your position	2	
An attention-grabbing beginning	2	
Includes at least 3 supporting reasons	6	
Provides evidence for each reason	6	
Includes AT LEAST 2 sources for all of your claims	2	
Includes a call to action at the conclusion	2	
Total	20	

Chapter 5
Blending Creative Drama and Computer-Supported Collaborative Learning for Socioscientific Argumentation

Aysegul Oguz Namdar
https://orcid.org/0000-0002-6853-8507
Recep Tayyip Erdogan University, Turkey

Bahadir Namdar
https://orcid.org/0000-0002-5076-6034
Ege University, Turkey

ABSTRACT

Argumentation has received a significant amount of attention from the science education community. To implement this important teaching method and scientific practice in science classrooms, curriculum reforms have recommended incorporating socioscientific issues. Research indicates that students have difficulty organizing information about complex socioscientific issues; students also tend to ignore counterclaims and only use evidence that supports their own views. Therefore, alternative teaching strategies should be used to overcome this barrier. This chapter advocates for blending creative drama and computer-supported collaborative learning in the context of socioscientific argumentation; it concludes with an instructional sequence for successful decision making about these issues.

DOI: 10.4018/978-1-7998-4558-4.ch005

INTRODUCTION

Argumentation, the process of justifying claims through evidence, has been receiving more attention than ever before—not only at the national level, but in the international science education community as well (Ministry of National Education, 2018; National Research Council, 2012). One of the core practices of science, argumentation has been highlighted in recent research because it promises to increase students' content knowledge, increase nature of science understandings, and improve motivation (e.g., Dawson & Venville, 2010; Eşkin & Ogan-Bekiroğlu, 2009; Uluçınar-Sağır & Kılıç, 2013; Zohar & Nemet, 2002).

Argumentation has been widely incorporated in science education in efforts to cultivate scientifically literate citizens. Scientifically literate citizens are expected to engage in evidence-based decision making about contemporary real-life issues (Hofstein, Eilks, & Bybee, 2011). These socioscientific issues, so-called because of their frequent ties to science and society, are dilemmas that society faces alongside advancements in science and technology (Sadler, 2004). The loosely structured assignments in SSI curriculum deal with issues that directly or indirectly affect the lives of students. Therefore, engaging students in evidence-based discussions and decision making about these issues is an important method for educating scientifically literate citizens (Lin & Mintzes, 2010).

When making decisions about socioscientific issues, students engage in informal reasoning processes; they weigh the advantages and disadvantages of these complex issues (Means & Voss, 1996). Unlike formal reasoning—in which premises are fixed and problems are solved through mathematics, logic, and the use of the available data alone—informal reasoning involves alterations to arguments as new evidence becomes available. Argumentation about socioscientific issues involves the verbal or written use of informal reasoning and could be used effectively to assess students' informal reasoning skills (Topçu, Yılmaz-Tüzün, & Sadler, 2011). Furthermore, researchers have indicated that argumentation makes students aware of the complexity of socioscientific issues (Chang-Rundgren & Rundgren, 2010) and helps them to develop informal reasoning skills and higher order thinking skills (Lindahl, Folkesson, & Zeidler, 2019). This occurs because in socioscientific argumentation, students engage in discursive practices such as evaluating evidence, assessing alternative perspectives, and addressing counterclaims (National Research Council, 2012).

In this research, we identified two major challenges that students encounter when engaging in socioscientific argumentation. First, research in science education indicates that students have difficulties using balanced reasoning because students tend to reason in biased ways—ignoring the counter-evidence and only using evidence that supports their own views (Sampson, Enderle, Grooms, & Witte, 2013). Research has also indicated that students tend to search for data that could potentially confirm their initial claim, instead of considering alternative information sources. They rarely realize that incorporating multiple perspectives and rebuttals increases the persuasiveness of arguments (Nussbaum & Kardash, 2005).

Second, the advent of new communication technologies offers access to a vast amount of information about these complex issues online, distributed via multiple data sources and presented in multiple formats. Students are asked to read and make sense of data and arguments in a "critical manner so as to identify their strengths and weaknesses" (National Research Council, 2012). However, research shows that when working on socioscientific argumentation, students grapple with identifying and organizing information from multiple resources in these multiple formats (Namdar, 2015).

Therefore, in this chapter we suggest that there is a need for blended learning opportunities that could provide students with a space to (a) organize knowledge in multiple modalities and formats and (b) discover counterarguments, which could potentially increase students' socioscientific argumentation quality by introducing multiple sides of an issue and encouraging deeper processing (Nussbaum & Schraw, 2007). We propose blending creative drama and computer-supported collaborative learning (CSCL) to overcome the two challenges identified above.

BACKGROUND

Socioscientific Argumentation

An argument is an assertion supported by a justification or reason (Zohar & Nemet, 2002). Toulmin (1958) defined the components of an argument as follows: claim, data, warrant, backing, rebuttal, and qualifier. The claims are a conclusion, or the answer to a question. The data are the facts and information that support the claim. The warrants are the statements that connect data to claim; backings justify or explain the warrant. The rebuttals are the arguments that contradict the counterclaims presented. Finally, qualifiers are statements that explain the conditions under which the claims are valid. Argumentation is the process of creating arguments by connecting claims and data through justification (Jiménez-Aleixandre & Erduran, 2008). Argumentation related to SSI (socioscientific argumentation) is the process of discussing these open-ended and ill-structured problems (Sadler & Donnelly, 2006).

Research shows that strong socioscientific argumentation is tied to strong informal reasoning (Topçu et al., 2011). Using multiple perspectives to reason about an SSI could be an indication of high-quality argumentation; drawing on different perspectives allows learners to construct counterarguments from different angles. Several studies identify the quality of participants' informal reasoning by assessing the number of reasoning modes or patterns used put forward in argumentation. For instance, Tal and Kedmi (2006) identified the number of aspects that students used in their arguments about a local SSI to determine quality of arguments. Wu and Tsai (2007) investigated high school students' informal reasoning about nuclear energy by using qualitative and quantitative reasoning indicators. They assessed the quality of argumentation by identifying the number of reasoning modes and justifications provided in each of the argumentation components. Similarly, in their study with college students, Topçu et al. (2011) identify preservice science teachers' informal reasoning about gene therapy, human cloning, and global warming by identifying claims, justifications, counter-positions, and rebuttals.

Role of Creative Drama in Socioscientific Argumentation

Although diverse points of view are pervasive, and their use in socioscientific argumentation has been proven to be important, research indicates that learners across age levels lack the critical argumentation skills needed to reason from multiple perspectives during socioscientific argumentation. In their study with Turkish preservice science teachers, Öztürk and Yılmaz-Tüzün (2017)found that the participants could not use multiple perspectives when arguing about nuclear energy issues. Similarly, Atasoy (2018) also found that preservice teachers only reasoned from a single perspective when they argued about hydroelectric power plants. To overcome this challenge creative drama could be incorporated in socioscientific argumentation instruction.

There has been an increasing interest in drama in science education as it promotes active participation of students in their own learning, increasing motivation, and student perception of empowerment and ownership (Odegaard, 2003). Dorion (2009) argued that in different fields of drama the description of the activities are so incomplete that there is a dispute between the definitions of it, and provided a more comprehensive definition. According to Dorion "drama could be seen to be the combination of three features: *role play* within an *imagined situation*, and enacted within the *human dimension*" (p. 2249).

Major reasons for incorporating drama in science education is to engage students in moral and ethical reasoning processes through empathy, as well as increasing students' own metacognitive awareness of their morality (Dorion, 2009). In her extensive literature review on drama in science education, Odegaard (2003) argues that drama can be used to highlight the connection between science and society. That is, drama can allow students to explore complex situations that create empathy and reflect diverse interests and ethical dilemmas. In line with this review, the literature suggests that drama can increase students' science knowledge and supporting argumentation skills by representing scientific debate (Braund, 2015) and student centered discourse (Somers, 1994).

In science education literature role plays are usually used to describe drama activities (Dorion, 2009). Role plays, a theatre technique is accepting and adopting the persona of another individual through impersonation (Adıgüzel, 2015), are suitable for teaching moral ethical issues and make students explore their place in the environment (McSharry & Jones, 2000). It has been frequently used in socioscientific decision making situations. Through role play, participants may develop empathy toward the situations and parties in question—situations that could be difficult or unlikely in the student's daily life. Using this technique learners freely express themselves and improve their communication and decision-making skills (McNaughton, 2004). Researchers incorporate role plays as reading role cards in small groups and engaging decision making through discourse without impersonation (Agell, Soria, & Carrio, 2015), assigning different roles of stakeholders coupled with classroom debates (Simonneaux, 2001), using court trials based on the history of science and historical empathy on evolution (Duveen & Solomon, 1994). A recent study by Archila (2017) suggests a teaching-learning sequence combining argumentation and drama to teach an ethical science topic in five steps as shown in Table 1.

Table 1. A teaching-learning sequence suggested by Archila (2017)

Steps	Explanations
(a) Initial argumentation	Before a classroom debate, students are encouraged to build their initial arguments about a given problem.
(b)Argumentation before debating	Students are provided the scenario and asked to reenact scenes from a play. They reflect on their initial arguments, considering the roles of the characters in the scenes. Meanwhile, students are also given a questionnaire to reflect on both the arguments put forward by the characters and on their own opinions
(c)Small group debate	Students work in groups of four to make decisions about the positions enacted in the reenactment. Students reflect on the arguments of the characters and come to a consensus in their small groups
(d)Whole class debate	Groups presents their decisions about the problem and teachers facilitate the discussion in the classroom
(e)Final argumentation	Students construct their final argument based on the previous steps

As can be seen from the literature students are expected to get into their parts by simply reading role (information) cards and in some situations associated data. Similarly, Odegaard (2003) argues that most interventions incorporate improvisational role plays rather than scripted plays. Furthermore, these studies do not sufficiently engage students in the role that is being played or giving proper time and environment for scripted role plays. Yet the relationship between drama and argumentation is not a well-established area of research and instruction; there is little empirical evidence on the effectiveness of interventions combining these two.

Paul and Elder (2009) argue that there is a relationship between creativity an ethical reasoning.

The world we live in is a world of interdependence, of impacting the lives of many other persons and sentient creatures. As the world becomes increasingly more complex and humans become increasingly more interdependent, our ability to reason ethically becomes ever more important. And our ability to reason ethically directly depends on our skills in criticality and creativity. For example, ethical reasoning entails thinking critically about (i.e. taking command of) our native egocentric and sociocentric tendencies. It depends upon our ability to find creative means for dealing with these destructive predispositions. (p.123)

Therefore, in this study we suggest using creative drama instead of solely using role plays to provide spaces for students so that they can find "creative means" for making decisions about these issues.

Creative drama can be defined as the meaning making of an event, an abstract concept, or a behavior through the representation/reenactment of previous experiences or new observations as a group. Creative drama involves individuals' own creativity, memories and experiences. Through these processes and improvisations, individuals explore their lived experiences, emotions, and attitudes (San, 1990).

Lessons structured based on creative drama method consists of three stages (Adıgüzel, 2006)

- Preparation/Warm Up stage: The purpose of this stage is to create a group dynamic and prepare students to the following stage. In this stage drama leader (the teachers) makes sure that the students are focused on the task and kinesthetically prepares students to task. In this stage harmony and trust studies are conducted to make participants work together in the upcoming stages.
- Impersonation Stage: Improvisation and role plays are two significant drama techniques that are used in impersonation stage. As the process of drama is not precisely laid out in improvisation, this technique allows the use of creativity. Improvisation can be conducted both individually and as a collective group effort. Role plays are activities in which individuals take a role of an identity or another person by accepting their place and their functions.

- Evaluation/Discussion Stage: In this stage, participants' feelings and their ideas about the process is explored. Both verbal and written techniques are used as well as drama techniques for evaluating the whole dramatic experience.

Role of CSCL for Knowledge Organization in Socioscientific Argumentation

Another challenge in socioscientific argumentation is the utilization of multiple representations in argumentation processes. Representations are fundamental tools of scientific discourse. These tools visualize and conceptualize entities or processes in another format and/or mode. Scientists use more than one representation to design experiments, explain findings, and disseminate claims. They also use representations to organize knowledge and to justify their claims and explanations (Lemke, 1998). Therefore, science can be defined as a form of multimodal discourse.

In K12 science education, teachers widely and frequently use multiple representations (Ainsworth, 2008). Similarly, in argumentation processes, multiple representations are used to define claims, support those claims, and evaluate ideas (Hand & Choi, 2010; Pallant & Lee, 2015). Students are also capable of creating and presenting their own representations (DiSessa, 2004). However, they have difficulties organizing these representations into an accurate whole that can be used to support and refute claims and ultimately to reach a decision about complex socioscientific issues.

CSCL environments provide contexts and opportunities for students to meaningfully organize and share representations in multiple modalities together. In these environments, learners construct intersubjective knowledge and make that knowledge visible to other learners (G. Stahl, Ludvigsen, Law, & Cress, 2014). In this process, learners can work both individually and collaboratively to construct arguments about complex socioscientific issues (Namdar & Shen, 2018). As CSCL expose students to alternative ideas and perspectives about complex issues and prompt them to consider the reasoning of others, a challenge becomes to navigate multiple representations for an informed decision making.

The current chapter is a conceptualization of how teachers can engage students in socioscientific argumentation based on a revised version of the SSI-Online-Argumentation Pattern (SOAP) suggested by Tsai (2018). The SOAP strategy engages students in controversial dilemmas so that they have an opportunity to express their ideas and explore ideas that differ from those of their peers. Regarding the online component of this strategy, Tsai discusses the benefits of online learning communities in terms of time and space, bidirectional communication, learning, and collecting data.

The SOAP strategy also advocates for engaging students in evidence-based reasoning and justification of opinions. When engaging in this kind of socioscientific argumentation, students are exposed to extensive information presented in multiple modalities. The literature indicates that they should organize their knowledge to engage in informed decision-making processes, reflected by sound arguments. Knowledge organization is defined as the externalization of knowledge via selecting, creating, clustering, tagging, sharing, and archiving in external representations (in this case, presented in CSCL environments) (Namdar & Shen, 2018). Hence, for the current investigation, knowledge organization was added as an additional step to the previous SOAP strategy. We named our strategy SOKAP: SSI Online Learning Knowledge Organization and Argumentation Pattern. It is important to note that we do not envision this strategy following a linear sequence—knowledge organization can be revisited after online knowledge sharing and argumentation processes (Figure 1).

We approach socioscientific argumentation as a collaborative process of building arguments and creating informed opinions. Small-group and whole-group discussions are indispensable tools for argumentative classroom interactions (Archila, 2017). Because research indicates that drama can be used to engage students in argumentation (Archila, 2017), we hope to combine drama and argumentation in line with the idea that both could enhance students' motivation for decision making about controversial socioscientific issues.

Figure 1. SOKAP strategy

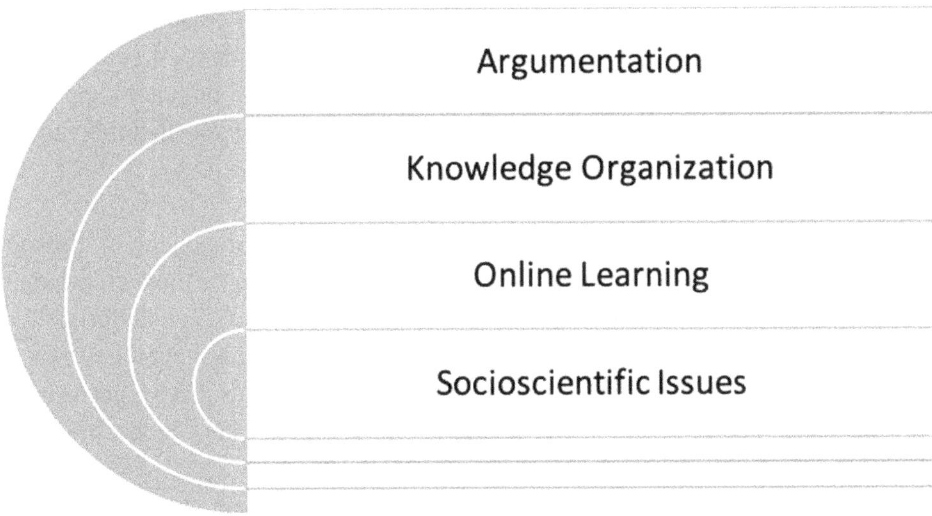

PROPOSAL

Blending Drama and SOKAP as a Flipped Classroom Strategy

Blended learning is a descriptor for formal education programs in which a course takes place partially online and partially face-to-face. Blended learning provides opportunities for students to control their learning pace, path, and time (Staker & Horn, 2012). It combines the strengths of in-person learning with online learning opportunities. These information and communication technologies increase student motivation, flexibility, reflection, and content knowledge learning experiences. Flipped classroom, as a model of blended learning, is based on the idea that the teachers should act as guides, instead of the sole source of knowledge (King, 1993). This model advocates for content instruction to take place at home, while students use in-class time to practice, elaborate, and reflect. Traditionally, most classroom time in colleges and universities is devoted to learning content knowledge. This gives learners limited time and space for reasoning (Du, Fu, & Wang, 2014). Out-of-class time is devoted to learning strategies such as projects, problem-solving contexts, and homework (Teo et al., 2014). Flipped classrooms have the opposite situation: students learn the subject matter mainly by means of video, with online course materials and tasks prepared by the teachers. In the classroom, students actively participate in their own learning processes (Lasry, Dugdale, & Charles, 2014). The flipped classroom model provides more flexible learning spaces because students are able to make decisions as to when, where, and how often they want to learn the course content. Therefore, students have the main responsibility in their learning processes. In-class time, then, is devoted to inquiry processes and laboratory activities.

To overcome the challenges associated with students' socioscientific argumentation skills identified above, we propose blending creative drama and computer-supported collaborative learning via a flipped classroom strategy. We believe that this model is also innovative in three ways.

- Providing a CSCL environment during an online component of flipped classroom provides opportunities for students to organize their knowledge through representations. As CSCL environments also has platforms for comment, share, discuss this could support knowledge building processes at a collective level (Cress & Kimmerle, 2008). In other words, CSCL potentially provides equal opportunities for students to explore knowledge base for increased knowledge about a given socioscientific issue.
- Second, as we discussed previously role play activities are important but yet could be considered insufficient to support role taking. In creative drama, role plays are considered as a part of impersonation stage. However, interventions

in the socioscientific argumentation literature points out a lack of proper alignment of drama stages. Through this model we provide preparation activities for students' kinesthetic and cognitive engagement to the dramatic activity, create a group dynamic, provide a space to better prepare themselves.

- Third, literature also suggests that students usually do not know the discussion rules (Sampson, Enderle, & Grooms, 2013). Teachers also sees that a students' domination of group discussion as one of the obstacles (Namdar & Shen, 2018). Therefore, creative drama can provide equal participation opportunities in the discussion.

The teaching strategy follows seven steps. We propose that four steps of this strategy take place online, while the remaining three steps occur face-to-face. In the following section we define and outline each step. Figure 2 illustrates the SOKAP strategy in a flipped classroom model (Figure 2).

Figure 2. SOKAP in a flipped learning model

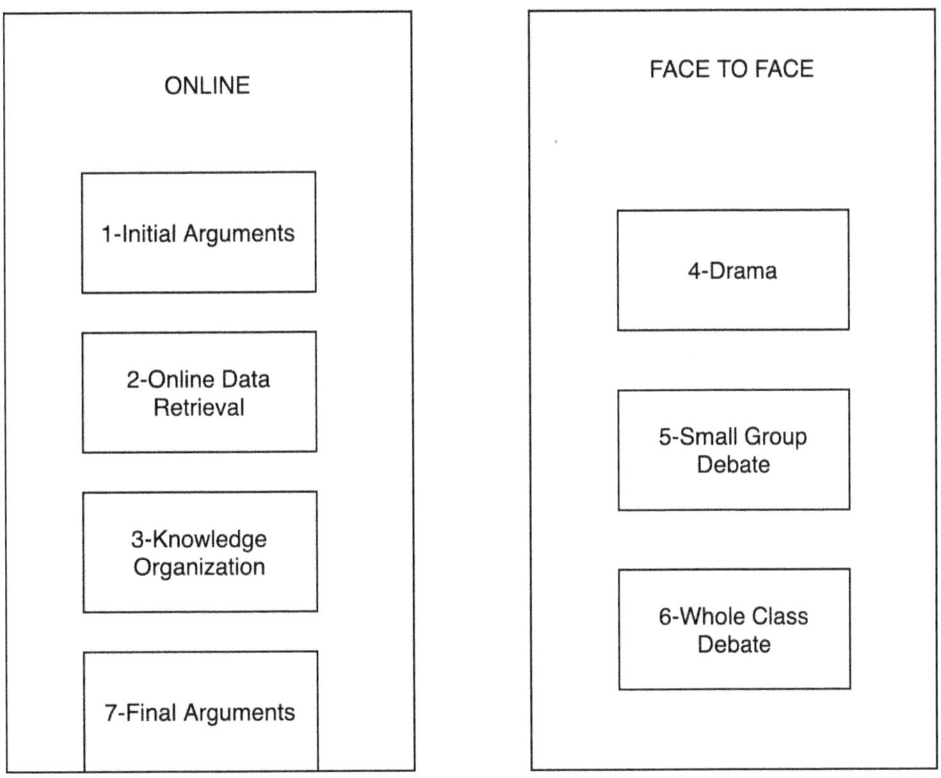

Teaching Sequence

Step 1: Initial Arguments

This step aims at getting students interested in the issue. Teachers create a reading or video covering two sides of the issue in question and share it via a CSCL environment. After students' curiosity about the issue has been piqued, they are asked to write down their initial arguments in the CSCL environment, using a synchronous or asynchronous discussion board. This step is important to make further judgments about students' reasoning. The written arguments provide opportunities for teachers to access students' reasoning and value-based judgments about these issues (Jiménez-Aleixandre & Erduran, 2008). Choi, Lee, Shin, Kim and Krajcik (2011) indicated 21st century citizenry requires the use of three values to reason about socioscientific issues. These include ecological worldview, social and moral compassion, and socioscientific accountability. By asking students to write their arguments teachers can see which values are expressed and which ones are missing so that they could use scaffolds such as explicit role-cards or information cards focusing on the missing value.

Step 2: Online Data Retrieval

After collecting students' initial arguments about a given socioscientific issue, instruction should promote learners' active participation in retrieval of evidence (Sadler, 2011).The internet can be used to collect and retrieve data and to access local and global resources easily. In the second step, students should be prompted to search about the different aspects of the given socioscientific issues. This could potentially increase the quality of students' informal reasoning; the literature states that students often argue about complex socioscientific issues considering single aspect of an issue.

Students gather information from different sources such as peers (Garcia-Mila & Andersen, 2007) and mass media, including newspapers, books, television, and online information (Maloney, 2007; Osborne, Erduran, & Simon, 2004). The ability to obtain and analyze information is an essential aspect of media literacy. Media literacy, which is generally defined as "the ability to access, analyze, evaluate and communicate messages in a wide variety of forms" (Aufderheide, 1993), focuses on the process skills needed to utilize media. Livingstone (2004) suggests that each skill supports the others in a non-linear process: "learning to create content helps one to analyze that produced professionally by others; skills in analysis and evaluation open the doors to new uses of the internet, expanding access, and so forth" (p. 3).

Although the traditional definition of literacy still prevails in K-12 education, teachers and researchers have begun to expand the definition to include media literacy. Media is more immediately available and accessible than ever. Importantly, media literacy is also inherently connected to both scientific literacy—which is defined in part as being able to engage in decision-making processes about scientific issues related to politics, economics, morality, and ethics—and decision-making about SSI, which includes being able to assess conflicting views and evaluate the trustworthiness of sources of data (Dani, Wan, & Henning, 2010; Klosterman, Sadler, & Brown, 2012). Media is at the heart of both of these concepts. Therefore, we consider media literacy when assessing students' scientific literacy and decision-making abilities about SSI. This becomes particularly important in the context of socioscientific decision making as children rely more heavily on the first information they encounter, and they get most information about socioscientific issues from media (S. Stahl, Hynd, Britton, McNish, & Bosquet, 1996)

Step 3: Knowledge Organization

Online data is presented in multiple formats, including tables, graphs, audio, video, and pictures. Considering that the socioscientific issues are inherently complex and include multiple aspects, data regarding these issues is distributed across different platforms in multiple modalities. Therefore, learners must meaningfully cluster the data to make decisions. We suggest that in the third step, *knowledge organization* (Namdar & Shen, 2018), students should be provided with a CSCL environment that can engage them in knowledge organization processes.

Knowledge organization process starts with externalizing knowledge in the CSCL platform via multiple external representations. Knowledge of course, not necessarily mapped onto the representations as it is in an individual's mind. CSCL allows students to cluster information such as assigning keywords to representations with similar content, and annotating pictorial representations. This process allow students for extending their knowledge due to deep processing (Matuk & Linn, 2012).

This individualized process will also have collective effect as the representations will be clustered based on the similarities and shared through a CSCL environment. Representations with similar content will be connected by CSCL platforms and then the representations becomes an epistemic artifact of a jointly created knowledge base. The knowledge then becomes part of a joint artifact and others in the learning environment can access it. Through this process, contributors can learn more from others' external representations and in return contribute to the knowledge base by creating more representations based on that new information. For this reason, we advocate for using CSCL technology that could potentially turn individual efforts into collective products.

Results of previous studies indicated that the learners mostly relied on textual representations rather than visual representations when organizing knowledge about complex socioscientific issues such as nuclear energy (Namdar & Shen, 2016). Furthermore, literature also illustrated that students' knowledge organization was important means in socioscientific argumentation practices as they were drawing information from organized representations (Namdar & Shen, 2016). Therefore, it is important to explicitly teach how to organize knowledge and also use it in socioscientific argumentation for informed decision making.

Step 4: Creative Drama

In the fourth step, *creative drama*, students engage in drama activities that should allow them to express the opinions (in)formed through the previous steps in a classroom environment. During this step, drama participants also act as the audience, hearing their peers' positions about the given SSI. Drama gives opportunities for students to consider not only their own positions about an issue, but also the counter-evidence and other evidence that does not support their views.

This section could be divided into three sub-steps. First, in the warm-up stage, the participants are prepared both physically and mentally for the dramatic exercise through games and exercises. This section is particularly important as to prepare students for the impersonation stage, which is usually missing in the socioscientific argumentation studies. Drama suggests that students get into their parts by first forming a group dynamic. Specifically, students who are not familiar with creative drama activities should be engaged in a warm-up activity so that they get use to each other and further empathize with others in the group (Akkocaoğlu Çayır & Erdoğan, 2017).

Second, students should discuss and review what they have found in their online learning and how they organized their knowledge about the issue in small groups. The impersonation stage includes the role-playing. Belova, Feierabend, and Eilks (2015) identified four type of role plays: (1) completely directed by a group of student moderators, or (2) by the teacher, (3) medium-quality role-plays with a certain amount of interactivity and free argumentation, and (4) real, spirited debates (p. 165). Their results further indicated that high quality argumentation rarely occurred during role-plays. Students are given role cards that include multiple personas with varying perspectives on the SSI. Through role-play, the teacher should remind students that they should create arguments and counter-arguments related to the issue for higher quality argumentation.

Step 5: Small Group Debate

The fifth and sixth step could be considered an evaluation stage of creative drama. After the role-playing, teachers should prompt students to make note of alternative explanations and counterclaims from other groups so that the students can discuss them in small groups and come to a consensus. Debates are important mediums for fostering critical thinking skills. In this step, students should be prompted to look for counterarguments and how to rebut them—i.e., high quality argumentation processes. Furthermore, students should also be encouraged to consider different aspects of the issue that did not come up during the impersonations. Debates allow students to understand the complexity of the issue, scientific principles, defend or criticize viewpoints, and evaluating facts surrounding the issues (Simonneaux, 2001). Particularly, small group debates gives students an active role in the discussion (Armstrong & Weber, 1991). A small group format gives students chances to explore multiple ideas and also change their position. Small group debate is also important for active involvement of more reserved students (Simonneaux, 2002).

Step 6: Whole Class Discussion

In the sixth step, the teacher initiates a *whole class discussion* and exposes students one last time to alternative positions about the SSI. There have been many studies that follow an instructional sequence starting with a small group debate followed by a whole class discussion (e.g. Namdar & Shen 2016). Class discussion following small group debates give opportunities for assessing the complexity of the issue (Shaw, 2012). During the discussion, the teacher can provide student groups with opportunities to present their arguments. This could be achieved through Web 2.0 technologies, such as encouraging students to create an infographic or a basic poster. As a guide during the discussion, a teacher should be mindful that the discussion covers different aspects of the SSI and the groups do not focus on a single aspect of an issue.

Step 7: Final Arguments

Finally, after leaving the classroom, students are asked to write their *individual final arguments* in the CSCL environment based on the previous steps. Dramatic influences are expected to manifest themselves in this stage as well. As learners are exposed to multiple perspectives and multiple values about the socioscientific issues, they are expected to transfer their reasoning in written arguments. Infact, the literature suggests that students can transfer their collaborative reasoning into writing (Dong, Anderson, Kim, & Li, 2008; Dong, Anderson, Lin, & Wu, 2009).

Table 2. Standards connections

MS-LS4-5. Gather and synthesize information about technologies that have changed the way humans influence the inheritance of desired traits in organisms. [Clarification Statement: Emphasis is on synthesizing information from reliable sources about the influence of humans on genetic outcomes in artificial selection (such as genetic modification, animal husbandry, gene therapy); and, on the impacts these technologies have on society as well as the technologies leading to these scientific discoveries.]		
Science and Engineering Practices: Obtaining, Evaluating, and Communicating Information Obtaining, evaluating, and communicating information in 6–8 builds on K–5 experiences and progresses to evaluating the merit and validity of ideas and methods. Gather, read, and synthesize information from multiple appropriate sources and assess the credibility, accuracy, and possible bias of each publication and methods used, and describe how they are supported or not supported by evidence.	**Disciplinary Core Idea: Obtaining, Evaluating, and Communicating Information** Obtaining, evaluating, and communicating information in 6–8 builds on K–5 experiences and progresses to evaluating the merit and validity of ideas and methods. Gather, read, and synthesize information from multiple appropriate sources and assess the credibility, accuracy, and possible bias of each publication and methods used, and describe how they are supported or not supported by evidence.	**Cross Cutting Concepts: Cause and Effect** · Phenomena may have more than one cause, and some cause and effect relationships in systems can only be described using probability. - ***Connections to Engineering, Technology, and Applications of Science*** **Interdependence of Science, Engineering, and Technology** · Engineering advances have led to important discoveries in virtually every field of science, and scientific discoveries have led to the development of entire industries and engineered systems. - ***Connections to Nature of Science*** **Science Addresses Questions About the Natural and Material World** · Scientific knowledge can describe the consequences of actions but does not necessarily prescribe the decisions that society takes.

This step could potentially give teachers and students an opportunity to see the progress in their arguments. Teachers could use several argumentation frameworks to assess this progress (for a review see Sampson & Clark, 2008).

Sample Lesson Plan

This lesson plan illustrates how we blend two important key ideas in socioscientific argumentation: drama for multiple perspective taking and computer-supported collaborative learning for utilizing multiple representations. We envision that three of the seven steps, namely creative drama, small group debate, and whole class discussion take place face-to-face while remaining steps take place online in a CSCL platform. We construct our lesson plan based on internationally accepted Next Generation Science Standards published in the United States (NGSS Leads States, 2013; https://www.nextgenscience.org/pe/ms-ls4-5-biological-evolution-unity-and-diversity)

Figure 3. iKOS PicTag Mode

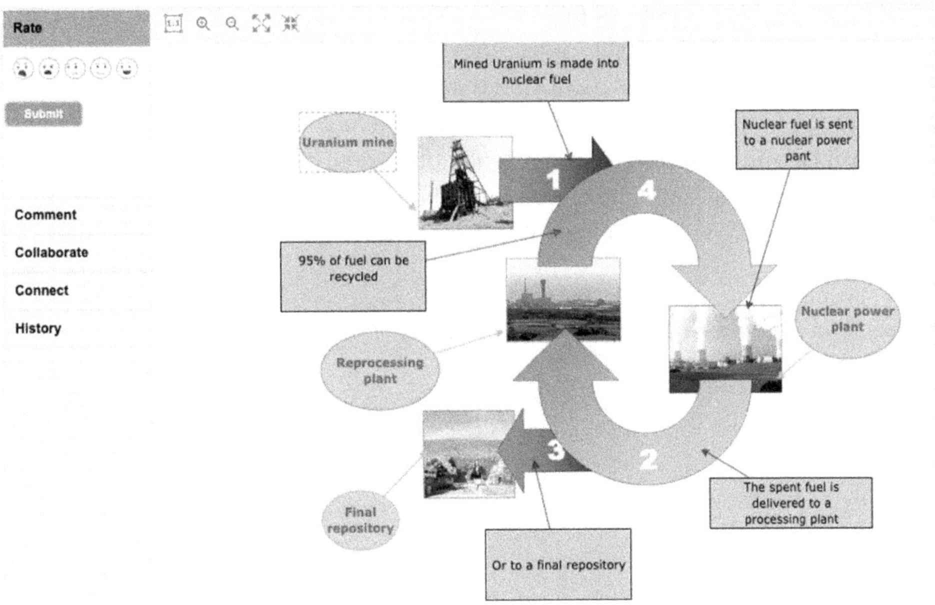

Timeline:

Step 1: Initial arguments: Online- 15 minutes
Step 2: Online data retrieval: Online- 60 minutes
Step 3: Knowledge organization: Online- 45 minutes
Step 4: Creative drama: Face-to-face- 45+45+45 minutes
Step 5: Small group debate: Face-to-face-45 minutes
Step 6: Whole class discussion: Face-to-face-20 minutes
Step 7: Final arguments: Online- 15 minutes
Materials: Computers with internet connection, role cards, stationary, pillows.

Step 1: Initial Arguments

To begin, the teacher opens an account in an open access, online platform called innovative Knowledge Organization system, or iKOS (www.ikos.miami.edu). This hypertext platform incorporates four representational modalities: PicTag, where students insert and annotate pictures (Figure 3); Wiki, where students can create textual representations similar to Wikipedia pages; Concept Map, where students can create graphical maps of their ideas; and Book, which allows students to create eBooks based on the representations that they created in the previous steps.

The platform also allows representations that share similar keywords, which are assigned by the creators, to create a knowledge web. When students utilize this knowledge web or "knowledge tree," they can explore the representations other students have created in the platform (Figure 4). The platform also incorporates collaboration tools such as commenting, rating, and co-creating a representation. Each box in Figure 4 represents a representation created by a student. When a student clicks on a box the system opens a dialogue box and asks if you wish to review the representation created by another student.

Figure 4. iKOS Knowledge Tree Mode

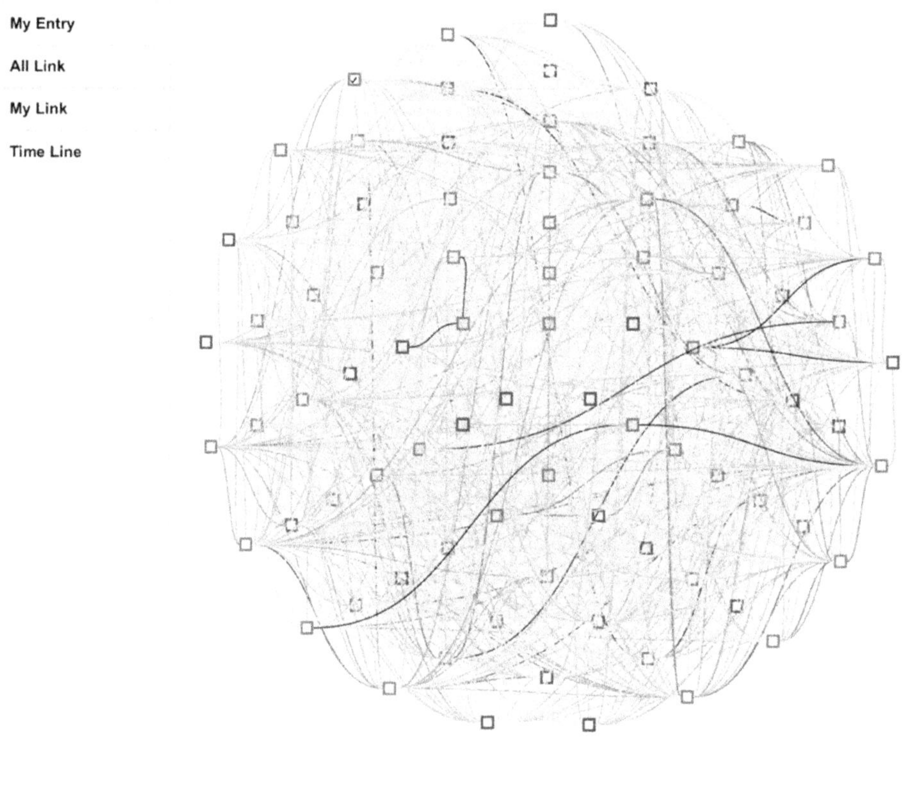

Within the software, the teacher first creates two Wiki entries about the benefits and challenges of GMOs. (This could also be done via videos.) The entries ask students: "Should the government ban the production of genetically modified food?" This step could be structured in two ways. First, the teacher can only post a generic question and asks student to write their arguments without giving any specific scaffolds. Second, the teacher prompts students to structure their argument by asking the following questions (c.f. Wu & Tsai, 2007): (a) Do you agree with the banning genetically modified food production? Why?, (b) If you want to convince someone else what evidence would you suggest to convince them? (c) Consider that there is someone else opposing your view. What arguments they might have? (d) Based on your hypothetical opponents' arguments, how would you justify your view? These four questions respectively address student's own position on the issue, supportive argument construction, counter-argument construction, and rebuttal construction. This could give students opportunities to explore multiple perspectives of the issue.

Step 2: Online Data Retrieval

Students are asked to freely explore the internet and find as much relevant data about the issue as possible. Throughout, students are encouraged to find and save data for later use. Students tend to ignore data that contradicts their ideas about complex issues (Sampson, Enderle, & Grooms, 2013). Therefore, in this step, teachers should remind students to search for data that both support and oppose their views. To better structure the online data retrieval process and to prevent cognitive load, students can be prompted to find certain number of data sources. Students are provided with an online worksheet and asked to critically evaluate the data they find. Some of the issues that a student should consider when evaluating online sources include: (a) source of the data, (b) intended audience, (c) accuracy of scientific information, (d) stakeholders involved, (e) writer's use of language, (f) author details, (g) whether biased opinions are expressed, (h) bibliography included, (i) other viewpoints expressed, and (j) whether it is up-to-date information.

Step 3. Knowledge Organization

In this step, learners are encouraged to find data in different modalities. The teacher reminds students to create a limited number of iKOS representations in the platform—this should ensure that the students mindfully select the most convincing data and not be overwhelmed by too many representations. During knowledge organization, students are asked to create keywords for their entries so that the iKOS environment could interlink other representations that share the same keywords. The teacher limits the creation of representations by one of each representation type.

Students are given one class period to complete the creation of representations in the iKOS platform. Students can invite others to co-create representations. Then, they are given time to investigate other representations in the platform and comment and rate the representations. In this way, students receive feedback and questions regarding their representations. This allows students to collaboratively learn about different aspects of the given issue. Overall, the knowledge organization process in the CSCL environment will allow students to both navigate through different pieces of knowledge and construct their own understanding and positions about the issue.

Step 4: Creative Drama

This step includes face-to-face activities. First, to prepare students mentally and physically to the role-play, students are engaged in two activities in the warm up stage. Activity 1 starts by forming three circles. Students count from 1 to 8 and forms 3 groups. Each participant takes their pillow and forms a circle by sitting on the floor. The leader distributes the puzzle pieces that she prepared before the class. Each puzzle will have a genetically modified food picture. After the groups finishes the puzzles they hang them on the wall. The leader initiates a discussion and asks students what these pictures represent. This activity aims at engaging students in the subject.

In warm up activity 2 students form a big circle in the class. A volunteer student comes to the middle of the circle. Spots, where each student standing, is marked with a piece of chalk. Each student is named as DNA, modification, and genetics. When the volunteer student says, for example, DNA, all the students named DNA switch spots. The student who stays in the middle becomes the next volunteer. When he says GMO all the students try to change their spots. Leader should also actively participate in the activities so that he/she contributes to the formation of a group dynamic.

Second, students are divided into groups of 6 randomly using an online tool, classtools (www.classtools.net).The groups are given role cards that define briefly the roles that will be played. Among the cards, students choose the role that they want to role-play. Because the SSI in question is GMOs, students are given role cards including (a) a parent who is concerned about GMOs, (b) a factory owner that produces GMOs, (c) a UN Food and Agriculture Council representative, (d) a scientist that produces transgenic foods, (e) an organic farm owner, and (f) a parent with seven children who lives in a metropolitan city area. Then, the groups are given a class period to prepare for their dramatization, noting that their scenario should clearly present their group's argument about GMOs. After the groups complete their scenarios, each group presents. When one group role-plays, other groups in the audience take notes about the strengths and weaknesses of the presenting group's argument, as well as the aspect of the issue that they tackle using the handout shown in Table 2. As discussed in background section, this will give students opportunities to explore alternative ideas and potentially will serve as an environment for high quality argumentation because students will be challenged by counterarguments and will create rebuttals.

For the evaluation stage, each group forms gathers around the learning stations namely painting, poem, slogan, and song. The leader distributes colored papers and pens. Then each group of students rotates through each station and continues to work on the station where the prior group left off. The purpose of this activity is to evaluate the drama activities. When all the groups rotate through all stations, the works are presented and the leader asks students what they have learned, their experiences regarding the drama, and how they felt about the process.

Step 5: Small Group Debate

In this part of the activity, students come back to their original groups and discuss the worksheets that they filled out during Step 4 for each group (Table 3). In this debate, students should focus on supporting or rebutting the claims put forward by the other groups. At the end of the debate, students create a poster or eBook (depending on time) via iKOS with multiple representations that reflect their argument about the GMO. The teacher should make sure that students present their arguments by focusing on multiple aspects of the issue.

Table 3. Small group debate handout

What is group #....'s claim about the GMO issue?		
What are the most convincing data that the group presented in their role-play?		
What are the weakest data points that you think do not support their claim? Why?		
Please circle the aspects of the issues mentioned:	Economy Ecology Political Environmental Science/Technology	

Step 6: Whole Group Discussion

The teacher creates a gallery walk, where the groups display their posters on the walls, asking one or two students to present their poster while other group members walk around and observe other posters. Finally, the teacher initiates the discussion with the entire class. This discussion provides an opportunity to check whether all aspects of the issue is covered during the presentations. The discussion ends with noting that these issues are controversial; decisions could change based on several factors, such as new health research and technological advancements.

Step 7: Final Arguments

The teacher asks students to write their individual final arguments on the iKOS platform. To see the progress of an argument, a framework for assessing socioscientific argumentation can be used: for instance, the socioscientific argumentation analysis rubric suggested by Tal and Kedmi (2006). This framework assesses the quality of socioscientific arguments based on number of justifications, use of scientific knowledge, number of aspects incorporated in the argument, and synthesis of counterargument and rebuttal. Thus, the teacher can see the progress of their students' socioscientific arguments by comparing initial and final arguments posted on iKOS platform.

SUMMARY

In this chapter, we provide a teaching sequence for socioscientific argumentation, incorporating reading, role-play, and writing. Integrating multiple representations into the student explanations and decision-making processes is a tool for fostering scientific literacy (Krajcik & Sutherland, 2010). This chapter also illustrates our rationale and methodology for the proposed sequence. First, we identify the importance of socioscientific issues and argumentation in science education. Second, we note two challenges in socioscientific argumentation and propose creative drama and computer-supported collaborative learning as two venues for overcoming these challenges. Third, we introduce the teaching and learning sequence blending these tools. Finally, we introduce a CSCL environment called "innovative Knowledge Organization System" (www.ikos.miami.edu) and give an example lesson plan incorporating this sequence in the context of using socioscientific argumentation to teach genetically modified organisms.

ACKNOWLEDGMENT

An earlier version of this chapter was accepted to be presented as a stand-alone paper at NARST 2020 Annual International Conference, Portland, OR, March 15-18, 2020. This research received no specific grant from any funding agency in the public, commercial, or not-for-profit sectors.

REFERENCES

Adıgüzel, A. (2015). *Eğitimde yaratıcı drama* [Creative drama in education]. Pegem Akademi.

Adıgüzel, H. Ö. (2006). Yaratıcı drama kavramı, bileşenleri ve aşamaları [The concept, components and stages of creative drama]. *Yaratıcı Drama Dergisi, 1*(1), 17–27.

Agell, L., Soria, V., & Carrio, M. (2015). Using role play to debate animal testing. *Journal of Biological Education, 49*(3), 309–321. doi:10.1080/00219266.2014.943788

Ainsworth, S. (2008). The educational value of multiple-representations when learning complex scientific concepts. In J. K. Gilbert, M. Reiner, & M. Nakhleh (Eds.), *Visualization: Theory and practice in science education* (pp. 191–208). Springer. doi:10.1007/978-1-4020-5267-5_9

Akkocaoğlu Çayır, N., & Erdoğan, T. (2017). Dramada planlama [Planning in drama]. In T. Erdoğan (Ed.), *Okul öncesinden ilköğretime kuramdan uygulamaya drama* [Drama from preschool to elementary, theory to practice]. Eğiten Kitap.

Archila, P. (2017). Using drama to promote argumentation in science education. *Science & Education, 26*(3–4), 345–375. doi:10.100711191-017-9901-7

Armstrong, K., & Weber, K. (1991). Genetic engineering: A lesson on bioethics for the classroom. *The American Biology Teacher, 53*(5), 294–297. doi:10.2307/4449297 PMID:11655276

Atasoy, Ş. (2018). Öğretmen adaylarının yaşam alanlarına göre yerel sosyobilimsel konularla ilgili informal muhakemeleri [Student teachers' informal reasoning of local socioscientific issues according to living places]. *Fen Bilimleri Öğretimi Dergisi, 6*(1), 60–72.

Aufderheide, P. (1993). *Media Literacy: A report of the national leadership conference on media literacy*. Aspen Institute.

Belova, N., Feierabend, T., & Eilks, I. (2015). The evaluation of role-playing in the context of teaching climate change. *International Journal of Science and Mathematics Education, 13*(S1), 165–190. doi:10.100710763-013-9477-x

Braund, M. (2015). No Drama and learning science: An empty space? *British Educational Research Journal, 41*(1), 102–121. doi:10.1002/berj.3130

Chang-Rundgren, S. N., & Rundgren, C.-J. (2010). SEE-SEP: From a separate to a holistic view of socioscientific issues. *Asia-Pacific Forum on Science Learning and Teaching, 11*(1), Article 2.

Choi, K., Lee, H., Shin, N., Kim, S., & Krajcik, J. (2011). Re-conceptualization ofscientific literacy in South Korea for the 21st Century. *Journal of Research in Science Teaching*, *48*(6), 670–679. doi:10.1002/tea.20424

Cress, U., & Kimmerle, J. (2008). A systemic and cognitive view on collaborative knowledge building with wikis. *International Journal of Computer-Supported Collaborative Learning*, *3*(2), 105–122. doi:10.100711412-007-9035-z

Dani, D., Wan, G., & Henning, J. E. (2010). A case for media literacy in the context of socioscientific issues. *New Horizons in Education*, *58*(3), 85.

Dawson, V. M., & Venville, G. (2010). Teaching strategies for developing students' argumentation skills about socioscientific issues in high school genetics. *Research in Science Education*, *40*(2), 133–148. doi:10.100711165-008-9104-y

DiSessa, A. (2004). Metarepresentation: Native competence and targets for instruction. *Cognition and Instruction*, *22*(3), 293–331. doi:10.12071532690xci2203_2

Dong, T., Anderson, R. C., Kim, I., & Li, Y. (2008). Collaborative reasoning in China and Korea. *Reading Research Quarterly*, *43*(4), 400–424. doi:10.1598/RRQ.43.4.5

Dong, T., Anderson, R. C., Lin, T., & Wu, X. (2009). Concurrent student-managed discussions in a large class. *International Journal of Educational Research*, *48*(5), 352–367. doi:10.1016/j.ijer.2010.03.005

Dorion, K. R. (2009). Science through Drama: A multiple case exploration of the characteristics of drama activities used in secondary science lessons. *International Journal of Science Education*, *31*(16), 2247–2270. doi:10.1080/09500690802712699

Du, S., Fu, Z., & Wang, Y. (2014). The flipped classroom – advantages and challenges. In *International Conference on Economic Management and Trade Cooperation* (pp. 17–20). Atlantis Press.

Duveen, J., & Solomon, J. (1994). The great evolution trial: Use of role play in the classroom. *Journal of Research in Science Teaching*, *31*(5), 575–582. doi:10.1002/tea.3660310510

Eşkin, H., & Ogan-Bekiroğlu, F. (2009). Investigation of a pattern between students' engagement in argumentation and their science content knowledge: A Case Study. *Eurasia Journal of Mathematics. Science and Technology Education*, *5*(1), 63–70.

Garcia-Mila, M., & Andersen, C. (2007). Cognitive foundations of learning argumentation. In S. Erduran & M. P. Jimenez-Aleixandre (Eds.), *Argumentation in science education: Perspectives from classroom-based research* (pp. 29–45). Springer. doi:10.1007/978-1-4020-6670-2_2

Hand, B., & Choi, A. (2010). Examining the impact of student use of multiple modal representations in constructing arguments in organic chemistry laboratory classes. *Research in Science Education*, *40*(1), 29–44. doi:10.100711165-009-9155-8

Hofstein, A., Eilks, I., & Bybee, R. (2011). Societal issues and their importance for contemporary science education—A pedagogical justification and the state-of-the-art in Israel, Germany, and the USA. *International Journal of Science and Mathematics Education*, *9*(6), 1459–1483. doi:10.100710763-010-9273-9

Jiménez-Aleixandre, M. P., & Erduran, S. (2008). Argumentation in science education: An overview. In S. Erduran & M. P. Jimenez-Aleixandre (Eds.), *Argumentation in science education: Perspectives from classroom-based research* (pp. 3–27). Springer.

King, A. (1993). From sage on the stage to guide on the side. *College Teaching*, *41*(1), 30–35. doi:10.1080/87567555.1993.9926781

Klosterman, M. L., Sadler, T. D., & Brown, J. (2012). Science teachers' use of mass media to address socio-scientific and sustainability issues. *Research in Science Education*, *42*(1), 51–74. doi:10.100711165-011-9256-z

Krajcik, J. S., & Sutherland, L. (2010). Supporting students in developing literacy in science. *Science*, *328*(5977), 456–459. doi:10.1126cience.1182593 PMID:20413490

Lasry, N., Dugdale, M., & Charles, E. (2014). Just in time to flip your classroom. *The Physics Teacher*, *52*(1), 34–37. doi:10.1119/1.4849151

Lemke, J. L. (1998). Multiplying meaning: visual and verbal semiotics in scientific text. In J. R. Martin & R. Vell (Eds.), *Reading science: critical and functional perspectives on discourses of science* (pp. 87–113). Routledge.

Lin, S., & Mintzes, J. J. (2010). Leaming argumentation skills through instruction in socioscientific issues: The effect of ability level. *International Journal of Science and Mathematics Education*, *8*(6), 993–1017. doi:10.100710763-010-9215-6

Lindahl, M. G., Folkesson, A.-M., & Zeidler, D. L. (2019). Students ' recognition of educational demands in the context of a socioscientific issues curriculum. *Journal of Research in Science Teaching*, 1–28. doi:10.1002/tea.21548

Livingstone, S. (2004). Media literacy and the challenge of new information and communication technologies. *Communication Review, 1*(7), 3–14. doi:10.1080/10714420490280152

Maloney, J. (2007). Children's role and use of evidence on science: An analysis of decision-making in small groups. *British Educational Research Journal, 33*(3), 371–401. doi:10.1080/01411920701243636

Matuk, C., & Linn, M. (2012). Technology integration to scaffold and assess students ' use of visual evidence in science inquiry. In American Educational Research Association (AERA) Conference (pp. 1–8). Academic Press.

McNaughton, M. J. (2004). Educational drama in the teaching of education for sustainability. *Environmental Education Research, 10*(2), 139–155. doi:10.1080/13504620242000198140

McSharry, G., & Jones, S. (2000). Role-play in science teaching and learning. *The School Science Review, 82*, 73–82.

Means, M. L., & Voss, J. F. (1996). Who reasons well? Two studies of informal reasoning among children of different grade, ability, and knowledge levels. *Cognition and Instruction, 14*(2), 139–178. doi:10.12071532690xci1402_1

Ministry of National Education. (2018). *Fen bilimleri dersi öğretim programı (İlkokul ve Ortaokul 3,4,5,6,7 ve 8. sınıflar)* [Science course instruction program (Elementary and middle school 3,4,5,6,7, and 8 grades]. Ankara, Turkey: Talim Terbiye Kurulu Başkanlığı.

Namdar, B. (2015). An examination of preservice science teachers' representational modality preferences during computer-supported knowledge organization. *Journal of Theory and Practice in Education, 11*(3), 949–970.

Namdar, B., & Shen, J. (2016). Intersection of argumentation and the use of multiple representations in the context of socioscientific issues. *International Journal of Science Education, 38*(7), 1100–1132. doi:10.1080/09500693.2016.1183265

Namdar, B., & Shen, J. (2018). Knowledge organization through multiple representations in a computer-supported collaborative learning environment. *Interactive Learning Environments, 26*(5), 638–653. doi:10.1080/10494820.2017.1376337

National Research Council. (2012). *A framework for K-12 science education: Practices, crosscutting concepts, and core ideas. Committee on conceptual framework for the new K-12 science education standards*. The National Academies Press.

NGSS Leads States. (2013). *Next generation science standards: For states, by states*. The National Academies Press.

Nussbaum, E. M., & Kardash, C. M. (2005). The effects of goal instructions and text on the genera- tion of counterarguments during writing. *Journal of Educational Psychology*, *97*(2), 157–169. doi:10.1037/0022-0663.97.2.157

Nussbaum, E. M., & Schraw, G. (2007). Promoting argument-counterargument integration in students' writing. *Journal of Experimental Education*, *76*(1), 59–92. doi:10.3200/JEXE.76.1.59-92

Odegaard, M. (2003). Dramatic science: A critical review of drama in science education. *Studies in Science Education*, *39*(1), 75–102. doi:10.1080/03057260308560196

Osborne, J., Erduran, S., & Simon, S. (2004). Enhancing the quality of argumentation in school science. *Journal of Research in Science Teaching*, *41*(10), 994–1020. doi:10.1002/tea.20035

Öztürk, N., & Yılmaz-Tüzün, Ö. (2017). Preservice science teachers' epistemological beliefs and informal reasoning regarding socioscientific issues. *Research in Science Education*, *47*(6), 1275–1304. doi:10.100711165-016-9548-4

Pallant, A., & Lee, H.-S. (2015). Constructing scientific arguments using evidence from dynamic computational climate models. *Journal of Science Education and Technology*, *24*(2), 378–395. doi:10.100710956-014-9499-3

Paul, R., & Elder, L. (2009). Critical thinking, creativity, ethical reasoning: A unity of opposites. In D. Ambrose & T. Cross (Eds.), *Morality, Ethics, and Gifted Minds* (pp. 117–132). Springer. doi:10.1007/978-0-387-89368-6_8

Sadler, T. D. (2004). Informal reasoning regarding socioscientific issues: A critical review of research. *Journal of Research in Science Teaching*, *41*(5), 513–536. doi:10.1002/tea.20009

Sadler, T. D. (2011). Situating socio-scientific issues in classrooms as a means of achieving goals of science education. In T. D. Sadler (Ed.), *Socio-scientific Issues in the Classroom Teaching, Learning and Research* (Vol. 39, pp. 1–9). Springer Netherlands., doi:10.1007/978-94-007-1159-4_1

Sadler, T. D., & Donnelly, L. A. (2006). Socioscientific argumentation: The effects of content knowledge and morality. *International Journal of Science Education*, *28*(12), 1463–1488. doi:10.1080/09500690600708717

Sampson, V., & Clark, D. B. (2008). Assessment of the ways students generate arguments in science education: Current perspectives and recommendations for future directions. *Science Education, 92*(3), 447–472. doi:10.1002ce.20276

Sampson, V., Enderle, P., & Grooms, J. (2013). Argumentation in science education: Helping students understand the nature of scientific argumentation so they can meet the new science standards. *Science Teacher (Normal, Ill.), 80*(5), 30–33. doi:10.2505/4/tst13_080_05_30

Sampson, V., Enderle, P., Grooms, J., & Witte, S. (2013). Writing to learn by learning to write during the school science laboratory: Helping middle and high school students develop argumentative writing skills as they learn core ideas. *Science Education, 97*(5), 643–670. doi:10.1002ce.21069

San, İ. (1990). Eğitimde yaratıcı drama [Creative drama in education]. *Ankara Üniversitesi Eğitim Bilimleri Fakültesi Dergisi, 23*(2), 573–583.

Shaw, J. A. (2012). Using small group debates to actively engage students in an introductory microbiology course. *Journal of Microbiology & Biology Education, 13*(2), 155–160. doi:10.1128/jmbe.v13i2.420 PMID:23653803

Simonneaux, L. (2001). Role-play or debate to promote students' argumentation and justification on an issue in animal transgenesis. *International Journal of Science Education, 23*(9), 903–927. doi:10.1080/09500690010016076

Simonneaux, L. (2002). Analysis of classroom debating strategies in the field of biotechnology. *Journal of Biological Education, 37*(1), 9–12. doi:10.1080/00219266.2002.9655839

Somers, J. (1994). *Drama in the curriculum*. Cassell Education.

Stahl, G., Ludvigsen, S., Law, N., & Cress, U. (2014). CSCL artifacts. *International Journal of Computer-Supported Collaborative Learning, 9*(3), 237–245. doi:10.100711412-014-9200-0

Stahl, S., Hynd, C., Britton, B., McNish, M., & Bosquet, D. (1996). What happens when students read multiple source documents in history? *Reading Research Quarterly, 31*(4), 430–456. doi:10.1598/RRQ.31.4.5

Staker, B. H., & Horn, M. B. (2012). *Classifying K – 12 Blended Learning*. Academic Press.

Tal, R. T., & Kedmi, Y. (2006). Teaching socioscientific issues: Classroom culture and students' performances. *Cultural Studies of Science Education, 1*(4), 615–644. doi:10.100711422-006-9026-9

Teo, T. W., Tan, K. C. D., Yan, Y. K., Teo, Y. C., & Yeo, L. W. (2014). How flip teaching supports undergraduate chemistry laboratory learning. *Chemistry Education Research and Practice, 15*(4), 550–567. doi:10.1039/C4RP00003J

Topçu, M. S., Yılmaz-Tüzün, Ö., & Sadler, T. D. (2011). Turkish preservice science teachers' informal reasoning regarding socioscientific issues and the factors influencing their informal reasoning. *Journal of Science Teacher Education, 22*(4), 313–332. doi:10.100710972-010-9221-0

Toulmin, S. (1958). *The uses of argument.* Cambridge University Press.

Tsai, C. Y. (2018). The effect of online argumentation of socio-scientific issues on students' scientific competencies and sustainability attitudes. *Computers & Education, 116*, 14–27. doi:10.1016/j.compedu.2017.08.009

Uluçınar-Sağır, Ş., & Kılıç, Z. (2013). İlköğretim öğrencilerinin bilimin doğasını anlama düzeylerine bilimsel tartışma odaklı öğretimin etkisi. *Hacettepe Üniversitesi Eğitim Fakültesi Dergisi, 44*, 308–318.

Wu, Y.-T., & Tsai, C.-C. (2007). High school students' informal reasoning on a socio-scientific issue: Qualitative and quantitative analyses. *International Journal of Science Education, 29*(9), 1163–1187. doi:10.1080/09500690601083375

Zohar, A., & Nemet, F. (2002). Fostering students' knowledge and argumentation skills through dilemmas in human genetics. *Journal of Research in Science Teaching, 39*(1), 35–62. doi:10.1002/tea.10008

ADDITIONAL READING

Allan, C. N., Shurety, W., & Crough, J. (2019). *Blended learning designs in STEM higher education.* Springer Singapore. doi:10.1007/978-981-13-6982-7

Baldwin, P. (2009). *School improvement through drama. A creative whole class, whole school approach.* Continuum International Publishing Group.

Erdoğan, T. (2017). *Okul öncesinden ilköğretime kuramdan uygulamaya drama* [Drama from preschool to primary education, theory to practice]. Eğiten Kitap.

Erduran, S., & Jiménez-Aleixandre, M.-P. (2007). *Argumentation in science education: Perspectives from classroom-based research.* Springer. doi:10.1007/978-1-4020-6670-2

Khine, M. S. (2012). *Perspectives on scientific argumentation. Theory, Practice and Research.* Springer Netherlands. doi:10.1007/978-94-007-2470-9

Kuhn, D. (1991). *The skills of argument.* Cambridge University Press. doi:10.1017/CBO9780511571350

McCaslin, N. (2000). *Creative drama in the classroom and beyond* (7th ed.). Longman.

Reidsema, C., Kavanagh, L., Hadgraft, R., & Smith, N. (2017). *The flipped classroom.* Springer Singapore. doi:10.1007/978-981-10-3413-8

Sawyer, R. K. (2008). *The Cambridge handbook of the learning sciences.* Cambridge University Press.

KEY TERMS AND DEFINITIONS

Argumentation: The process of justifying claims through the use of evidence.

Creative Drama: The process of using drama techniques to improvise and dramatize an event, a situation or a process.

Flipped Classroom: A blended teaching/learning approach where students learn course content online and practice it in a face-to-face learning environment.

Knowledge Organization: The process of creating, sorting, sharing and clustering epistemic artifacts.

Representation: An artifact that symbolize an idea, a process, or a phenomenon.

Socioscientific Issues: Complex issues with ties to science and society, which create dilemmas and prone to multiple solutions.

SSI-Online-Knowledge Organization-Argumentation Pattern (SOKAP): A teaching strategy that incorporates socioscientific issues, online learning, knowledge organization and argumentation.

Section 3

Socioscientific Issues for Scientific Literacy in the Undergraduate Classroom

Chapter 6
Enhancing Socioscientific Reasoning Through Nature of Technology

Hyunok Lee
https://orcid.org/0000-0001-8981-9263
Seoul National University, South Korea

Hyunju Lee
https://orcid.org/0000-0003-4976-6544
Ewha Womans University, South Korea

ABSTRACT

Fostering informed Socioscientific Reasoning (SSR) is an essential component of developing scientific literacy. In this chapter, the authors suggest that enhancing Nature of Technology (NOT) understanding can be one way to leverage students' informed socioscientific reasoning. The authors describe a proposed NOT conceptual framework with four dimensions and detailed components, and present an analysis of students' reasoning of various socioscientific issues using this framework. Finally, the authors present the finding that NOT components were present in student discussions with varying levels of understanding. The SSR analysis reveals that students with NOT informed understanding can appreciate the integrated characteristics of technology, so as to make sophisticated decisions about science and technology that will change society in fundamental ways, for both better and worse.

DOI: 10.4018/978-1-7998-4558-4.ch006

INTRODUCTION

Fostering informed Socioscientific Reasoning (SSR) is an essential component of developing scientific literacy (Davies, 2004; Kolstø, 2001; Romine, Sadler, & Kinslow, 2017; Sadler, Barab, & Scott, 2007; Sadler, & Zeidler, 2009; Zeidler et al., 2005). Science educators have implemented various strategies to enhance students' SSR. One of the conventional approaches was the Nature of Science (NOS) instruction in the context of socioscientific issues (SSI), which argues that understanding of NOS – an epistemology of scientific knowledge and its development – would influence the way students apply scientific evidence to SSIs. However, empirical findings in previous studies did not sufficiently support the relationship between NOS and SSR. For example, Bell and Lederman (2003) found that two groups of people who had different NOS views did not show significant differences in their SSR. Although Walker and Zeidler (2007) reported a significant increase in students' NOS understanding after implementing the NOS program in the context of SSI, students' actual discussions did not represent their NOS understanding.

In this chapter, the authors suggest that utilizing Nature of Technology (NOT) would be a promising approach to enhance students' SSR. One reason for considering NOT is that many cases of SSIs have been derived from cutting-edge technology and engineering with natural science. Examples of SSI are not limited to science, but also include technology and engineering; biotechnology (e.g., GMOs, vaccines, etc.), energy technology (e.g., nuclear power plants, renewable energy resources, etc.), and environmental engineering (e.g., pollution, reclamation, etc.). Indeed, specific examples of SSI often encompass aspects of technology. Another reason is that students spontaneously consider the impacts of technical artifacts in their lives and society to negotiate SSIs. They often regard technology as, "a solution for real-world problems," compare the advantages and drawbacks in the process of technology development, and consider "multiple stakeholders," including consumers and producers. It indicates that students might apply NOT aspects in SSI contexts. Moreover, NOT instruction would assist students in understanding what technology is, how technology is developed with social, cultural, economic, and other factors, how individuals and society interact with technology, and a Faustian bargain of technology. Effectively addressing these and multifaceted aspects of NOT would cultivate habits of thought and SSR for citizenship in society with advanced science and technology (Clough, 2013).

Understanding NOT, like NOS, has long been regarded as an integral part of scientific literacy (AAAS, 1990). However, there has been little research exploring the features of technology or how students and teachers in the fields of technology, engineering, or science education approach technology in the context of SSIs. This has not been the case for NOS, which has had its aspects and assessments of students'

NOS understandings according to various groups and implementations thoroughly researched (Lederman, 1992, 2007; Lederman, Abd-El-Khalick, Bell and Schwartz, 2002). The authors believe that science educators need to pay more attention to NOT and provide practical guidance on how NOT may be better articulated and applied in the context of science education.

Thus, in this chapter, the authors first propose our conceptual framework for NOT, which was developed based on extensive literature studies from the fields of history, philosophy, and sociology of technology, peer-reviews, and empirical findings on students' reasoning (Lee, 2015). Second, the authors present how students conceptualize NOT in various SSI contexts (Lee & Lee, 2016). Third, the authors illustrate an intimate link between NOT understanding and SSR, displaying that the more sophisticated NOT understanding in SSI decision-making, the higher level in some aspects of SSR. Finally, the authors suggest some implications for SSI instruction focused on NOT understanding.

CONCEPTUAL FRAMEWORK OF NATURE OF TECHNOLOGY

Definition of Technology

A broad spectrum of ideas about technology is found in the literature of the field of science education, history of technology, and philosophy of technology (AAAS, 1990, 1993; ITEA, 2007; Hughes, 2004; Layton, 1993; Mitcham, 1994; NRC, 1996, 2006, 2012; Nye, 2006; Pacey, 1983; Volti, 2010). In the document of science education standards, technology embraces even "all types of human-made systems and processes" (NRC, 2012, p. 11-12). In addition, Hughes (2004), a renowned historian of technology, mentioned that attempts to define technology have rarely been made; rather, definitions have been inferred through lists that encompass examples of technology.

The meaning of the term "technology" varied according to period and fields. First, the meaning of technology has been changed through period. Formerly, technology was related to physical labor, which was performed by mostly illiterate craftsmen and artisans. Latterly, technology has extended a more abstract and inclusive ideas including social enterprises such as manufacturing, management, maintenance, marketing, and finance (Mitcham, 1994; Nye, 2006; Pacey, 1983). Second, there are two fields, engineering and humanities, in which ideas about technology have been distinctly articulated (Mitcham, 1994; Pacey, 1983). Engineers typically define technology narrowly, including skill, knowledge, technique, tools, machines, and products. Their propensity has historical origins. In the 19th century, the engineer groups began to appear. They generally received a formal education in a university, including science. They had a desire to differentiate themselves from artisans or technicians who had lower social status than scientists (Kim et al., 2013). Thus, they would have a benefit by defining technology narrowly and relating engineering with science. In contrast, scholars in the field of humanities such as history, philosophy, and sociology of technology commonly define technology as broad ideas extending to its social and cultural aspects, which from its narrow definition, are regarded as external factors.

Despite these potential definitions, many researchers found that most people showed a limited understanding of technology (Burns, 1992; De Klerk Wolters, 1989; De Vries, 2005; DiGironimo, 2011; Jarvis & Rennie, 1996; Scherz & Oren, 2006). These researchers found that most people have a tendency to define technology as modern high tech artifacts, such as computers, electronics, or machines, while ignoring relatively simple and familiar objects, such as pencils or reading glasses. Furthermore, people tend to restrict technology to its final product and fail to recognize the process or activities in which technological creativity is crucial.

Despite technology's diverse notions and its rather 'narrow' public perceptions, the authors chose a more inclusive definition of technology for NOT educational purposes. The authors operationally defined it as the sum of various tools, machines, instruments, and immaterial entities, including engineers' techniques and knowledge, as well as social and cultural resources that work to transform the world to meet humans' needs and desires. The followings are educational grounds to implement the inclusive definition of technology. First, this inclusive definition of technology provides a rich foundation for a NOT framework that extends the public's limited understanding of technology and encompasses a variety of students' conceptualizations of technology. Second, documents for the standards of science and technology education show that technology is not limited to artifacts or practical skills in its narrow meaning, but that broader social and cultural factors need to be included to foster scientific and technological literacy (ITEA, 2007; NRC, 2006, 2012). Third, the fields of the history, philosophy, and sociology of technology, which provide primary contents for the NOT framework, assume that technology is more than skill, knowledge, technique, and final technological products. Finally, the broad definition of technology, including social, cultural, and ethical aspects, is crucial to promote scientific literacy. Students need to be prepared to engage in critical conversations and make informed decisions on contemporary issues like climate change, human health, clean water, energy shortage, and environmental pollution in a technically developed society. These issues have raised ethical, social, and cultural concerns that are interwoven with scientific and technological considerations.

Nature of Technology Framework

The NOT conceptual framework was constructed from extensive literature reviews in the fields of history, philosophy, and sociology of technology (Lee, 2015). In its early stages, technology was first operationally defined by surveying and analyzing encyclopedias or books of introductory and liberal arts courses (e.g., De Vries, 2005; Nye, 2006; Pacey, 1983; Volti, 2010). Then, specialized subjects and journal articles were scrutinized to derive the subcomponents of the NOT (e.g., Friedel, 2010; Gardner, 1992; Hughes, 1983, 1986, 2004, 2012; Kroes, 2012; Mitcham, 1994; Polanyi, 1958; Vincenti, 1990). After several revisions and discussions among the authors, a temporary NOT framework was conceptualized. Then this framework was subjected to a peer-review process, which included a philosopher of technology, a historian of technology, and an engineer, who reviewed its contents and confirmed that the framework had representative and balanced components across fields in order to prevent distortion of ideas about technology when applied to science education. This peer-review panel was selected according to the scholars' publication history, longevity and reputability in the field, and experience in research and teaching ideas about technology. Once the peer-review panel made its recommendations and corrections were made, the authors focused on testing the pedagogical feasibility of the NOT framework. To investigate this, the authors developed a questionnaire and administered it to students to understand perceptions of NOT as a framework to teach socioscientific reasoning among secondary and general education levels.

The NOT conceptual framework is composed of four dimensions. Each of these dimensions includes three components, summing up to 12 components in total (Fig 1). The four dimensions of NOT - *artifacts*, *practices*, *knowledge*, and *systems* - explicitly take into account broad notions of technology. Scholars in the past have attempted to define and conceptualize technology with various dimensions in order to be inclusive of broad meanings of technology (AAAS, 1990; DiGironimo, 2011; ITEA, 2007; Mitcham, 1994). Specifically, Mitcham (1994)'s four dimensions: *artifacts*, *activity*, *knowledge*, and *volition,* were revised and adapted. The *volition* dimension was perceived to be too abstract or narrow for K-12 education or general public education, even though the dimension itself was theoretically reasonable in the field of philosophy of technology. Thus, technology as a *volition* was replaced by technology as *systems* because of its relative acceptance in the context of education and because understanding technology as systems contributes to overall educational interests. Specifically, technology as systems connotes a holistic view, where the whole is more significant than the sum of its parts. This holistic perspective is advocated to embrace various factors and the interactions between them, while also explaining the complex aspects of technological development such as artifacts with physical properties; human factors, including technicians, engineers, consumers, etc.; organizational and social aspects, such as law, regulation, and institutions; as well as scientific knowledge and research programs (Hughes, 2012). In addition, technology as an *activity* was replaced by technology as *practices* to emphasize engineers' repetition of an activity as usual behavior in their profession.

Figure 1. Conceptual framework of nature of technology

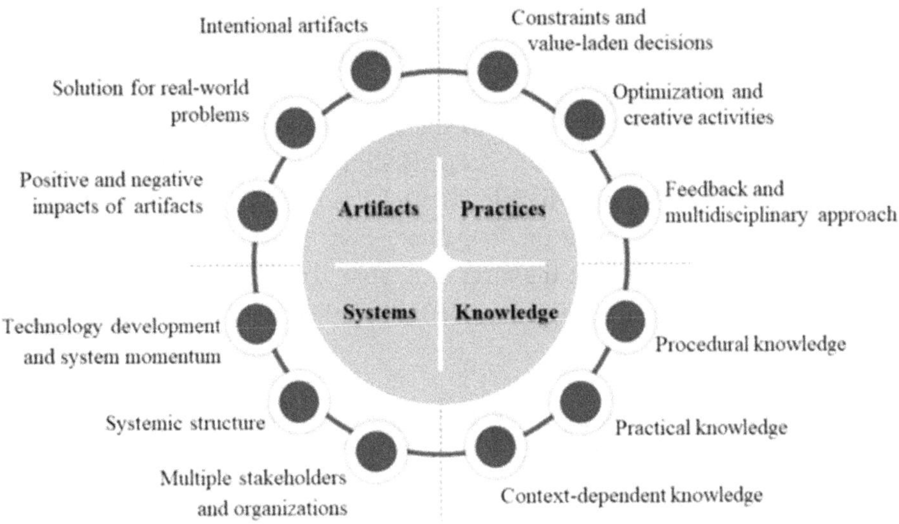

Artifacts refer to material products, such as the tools and machines that individuals generally think about when 'technology' is mentioned. *Artifacts* not only include products of up-to-date technology, like computers and airplanes, but also objects that people might consider basic tools to accomplish tasks, such as pencils, scissors, plates, and devices that humans have created throughout history (De Vries, 2005; DiGironimo, 2011; Mitcham, 1994). *Practices* are a series of activities that engineers conduct, such as designing, troubleshooting, developing, repairing, maintaining, and evaluating, among others. In particular, the authors compartmentalized *practices* by focusing on designing, among the various activities of engineers, because philosophers of technology and engineers themselves emphasize designing as the most distinctive activity among technological practices (De Vries, 2005; Son, 2011; Vincenti, 1990). *Knowledge* is defined as the information and skill sets that are practically used and produced by engineers in their realm of expertise. Even though engineers apply scientific knowledge to their tasks, and scientific knowledge is often transferred to engineering knowledge, this does not mean that technology is the same as applied scientific knowledge (Constantinou et al., 2010; Custer, 1995; Layton, D., 1993; Layton, E. T., 1974; De Vries & Tamir, 1997; Gardner, 1992, 1994, 1995; Hughes, 1986, 2012; Mitcham, 1994; Nye, 2006; Vincenti, 1990). For example, William J. M. Rankine, referred to as the founder of engineering, tried to apply Newton's law of motion as a scientific theory in building structures, but he later acknowledged the limitations of applying scientific principles to engineering tasks. Instead, Rankine introduced new concepts, such as *stress* and *strain*, and then developed the relevant field of structural engineering (Hong, 1999). Thus, engineering knowledge exists along a continuum ranging from tacit to symbolic knowledge (Custer, 1995; Frey, 1991; Vincenti, 1990). Finally, *systems* consist of various components, or sub-systems, and their interactions, as well as social and cultural factors of technology such as marketing, policy, and laws (AAAS, 1990, 1993; ITEA, 2007; NRC, 2012, Volti, 2010; Waight & Abd-El-Khalick, 2012).

The NOT components under the four dimensions are described in Table 1. It is important to note that despite the form of a table, the NOT components are not intended to be utilized as declarative knowledge or as lists for students to memorize.

CONTEXTUALIZED NATURE OF TECHNOLOGY IN SOCIOSCIENTIFIC ISSUES

In this section, the authors present how students conceptualized NOT in various SSIs (Lee & Lee, 2016). Undergraduate students developed SSI topics as a team project assignment and led a class discussion. The following SSI topics were analyzed: 1) Google Glass; 2) closed-circuit television; 3) nuclear power plants; 4) weapons of war; 5) insecticide DDT; 6) AIDS treatment; and 7) animal experimentation. For example, Google Glass resembles eyeglasses that have photo and video shooting and sharing capabilities, but with drawbacks such as the potential for invasion of privacy and various information leaks. Students were asked to discuss if they would purchase this product as a consumer and the potential social consequences of this product.

Table 1. Descriptions of NOT components (Lee, 2015; Lee & Lee, 2016, 2017)

Categories	NOT components	Contents
Artifacts	Intentional artifacts	Human intention is revealed through the functional and physical properties in artifacts. However, designs and plans to invent an artifact may not be completed according to engineers' intentions due to a variety of factors, such as the traits of its materials or a potential shortage of utility techniques. Beyond this, an engineer's plans for an artifact, such as its proper usage, may not be manifested when the final product is handed over to users, or is used in different contexts.
	Solution for real-world problems	Throughout history, human beings have created various technological products to solve real-world problems. Problems include not only technical ones in a narrow sense (e.g., a friction resistance device for mileage improvement, noise prevention devices, etc.) but also social problems conceptually related to science and technology (e.g., environmental pollution, energy shortage, food shortage, etc.). It is worth noting that the introduction of artifacts cannot guarantee the complete and successful resolution of real-world problems because the artifacts may cause other unexpected difficulties or may only partially address the original issue.
	Positive and negative impacts of artifacts	Artifacts are developed and produced based on the necessity and demands of individuals. Thus, they are generally regarded as beneficial. However, they could also create negative impacts when broken, used inappropriately, or when they create unintended side effects for its users. Some artifacts that have permeated our daily lives without presenting issues or creating damage during long-term usage go unnoticed, making it challenging to recognize their possible negative influence. In addition, the impact of an artifact may differ according to its context and how it is regarded by different stakeholders.
Practices	Constraints & value-laden decisions	In the design process, engineers take into account various constraints to create the needed product. These constraints fall into two categories. Constraints can be rigid and unchangeable, such as the laws of physics, properties of matter, or a reserved amount of resources, or they can be flexible and changeable, due to temporal and contextual elements such as culture, marketing, consumer demands, risk awareness of technical products, among others. Engineers' decision-making processes are value-laden in that they prioritize the weight of the constraints and proceed to incorporate them into their designs. For example, a civil engineer constructing a dam might use a conventional design model with reasonable costs as a point of departure, but if the budget is reduced, the engineer will need to make a decision that accounts for new constraints.
	Optimization & creative activities	In the real world, design constraints often contradict each other, so engineers need to find the most functional options to attempt to address each constraint and complete a task. Namely, engineers conduct trade-offs, which entail comparing several prototypes with different features to choose a final product that addresses the most constraints. Creative engineers implement optimal solutions or introduce novel and innovative approaches, develop new techniques, adjust the existing procedure, and interpret their constraints. In contrast to artists who are creative to achieve aesthetic value, engineers use their creativity to accomplish functional value.
	Feedback & multidisciplinary approach	Engineers attempt to estimate reactions of target users and predict defects or side effects when designing products. However, engineers may have difficulty estimating if a technological product will be successful or which technology becomes influential, regardless of their efforts. Because the public has diverse tastes and unexpected ways of using a product, engineers collaborate with experts in various fields to satisfy the demands of users and consumers. This multidisciplinary approach provides information on complicated technological development that engineers would not have access to otherwise.
Knowledge	Procedural knowledge	Engineering knowledge includes knowledge about procedures such as designing, developing, and troubleshooting, which seek to fulfill both human needs and practical goals. In particular, engineers have often proposed and utilized models which describe design processes, frequently in the form of flowcharts. Even though the models may not completely represent the details of different design processes, they can still be beneficial as pedagogical tools for students or as a way to increase efficiency when communicating the design process with others who may have a different understanding. Engineers may use a typical model according to a specific design task, but this does not mean that model is the only possible one.
	Practical knowledge	Practical knowledge is tacit knowledge. This kind of knowledge is derived from accumulated practices, which require several attempts and is difficult to be expressed in written or verbal forms of communication. Individuals must remain cautious and avoid misinterpreting it as effortless dexterity, arbitrary knowledge, or knowledge obtained through simple trial and error. Engineering knowledge ranges in a broad spectrum, from tacit to symbolic knowledge.
	Context-dependent knowledge	In contrast to scientists, who pursue universal knowledge like the law of universal gravitation, engineers seek knowledge that is context-specific and appropriate for a particular situation. Thus, engineering knowledge that has been successfully applied in one situation does not guarantee success in another, which forces engineers to consider numerous conditions and variables during the design process.
Systems	Multiple stakeholders and organizations	Beyond technical artifacts, technology also involves humans and organizations, including engineers, consumers, patent and legal professionals, business groups, among other participants with various social, cultural, economic, and political interests. *Technological closure* occurs when a particular technical product becomes the norm after historically competing with other artifacts. Although contemporary consumers attribute technological closure to physical or functional merits that create an extraordinarily efficient product, in reality, various stakeholders and organizations played a significant role in improving the product as the final form.
	Systemic structure	Technology is a systemic structure composed of sub-components. Although these sub-components have certain freedoms, they are interconnected, and when they do not function properly, problems can occur in the entire system. Thus, social and economic sub-components within a country can affect the *diffusion of technology*. For example, in Pakistan, the introduction of labor-saving tractors replaced the jobs of many farm laborers, causing poverty rates to rise. The dramatically different economic and social sub-components of Pakistan compared to other more developed nations caused the system not to function properly. The systemic problems in technology diffusion across different contexts cannot be understood nor estimated with the view that technical products and techniques are independent and isolated artifacts.
	Technology development and system momentum	While it is difficult to estimate the precise trajectory of technology development, adapting the perspective of technology as a system can help identify characteristics that appear through the development of a large technological system. New technology might develop in unanticipated directions or disappear entirely in its early stages. However, when a particular technology is successfully developed and grounded into society as a large technological system, *system momentum* takes place. System momentum occurs when a large system with its many stakeholders, organizations, and their complex interactions present resistance or *inertia* to keep the technological system growing in a fixed direction. Under these circumstances, individuals can often feel passive without the autonomy to decide whether or not to use the technology. Still, system momentum does not imply the perpetuity of the large technological system because it could still be transformed and replaced by other technology when collective demands of stakeholders and simultaneous changes of relevant components occur.

Participants were 45 first-year undergraduate students, 18 humanities majors and 27 engineering majors, who were enrolled in a liberal arts course on science and technology. The students participated in a team project with 4-5 people where they prepared and led a whole classroom discussion on SSI topics. Data was collected from the preparation process, including preliminary reports for selecting an SSI topic, interviews with each group, and a class discussion including audio-recorded dialogue among students, presentation materials, peer evaluation reports, and final team reports containing individual opinions on the selected topic. The project provided various data such as the individual and group opinions of students who majored in humanities and engineering in written and oral forms, which were expected to show a broad spectrum of SSI reasoning.

In order to identify NOT conceptions in the SSIs, the authors utilized the NOT framework. The two researchers were involved in the development of the NOT framework, and one of them has been conducting SSI research for over ten years. The authors identified potential NOT conceptions with students' excerpts while reading the data and compared them. Based on the first analysis results, the authors derived keywords and typical contents to re-analyze the relevant NOT conceptions through the whole data. The authors continued to analyze the data with this method until reaching an agreement (Miles, & Huberman, 1994).

The NOT conceptions were present in each of the seven topics developed, but the dimensions of *artifacts* and *systems* dominated the discussions. *Knowledge* and *practices* were implicitly shown in students' SSR. In the following sections, the authors introduce examples of contextualized NOT.

Contextualized NOT Under Artifacts Dimension

Intentional Artifacts

Students often mentioned *intentional artifacts* in discussions about the development of new technological products and their uses. For example, in discussions about Google Glass (SSI 1) and closed-circuit television (SSI 2), students focused on developers' intentions in maximizing these technologies' physical functionality like comfort, wearability, and style. Students also frequently addressed the intentions of scientists and engineers when discussing weapons of war (SSI 4) as they role-played a hypothetical trial in which the damages of a high-tech weapon were disputed. For example, the majority of students believed that the primary intention for developing weapons was to harm and kill people efficiently, and some students emphasized certain weapons when explaining their physical functions to damage the enemy. In their reasoning, some students understood that a plan to invent a weapon might not evolve according to developers' intentions due to material limitations or a potential shortage of needed technical skills. In addition, some students identified the material properties and functions of weapons and provided these as the underlying evidence to judge the responsibilities of the developers. The following vignette is an excerpt of a role-play where developers of the high-tech weapon, Electronic Magnetic Pulse (EMP), are put on trial.

Defendant Attorney: So far, we have talked about the advantages and disadvantages of the developed EMP weapons. In a preliminary investigation, I heard that you were offered two proposals for a weapon development project, right? One was the K-9 nuclear weapon, and the other was the EMP, which had no direct damage to lives but indiscriminately destroyed electronics, making the area a primitive society as a result. What were the criteria for you to choose to participate in the weapon development project?

Developer: I thought the bottom line should be to reduce casualties.

Defendant Attorney: The developer tried to minimize the damage and casualties caused by the weapon. Please, honorable judge and jury, refer to his intentions. Thank you.

Plaintiff Attorney: Once he developed the EMP weapon, it showed his willingness to contribute to the war and cause conflict. In addition, electromagnetic pulses from EMP continue to affect a wide range of areas for approximately 10 years. In the meantime, many people will suffer mental as well as physical difficulties because of a broken electronic system, which serves as the backbone of modern society. Also, EMP could cause some damages to neighboring countries because of its form of electromagnetic radiation, and we did not find evidence that EMP is in fact harmless to the human body.

Students took into account the intentions of scientists and engineers and attempted to judge their responsibility on the development of the advanced weapon of war, EMP. A student pointed out in the discussion that while weapons have similar purposes, the jury needed to judge EMP developers as less guilty given that the EMP only damages electronic devices; thus, it causes less harm, whereas nuclear bombs result in the loss of human lives. In this role-play, most students showed awareness of the intentions of engineers and developers, and some identified inconsistencies between those intentions and the final impact of the technology.

Solution for Real-World Problems

One of the typical SSI contexts is the debate over how to solve societal problems related to science and technology. Our analysis of students' reasoning using the NOT framework revealed that most students addressed the NOT component, *solutions for real-world problems*. For example, students frequently referred back to this NOT component during the closed-circuit television (SSI 2) discussion. Students proposed connected CCTV networks to address high crime rates in their community and compared the pros and cons of creating this advanced version of conventional CCTVs that could collect data across the country and share it in real-time. In the following excerpts, two students discussed solutions, the connected CCTV networks or alternatives, to address high crime rates.

A: I see no other solution. I wish we had another way, but there is no other solution, so maybe that's why the connected CCTV networks were developed?

B: In my opinion, if we focus on the underdeveloped areas and assign more workforces to patrol them, the crime rate will decrease. Saying that this is the only solution and relying on the device excludes other possibilities.

Many students considered CCTV networks to be a favorable alternative to lower crime rates. However, some critiqued this prevailing attitude toward the CCTV networks arguing that it would not reduce crime rates but only arrest rates of criminals. As shown, student B urged to consider other solutions like increasing patrol in underdeveloped areas instead of CCTV. Furthermore, some students mentioned that the fundamental cause of the crime rate increase needed to be analyzed in order to come up with appropriate solutions and consider potential drawbacks of relying on CCTV networks.

C: In my opinion, we need to find out the underlying reason why the crime rate is increasing. The CCTV networks could work to stop criminals temporarily. Thus, if the CCTV system fails to operate, the community will return to its original situation.

D: If you search your name on Google, you will find all kinds of information and social media posts. So, if the new connected CCTV networks are installed, we will likely reveal our information even more. Innocent citizens will be at risk to be inspected, and power authorities might take advantage of the information for their own gain.

In this excerpt, student C advised identifying the underlying cause of the rising crime rate and warned that the CCTV networks could serve only as a temporary and superficial solution. Student D stated that the connected CCTV networks at the national level could lead to an invasion of privacy and other consequences such as abuse of state power. Evidently, students debated the pros and cons of implementing a connected CCTV network as a method to reduce crime rates and engaged in discussions to unpack whether this technology would solve this real-world problem.

Positive and Negative Impacts of Artifacts

Most students knew that technical artifacts could create negative impacts, even though technology is generally regarded as beneficial. Some students, like the following, stated it explicitly, "technology benefits us while also putting us at risk."

The following excerpt depicts one of several students who articulated the positive and negative impacts of particular artifacts in the context of SSI.

I think technology has the power to protect us from natural threats. Some representative examples are tsunami warning devices and earthquake-resistant architectural designs. Medical equipment also improves our lives. However, technology not only benefits us but can also bring us danger. For example, plastic that releases environmental hormones and genetically modified foods, despite their controversy, have drawbacks.

Another example of students engaging in this conversation occurred with SSI 1. Google Glass was described as a high-tech product with many features that individuals generally want to purchase. Most students compared positive and negative features of the Google Glass to decide whether or not they would purchase and use it. Some students recognized that Google Glass could have social and cultural drawbacks, unlike advertisements that only highlighted its convenient features. The following excerpt depicts a student warning the other students about the negative impacts of Google Glass.

Google Glass is different from products such as smartphones and digital cameras. Because Google Glass is stealthier than others, it is much more likely to be exploited. ···Once it is commercialized, it is hard to predict the impacts of this product on society. You should be careful to buy it because it can infringe on many people's portrait rights

Students argued that technological artifacts are generally positive for us because they are ultimately invented to satisfy human needs and desires. Still, some of them suspected that they could cause negative impacts despite their convenient functions. Moreover, a few students recognized that over time or in certain situations and regions, the positive and negative impacts could be debated. For example, when the car was invented, it was considered environmentally sustainable technology. At that time, driving a car instead of a horse-drawn wagon improved street sanitation because the streets were not contaminated with horse manure. However, the fumes emitted by automobiles are now considered serious environmental pollutants in modern society. Students with sufficient understanding of this NOT could articulate the technology of which positive and negative impacts were debated over time or across different cultures.

Contextualized NOT Under Systems Dimension

Students recognized NOT components under the *systems* dimension in several SSIs. Specifically, students referred to the *multiple stakeholders and organizations* and the *systemic structure* components in their role-plays. However, most understandings of the *Technology development and system momentum* components remained superficial.

Multiple Stakeholders and Organizations

In many SSIs, students referred to stakeholders such as developers, engineers, consumers, or the general public. Specifically, students leading the class discussion described stakeholders' positions explicitly to explain the context of SSIs. In discussions about the insecticide DDT (SSI 5), students discussed DDT as an effective pesticide that can reduce malaria incidence but can simultaneously cause environmental pollution. They debated two potential positions: one of developed countries who privilege environmental issues versus another one of underdeveloped countries who privilege addressing the malaria epidemic. Other students played devil's advocate for both positions.

In AIDS treatment (SSI 6), students role-played as multinational pharmaceutical companies, AIDS patients, health authorities, and the international AIDS society. The following is an excerpt from a discussion that a student began about AIDS treatment policy. The student introduced related stakeholders and simultaneously critiqued and advocated on behalf of a multinational pharmaceutical company selling an AIDS vaccine.

AIDS patients, the international AIDS society, a multinational pharmaceutical company, and health authorities have conflicting priorities. What is your opinion about a multinational pharmaceutical company selling AIDS medicine at a high price? Do you think it is a natural right of a for-profit company to do this, or is it unethical corporate tyranny? I want you to think together about what kind of policy the government should take in response to this issue.

In their discussion, students presented each stakeholder's view in detail. For example, a student role-playing the multinational pharmaceutical firm argued that, "their fundamental attribute is seeking to make profit," and that the AIDS patient and the support institutions were more concerned about, "the lack of provisions for the expensive AIDS drugs in our country." This student warned of a widespread HIV infection by untreated patients that could have social consequences such as discrimination against AIDS patients and an eventual economic collapse. Evidently, the students were able to differentiate between the needs of stakeholders and argue in favor or against their individual interests.

Systemic Structure

Students frequently referenced the NOT component *systemic structure* in the SSIs where *multiple stakeholders and organizations* manifested. Many students were able to recognize the existence of and describe each of the various stakeholders' arguments, but only a few convincingly pointed out the interrelationships between them. The following excerpt shows the interconnected nature of relationships between stakeholders responsible for developing and disseminating the AIDS vaccine.

E (Playing the Role of Health Authorities): In the case of Korea, the government supports AIDS patients by contributing 300,000 to 600,000 won for one dose of the vaccine. Despite the government grant, the vaccine cost is way too expensive on the market, so AIDS patients have not been appropriately vaccinated. The government needs to consider an increase of the subsidy. Otherwise, multinational pharmaceutical companies will need to consider reducing the price of the AIDS vaccine. The number of AIDS patients is increasing and we have a risk of widespread of HIV infection. Because of this, I exhort multinational pharmaceutical companies to lower the vaccine cost because it benefit them in the long run, despite how counterintuitive this may seem.

F (As the Role of the Multinational Pharmaceutical Company): We can't just consider Korea's situation. If we lower the price in Korea, we, as a multinational company, need to reduce the price at all of our branches around the world. Therefore, the challenges the Korean government is facing are the same as the others. Controlling the spread of HIV infection in a country is each government's responsibility. I don't think it's appropriate to shift the blame onto a multinational pharmaceutical company.

E (As the Role of Health Authorities): Putting moral responsibility on a company may sound like idealism. However, a company can benefit from addressing conflicts between civil society and patients, which will ultimately improve their public image to be seen as socially conscious.

Students role-playing the multinational pharmaceutical company had differing interests with health authorities. In particular, student F conveyed that the multinational company had more complex interests because it needed to remain consistent in the global market. However, student E role-playing health authorities suggested that even though a company inherently pursues profit, it is not necessarily opposed to issues of public health and well-being as a whole. Student E argued that the company could benefit from being perceived as socially responsible and their revenue might increase. Namely, students considered stakeholders' interrelationships and various situations to resolve conflicts.

Technology Development and System Momentum

The NOT component *technology development and system momentum* was represented in a few SSIs with limited levels of depth. Students noticed this NOT component but did not apply it with sufficient detail. *System momentum* occurs when a large technological system with its many stakeholders and sub-elements present resistance to keep the system growing in a fixed direction. Under these circumstances, individuals often feel forced to follow the dominant trend. However, an individual with a profound understanding of *system momentum* is capable of explaining how the trend has been created by various stakeholders and organizing collective actions and simultaneous changes of the system sub-elements to transform or replace that technological system. For example, when deciding whether to buy Google Glass in SSI 1, students felt forced to buy it if most people had purchased the device and it had become the norm.

When Google Glass starts to sell, I'm not going to buy it because it has negative repercussions, like the leaking of private information. But, if everyone else uses it, I could not help but to buy it because I would be afraid that others may keep an eye on me through the glasses so I would have to do the same to them to keep the upper-hand.

This student felt that if Google Glass became commonplace, he would feel pressured to use it due to the anxiety of being monitored by others. However, this student did not correctly understand the concept of system momentum related to Google Glass as he attempted to analyze stakeholders and their dynamics. Beyond users' surveillance of others, the Google Corporation as the main stakeholder could collect massive information and monitor numerous people, and it might turn into a severe risk in society. The student did not refer to the main stakeholder and their interrelationship and reluctantly decided to buy the technical product, which would result in the acceleration of the trend.

In the discussion of the development of weapons (SSI 4), students presented similar responses. They felt helpless against technology development in the defense industry, claiming that, "weapons and military technology should be suppressed, but I cannot stop it." Even though it is difficult for individuals to change the large system due to its *system momentum*, it does not necessarily imply that the system cannot be transformed or changed. Collective demands and actions could influence the technological system. Still, students did not recognize nor mention this possibility.

Contextualized NOT Under Practices and Knowledge Dimension

The NOT components under *practices* and *knowledge* tend to appear indirectly and infrequently in students' reasoning. Presumably, since the *practices* NOT components were derived from a series of engineering activities, this dimension may not have been salient in students' reasoning since they conducted their SSI discussions in the role of consumers or citizens. However, in a few SSIs, students were asked to make decisions as if they were engineers or scientists, which led them to briefly mention engineers' positions and practices. The *practices* NOT component, *feedback and multidisciplinary approach*, was mentioned in the discussion of the insecticide DDT (SS5) as a human effort to end malaria. Students discussed DDT, which had been used as an effective means to kill malaria-causing mosquitoes, but eventually failed to exterminate mosquitoes and resulted in large-scale environmental repercussions.

To prevent malaria, people used a large amount of the chemical called DDT at once and it failed. This historical event taught us why we need to explore various solutions and discuss with individuals in different areas.… I would say that this is the most effective way to find a solution, by considering various areas rather than finding a solution in one field.

The student suggested that, based on his understanding of the history of DDT, new efforts to address malaria should not be limited to narrow areas but consulted with experts in various fields. Namely, the multidisciplinary approach was recommended.

The *knowledge* NOT components appeared mostly in discussions of the animal experiment (SS7). Some students justified animal experiments arguing that they were necessary to ensure the safety of chemicals and drugs used by humans, but some doubted the reliability of the knowledge obtained from animal experiments. In the following excerpt, a student mentioned the *knowledge* NOT component, *context-dependent knowledge*.

In some cases, harmful substances in animal experiments could be harmless to humans, but sometimes the same substances display different symptoms in animals and humans. This raises questions about animal experiments because even though it produces accurate results in animals, it may display different results on humans.

The students pointed out that knowledge from animal experiments could not necessarily be applied to humans due to their different biology. Other students in the discourse gave examples of animal experiment failures. One example is the thalidomide, the morning sickness inhibitor for pregnant women, which had been safe in animal experiments, but manifested significant side effects in humans.

LINK BETWEEN NATURE OF TECHNOLOGY UNDERSTANDING AND SOCIOSCIENTIFIC REASONING

In this section, the authors show that NOT understanding in SSI decision-making is linked to SSR. Sadler et al. (2007) conceptualized SSR as a conceptual resource and analytical means to promote scientific literacy, focusing on citizenship goals. They elaborated on its four aspects, *complexity*, *multiple perspectives*, *ongoing inquiry* and *skepticism*, and developed its emergent rubric to assess students' negotiation and decision-making in the context of SSI. The four aspects are not exhaustive features of SSR, and some scholars proposed the addition of different aspects (Simonneaux & Simonneaux, 2009). Here, the authors focus on the four aspects because Sadler and the colleagues provided a rubric that provides a basis to compare with levels of NOT understanding. In the following, the authors introduce some students' responses with varied NOT understanding in our previous work (Lee & Lee, 2017), and then display that the more informed NOT understanding, the more advanced level in the four aspects of SSR.

Varied Levels of NOT Understanding

The authors collected 58 college students' essays on golden rice scenario (a type of genetically modified food, GMF) to assess NOT understanding (Lee & Lee, 2017). The GMF issue is not only a frequent topic in previous SSI studies, but also precisely demonstrates the characteristics of SSI (e.g., multiple perspectives, controversial and ill-structured issues, etc.) (Albe, 2008; Khishfe, 2012; Walker & Zeidler, 2007). The golden rice scenario was adopted from Khishfe (2012) and involved the controversies of golden rice as a technical artifact that aims to address vitamin A deficiency, which can produce blindness in over 500,000 children in developing countries when untreated. The scenario also presents two disagreements among scientists: a safety concern about the consumption of golden rice by humans, and an environmental concern about the potential contamination of conventional rice if golden rice is grown in the same region. This scenario explicitly connects to NOT components under the *artifacts* dimension: *intentional artifacts*, *solution for real-world problems*, and *positive and negative impacts of artifacts*.

NOT understanding was assessed with two criteria: firstly, how many different components students recognized from the four dimensions, *artifacts*, *practices*, *knowledge*, and *systems*; and secondly, how informed and sophisticated students'

NOT application was in a given issue. Since NOT understanding was evaluated in a specific context, students needed to acquire detailed information about the SSI and utilize relevant knowledge appropriately to analyze the given scenario. Namely, NOT informed understanding in a particular context is not merely declaring or memorizing general features of technology, but more about identifying the various NOT components and providing them with detailed information in a given context.

The following table presents NOT understanding levels in the golden scenario. The context explicitly indicates NOT components under artifacts such as *solution for real-world problems*, and *positive and negative impacts of artifacts*. Thus, many students identified the artifacts NOT, and a few students demonstrated NOT components under *practices*, *knowledge*, and *systems*, and discussed them with relevant information.

Table 2. NOT understanding levels in the context of SSI

Level of NOT understanding	Features	Reasoning patterns in the golden rice / GMF issue
naïve understanding	• Non-recognition or misunderstanding of NOT components. • Briefing on the NOT components explicitly presented in the given scenario.	• Emphasizing the positive aspects of golden rice / GMF without considering negative aspects • Arguing that golden rice can completely solve vitamin A deficiency • Referring to the safety and environmental concerns of golden rice / GMF without additional information
informed understanding	• Elaborating and providing a variety of relevant information about the NOT components explicitly presented in the given SSI scenario. • Referring to NOT components other than *artifacts* that are not explicitly provided in the given SSI scenario. • Recognizing the multiple dimensions of NOT as well as its components and providing contextualized information	• Applying information or data to judge the safety issue of golden rice / GMF • Citing comparable examples to propose potential side effects of golden rice / GMF in the long term • Discussing specific alternatives to solve vitamin A deficiency from the ethical concerns of childhood blindness in developing countries • Analysing golden rice / GMF issues with multiple perspectives in consideration of a level of technology development as well as social, economic, and political aspects

Naïve Understanding of NOT

The students with naive understanding of NOT often only emphasized the positive aspects of golden rice or GMF, and expressed blind trust in that GM technology could solve real-world problems.

I hope it will be useful because genetically modified food is **the only solution to the food shortage problem**. *(S1)*

Even if GMF harms the human body, people could make harmless ones through more experiments and research **because science and technology are omnipotent.** *(S15)*

Student 1 and 15 showed a misunderstanding of the *artifacts* NOT components. Student 1 perceived that the technical artifact GMF was the only solution to food shortage, which qualifies as a misconception of *solution for real-world problems* because other alternatives may exist or be developed in the future. Student 15 was plainly uncritical of the potential effects of GMF and technology in general. Although certain artifacts could solve real-world problems, both of these students guaranteed that GMF would solve the real-world problem in question without considering potential unexpected difficulties.

Some students referred to the *artifacts* NOT components but failed to apply additional information or data to elaborate on the issues in the golden rice scenario. For example, the following two students briefly discussed that golden rice could solve the problem of vitamin A deficiency while also referring to safety and environmental concerns as mentioned in the scenario.

As noted in the scenario, a large number of people suffer from vitamin A deficiency. To help them promptly, we need to accept this technology and mass produce golden rice. However, some scientists and engineers are against the GMF. They claimed that the potential risks of GMF could emerge later, even if these do not occur rapidly. (S4)

There is no evidence that golden rice is harmful to the human body. If we do not use the crops, they get wasted. If golden rice possibly contaminates other rice, you can raise it separately from ordinary rice. Golden rice has the amazing effect of treating vitamin A deficiency for 500,000 children every year. Considering its benefits to humankind, it is necessary to grow golden rice. (S18)

Even though student 4 and 18 were aware of the NOT component, *solution for real-world problems*, the students favored golden rice only considering the information presented in the scenario, nor did they recall other NOT components other than *artifacts*.

Link With Lower SSR Aspects

The preceding excerpts are comparable to lower levels of SSR aspects, *complexity*, and *ongoing inquiry*. Sadler et al., (2007, p.380) presented that a student with a lower level of SSR aspects "offers a very simplistic or illogical solution without considering multiple factors"; "considers pros and cons but ultimately frames the issue as being relatively simple with a single solution"; "fails to recognize the need for an inquiry." For example, S1 and S15 are likely to be assessed at lower levels of complexity, with S1 offering GMF as a solution for food shortage without considering other alternatives; and S15 being aware of GMF harms but showing blind faith in science and technology to overcome any drawbacks. In addition, S4 and S18 responses seem to be equivalent to a lower level of *ongoing inquiry*.

The authors presume that NOT naive understanding is included at lower levels of SSR aspects. NOT components depict more specific features on technology and engineering rather than SSR aspects, which could be adapted in a broad context. For example, *complexity* in the context of SSI is assessed by a question such as "Is this a difficult problem?" so that students provide diverse concerns. In comparison, a NOT understanding of *solution for real-world problems* corresponds to the recognition that the introduction of artifacts cannot guarantee that the problem will be solved completely because the artifacts may cause other unexpected difficulties or may only partially address the original issue.

Informed Understanding of NOT

Students at NOT informed understanding recognized multiple dimensions with their components and provided contextualized and informed arguments. Students at this level analyzed the underlying causes of vitamin A deficiency in children from developing countries, judged if golden rice could practically address those needs, or considered scientific and technological evidence, as well as social, cultural, and political factors that would affect its success.

*In **developing countries**, the underlying cause of vitamin A deficiency is an economic structure that causes famine. How could people buy golden rice if they don't have the money to provide any food? I assume that **developer**'s intentions are good to nourish **poor children**, but the process of developing and producing technology costs money and requires the involvement of **multinational agricultural companies**, which are mostly for-profit corporations, not charities with a social mission. Unavoidably, the price of golden rice will be higher with **royalties**, making golden rice the pot of gold at the end of the rainbow. Thus, I think it is better to encourage economic development to solve widespread poverty rather than cultivating golden rice, which might have risks in the future. (S17)*

Student 17 referred to and explained additional NOT components under the *systems* dimension: *multiple stakeholders and organizations* and *systemic structure* that went beyond the material aspects of vitamin A deficiency to analyze a potential root cause of this issue, such as an economic structure that perpetuates poverty. The student deliberated the feasibility of golden rice as a solution by taking into account various stakeholders and their relationships as well as comprehensively accounting for social and economic factors in that context. In particular, he referenced a multinational company as a stakeholder and included the term, "royalties" as an additional and specific economic factor to support the economic structure. This response differed from other students at other levels who simply repeated that golden rice could solve vitamin A deficiency of children in developing countries.

The preceding excerpt shows the informed reasoning of a student with a pessimistic opinion on golden rice. The following presents another example of an informed understanding of a student who supported golden rice in order to clarify that NOT understanding differs from expressing particular views.

In my opinion, golden rice is worth producing and marketing. Firstly, golden rice is rich in vitamin A to help reduce childhood blindness. Eating just one meal of golden rice a day meets the daily vitamin A recommended allowance. Secondly, since golden rice was developed for the public benefit without being patented, commercializing golden rice does not bring significant disadvantages to local farmers. In addition, farmers below a certain income level are supported by the provisions of golden rice seeds. Lastly, ***recent tests have not shown any allergic or toxic reactions from golden rice, and rather large-scale field tests have proven its safety****. Thus, we have no reason to decline golden rice. (S54)*

Student 54 favored golden rice and provided several reasons that take into account specific circumstances in this scenario. He first estimated the amount of golden rice needed to address the vitamin A deficiency and judged it an efficient solution. He then considered various stakeholders and their perspectives with detailed information to decrease negative consequences such as a foreign company's monopoly on agriculture in developing countries. Moreover, the student introduced the results of an experiment of golden rice rather than inferring from previous results of GM soybean and corn to examine the safety concern. By acknowledging that the results of GM crops like soybean and corn did not guarantee the safety of golden rice, the student recognized the *knowledge* NOT component *context-dependent knowledge*.

Link with Higher SSR Aspects

The responses of students with NOT informed understanding are likely to be assessed at advanced levels of SSR aspects, *multiple perspectives*, *skepticism*, and *ongoing inquiry*. Sadler et al. (2007, p.380) described that a student with higher levels of SSR aspects "assesses the issue from multiple perspectives, describes differences and discusses the significance of conflicting interests"; and "suggests a plan for inquiry focused on the collection of scientific and social data." For example, S17 is probably equivalent to higher levels for SSR *multiple perspectives* and *skepticism* in consideration of his reference to stakeholders such as the golden rice developer, multinational agricultural companies, developing countries, and children suffering vitamin A deficiency regarding conflicting interests and economic concerns. S54's response is likely to be assessed at the advanced level for SSR *ongoing inquiry* because he made his decision after examining additional information or data such as a large-scale field test for safety issues and patent protection details regarding potential troubles after commercializing golden rice.

Some NOT components, *multiple stakeholders and organization* and *systemic structure*, intimately overlap with an SSR aspect, *multiple perspectives*. As a view of technology as systems, the NOT components depict that technology involves humans and organizations, including developers, consumers, patent and legal professionals, business groups among other stakeholders with various social, cultural, economic, and political interests, and further, these are interconnected despite some degree of freedom. In addition, a NOT component, *context-dependent knowledge*, could be related to *on-going inquiry*, which assumes that SSIs are ill-structured problems related to frontier science or science in progress in society so as to be characterized by uncertainty. A student assessed at advanced levels of on-going inquiry is able to pose questions and plan to collect information and data, including scientific and social dimensions of SSIs. The NOT component, *context-dependent knowledge*, indicates a feature of technology/engineering knowledge that is context-specific and appropriate for a particular situation so that one case of success does not guarantee the others are owing to different conditions and variables. In the golden rice scenario, a student with NOT understanding is able to require and search for additional information and test results of golden rice as a new type of GMF beyond the other conventional GMFs such as soybean or corn.

CONCLUSION

In this chapter, the authors proposed the Nature of Technology (NOT) conceptual framework with four dimensions and detailed components. The authors then applied this framework to analyze students' NOT conceptions, and found that NOT components were present recurrently in the most SSIs with varying levels of understanding. Furthermore, the authors inferred that NOT understanding was linked to SSR focused on its four aspects, *complexity*, *multiple perspectives*, *ongoing inquiry,* and *skepticism.* Based on the results, the authors propose that enhancing NOT understanding is one way to leverage students' advanced SSR.

Some of NOT components are directly linked to SSI contexts so that students examined them as key determinants during decision-making. This result contrasts with studies on NOS in the context of SSI decision-making. NOS aspects rarely emerged in students' spontaneous negotiation and decision-making (Bell & Lederman, 2003; Eastwood et al., 2012; Walker & Zeidler, 2007; Salder, Chambers, & Zeidler, 2004). Notwithstanding, Khishfe (2012) suggested that students would apply NOS aspects after its explicit instruction where they were guided to utilize them in SSI decision-making. As she hypothesized, it was found that students in the treatment groups referred to more NOS aspects, *tentative*, *empirical*, and *subjective*, in their decision-making. For example, students applied empirical NOS during decision-making on golden rice, such as the following: "we should look for data and what it tells us, just like scientists"; and "we need to have more data and evidence about the consequences before making our decision" (Khishfer, 2012, p.90). However, the students' excerpts did not demonstrate that decision-making illustrated with NOS aspects were more advanced SSR. The tendency of postponement of the decision while discussing the lack of evidence and data is prevalent in previous studies (e.g., Chang & Lee, 2010; Sadler, Chambers, & Zeidler, 2004). Despite the potential role of NOS in SSI, the authors argue that the NOT components are more direct and significant determinants in the student's spontaneous decision-making (Lee & Lee, 2015). Further, our analysis reveals that students with NOT informed understanding in the context of SSI can appreciate the integrated characteristics of technology, so as to make a well-reasoned decision to select technical products among the wide variety in their daily lives, and policies on technology that will change society in fundamental ways, for both better and worse.

The following are suggestions and implications for SSI instruction with NOT. Firstly, science educators could use the *artifacts* dimension as a starting point to reflect students' familiarity. SSI scenarios and discussions are often directly related to the NOT components, *solution for real-world problems* and *positive and negative impacts of artifacts*. Technology is often proposed to solve the societal problems we face (e.g., environmental issues, food problems, diseases, etc.). However, technology itself principally changes the phenomenon caused by the issues rather than resolving fundamental problems or tensions. Thus, a technical solution could cause unexpected problems or partially address the initial difficulties. Technology is developed based on the needs of humans so as to be generally regarded as beneficial, but it can create negative impacts. Students could have difficulties to recognize possible negative influences of technology in the context of SSIs. Clough (2013, p.380) suggested that educators utilize questions to provoke students' misconceptions about technology, such as the following: "what positive outcomes occur by employing this technology?" and "what negative outcomes occur by employing this technology?" Thus, teachers provide students with the opportunity to consider the benefits and positive impacts of technology, as well as its problems and side effects. Secondly, the history and philosophy of technology provide rich sources and historical cases to teach NOT as well as SSR aspects. For example, teachers can introduce the *technological closure* idea with historical cases to explain *multiple perspectives*. The concept is defined as the point when a particular technical product becomes the norm after competing with other alternatives (Volti, 2010). The historical cases exhibit various stakeholders who interpreted the success of technical products with different criteria. Thirdly, the *systems* NOT, *technology development and system momentum*, could provide students with the motivation to engage in discussion and take action to change problems related to SSI. The NOT points out that technology development is not equal to progress. Adopting advanced technologies do not necessarily lead to advances in humanity, even though new techniques and technical artifacts are easily associated with the ascendancy of beliefs that our lives will be better than they were. In addition, it implies that the direction of technology development can be changed by various factors, including cultural, social, and political ones. Thus, a student informed of the NOT could recognize that his/her own SSI decisions can play a role in changing the trend of technology through individuals' product purchases or relevant policies. However, it is worth noting that if students are aware of *system momentum* but do not fully understand it, they are likely to commit to discussions with helplessness, and be skeptical as to whether any changes within the consolidated phase of the technological system could be made. In the debate on whether to purchase Google Glass (see p.13), a student with a limited understanding of *system momentum* felt pressure to buy it despite awareness of its drawbacks. When a particular technology is developed and grounded in society as an extensive technological system, *system momentum* takes place. It exhibits resistance or inertia to keep the technological system growing in a fixed direction. Under these circumstances, individuals can often feel passive without the autonomy to decide whether or not to use the technology. Therefore, it is important to note that the high value of *system momentum* does not imply its inherency or perpetuity. Technological systems could be transformed and replaced by others when there are collective demands or simultaneous changes of relevant factors. Educators need to provide not only NOT contents but also opportunities to apply and analyze them to various cases to reach a sufficient understanding. Lastly, the authors suggest further research on the *knowledge* and *practices* NOT dimensions, which were rarely mentioned in students' negotiation and decision-making on SSIs. It does not mean that there are no roles of the NOT in SSIs. The authors anticipate that they might play a role in constructing solutions or alternatives to problems in SSIs.

REFERENCES

Albe, V. (2008). When scientific knowledge, daily life experience, epistemological and social considerations Intersect: Students' argumentation in group discussions on a socio-scientific issue. *Research in Science Education, 38*(1), 67–90. doi:10.100711165-007-9040-2

American Association for the Advancement of Science (AAAS). (1990). *Science for all Americans.* New York, NY: Oxford University Press.

American Association for the Advancement of Science (AAAS). (1993). *Benchmarks for science literacy.* New York, NY: Oxford University Press.

Bell, R. L., & Lederman, N. G. (2003). Understandings of the nature of science and decision making on science and technology based issues. *Science Education, 87*(3), 352–377. doi:10.1002ce.10063

Burns, J. (1992). Student perceptions of technology and implications for an empowering curriculum. *Research in Science Education, 22*(1), 72–80. doi:10.1007/BF02356881

Chang, H. S., & Lee, H. J. (2010). College students' decision-making tendencies in the context of socioscientific issues (SSI). *Journal of the Korean Association for Science Education, 30*(7), 887–900.

Clough, M. P. (2013). Teaching about the nature of technology: Issues and pedagogical practices. In M. P. Clough, J. K. Olson, & D. S. Niederhauser (Eds.), *The nature of technology: Implications for learning and teaching* (pp. 345–369). Sense. doi:10.1007/978-94-6209-269-3_19

Constantinou, C., Hadjilouca, R., & Papadouris, N. (2010). Students' epistemological awareness concerning the distinction between science and technology. *International Journal of Science Education, 32*(2), 143–172. doi:10.1080/09500690903229296

Custer, R. L. (1995). Examining the dimensions of technology. *International Journal of Technology and Design Education, 5*(3), 219–244. doi:10.1007/BF00769905

Davies, I. (2004). Science and citizenship education. *International Journal of Science Education, 26*(14), 1751–1763. doi:10.1080/0950069042000230785

De Klerk Wolters, F. (1989). A PATT study among 10 to 12-year-old students in the Netherlands. *Journal of Technology Education, 1*(1), 21–31.

De Vries, M. J. (2005). *Teaching about technology: An introduction to the philosophy of technology for non-philosophers* (Vol. 27). Springer.

De Vries, M. J., & Tamir, A. (1997). Shaping concepts of technology: What concepts and how to shape them. In M. J. De Vries & A. Tarmir (Eds.), *Shaping concepts of technology* (pp. 3–10). Springer. doi:10.1007/978-94-011-5598-4_1

DiGironimo, N. (2011). What is technology? Investigating student conceptions about the nature of technology. *International Journal of Science Education, 33*(10), 1337–1352. doi:10.1080/09500693.2010.495400

Eastwood, J. L., Sadler, T. D., Zeidler, D. L., Lewis, A., Amiri, L., & Applebaum, S. (2012). Contextualizing nature of science instruction in socioscientific issues. *International Journal of Science Education, 34*(15), 2289–2315. doi:10.1080/09500693.2012.667582

Frey, R. E. (1991). Another look at technology and science. *Journal of Technology Education, 3*(1), 16–29. doi:10.21061/jte.v3i1.a.2

Friedel, R. (2010). *A culture of improvement: Technology and the western millennium*. MIT Press.

Gardner, P. L. (1992). The application of science to technology. *Research in Science Education, 22*(1), 140–148. doi:10.1007/BF02356889

Gardner, P. L. (1994). The relationship between technology and science: Some historical and philosophical reflections. Part I. *International Journal of Technology and Design Education, 4*(2), 123–153. doi:10.1007/BF01204544

Gardner, P. L. (1995). The relationship between technology and science: Some historical and philosophical reflections. Part II. *International Journal of Technology and Design Education, 5*(1), 1–33. doi:10.1007/BF00763650

Hong, S. (1999). *Science and technology as productive power and culture*. Moonji Press.

Hughes, T. P. (1983). *Networks of power: Electrification in western society*. Johns Hopkins University Press.

Hughes, T. P. (1986). The seamless web: Technology, science, etcetera, etcetera. *Social Studies of Science, 16*(2), 281–292. doi:10.1177/030631278601600204

Hughes, T. P. (2004). *Human-built world: How to think about technology and culture*. University of Chicago Press. doi:10.7208/chicago/9780226120669.001.0001

Hughes, T. P. (2012). The evolution of large technological systems. In W. E. Bijker, T. P. Hughes, T. J. Pinch, & D. G. Douglas (Eds.), *The social construction of technological systems: New directions in the sociology and history of technology* (pp. 45–74). MIT Press.

International Technology Education Association (ITEA). (2007). *Standards for technological literacy: Content for the study of technology*. International Technology Education Association.

Jarvis, T., & Rennie, L. J. (1996). Understanding technology: The development of a concept. *International Journal of Science Education*, *18*(8), 977–992. doi:10.1080/0950069960180809

Khishfe, R. (2012). Nature of science and decision-making. *International Journal of Science Education*, *34*(1), 67–100. doi:10.1080/09500693.2011.559490

Kim, D., Moon, J., Lee, J., Song, C., & Park, J. (2013). *The birth of modern engineering*. Ecolivres Press.

Kolstø, S. D. (2001). Scientific literacy for citizenship: Tools for dealing with the science dimension of controversial socioscientific issues. *Science Education*, *85*(3), 291–310. doi:10.1002ce.1011

Kroes, P. (2012). *Technical artefacts: Creations of mind and matter: A philosophy of engineering design*. Springer. doi:10.1007/978-94-007-3940-6

Layton, D. (1993). *Technology's challenge to science education*. Open University Press.

Layton, E. T. (1974). Technology as knowledge. *Technology and Culture*, *15*(1), 31–41. doi:10.2307/3102759

Lederman, N. G. (1992). Students' and teachers' conceptions of the nature of science: A review of the research. *Journal of Research in Science Teaching*, *29*(4), 331–359. doi:10.1002/tea.3660290404

Lederman, N. G. (2007). Nature of science: Past, present, and future. In N. G. Lederman & S. K. Abell (Eds.), *Handbook of research on science education* (pp. 831–879). Lawrence Erlbaum Associates.

Lederman, N. G., Abd-El-Khalick, F., Bell, R. L., & Schwartz, R. S. (2002). Views of nature of science questionnaire: Toward valid and meaningful assessment of learners' conceptions of nature of science. *Journal of Research in Science Teaching*, *39*(6), 497–521. doi:10.1002/tea.10034

Lee, H. (2015). *Construction of nature of technology framework and its utilization for investigation of changes in college students' perception of nature of technology through SSI-based program* (Unpublished doctoral dissertation). Ewha Womans University, Korea.

Lee, H., & Lee, H. (2015). Analysis of students' socioscientific decision-making from the nature of technology perspectives. *Journal of The Korean Association For Science Education, 35*(1), 169–177. doi:10.14697/jkase.2015.35.1.0169

Lee, H., & Lee, H. (2016). Contextualized nature of technology in socioscientific issues. *Journal of the Korean Association for Research in Science Education, 36*(2), 303–315. doi:10.14697/jkase.2016.36.2.0303

Lee, H., & Lee, H. (2017). Development and application of rubric for assessing nature of technology in the context of socioscientific issues. *Journal of the Korean Association for Research in Science Education, 37*(2), 323–334.

Miles, M. B., & Huberman, A. M. (1994). *Qualitative data analysis: An expanded sourcebook*. Sage.

Mitcham, C. (1994). *Thinking through technology: The path between engineering and philosophy*. University of Chicago Press.

National Research Council (NRC). (1996). *National science education standards*. Washington, DC: National Academy Press.

National Research Council (NRC). (2006). *Tech tally: Approaches to assessing technological literacy*. Washington, DC: National Academy Press.

National Research Council (NRC). (2012). *A framework for K-12 science education: Practices, crosscutting concepts, and core ideas*. Washington, DC: National Academies Press.

Nye, D. E. (2006). *Technology matters: Questions to live with*. MIT Press.

Pacey, A. (1983). *The culture of technology*. MIT Press.

Polanyi, M. (1958). *Personal knowledge: Towards a post-critical philosophy*. University of Chicago Press.

Romine, W. L., Sadler, T. D., & Kinslow, A. T. (2017). Assessment of scientific literacy: Development and validation of the Quantitative Assessment of Socio-Scientific Reasoning (QuASSR). *Journal of Research in Science Teaching, 54*(2), 274–295. doi:10.1002/tea.21368

Sadler, T. D., Barab, S., & Scott, B. (2007). What do students gain by engaging in socioscientific inquiry? *Research in Science Education, 37*(4), 371–391. doi:10.100711165-006-9030-9

Sadler, T. D., Chambers, F. W., & Zeidler, D. L. (2004). Student conceptualizations of the nature of science in response to a socioscientific issue. *International Journal of Science Education, 26*(4), 387–409. doi:10.1080/0950069032000119456

Sadler, T. D., & Zeidler, D. (2009). Scientific literacy, PISA, and socioscientific discourse: Assessment for progressive aims of science education. *Journal of Research in Science Teaching, 46*(8), 909–921. doi:10.1002/tea.20327

Scherz, Z., & Oren, M. (2006). How to change students' images of science and technology. *Science Education, 90*(6), 965–985. doi:10.1002ce.20159

Simonneaux, L., & Simonneaux, J. (2009). Students' socio-scientific reasoning on controversies from the viewpoint of education for sustainable development. *Cultural Studies of Science Education, 4*(3), 657–687. doi:10.100711422-008-9141-x

Son, W. C. (2011). Engineering design and philosophy of technology. *Journal of the Society of Philosophical Studies, 94*, 107–136.

Vincenti, W. G. (1990). *What engineers know and how they know it*. John Hopkins University Press.

Volti, R. (2010). *Society and technological change*. Worth.

Waight, N., & Abd-El-Khalick, F. (2012). Nature of technology: Implications for design, development, and enactment of technological tools in school science classrooms. *International Journal of Science Education, 34*(18), 2875–2905. doi: 10.1080/09500693.2012.698763

Walker, K. A., & Zeidler, D. L. (2007). Promoting discourse about socioscientific issues through scaffolded inquiry. *International Journal of Science Education, 29*(11), 1387–1410. doi:10.1080/09500690601068095

Zeidler, D. L., Sadler, T. D., Simmons, M. L., & Howes, E. V. (2005). Beyond STS: A research-based framework for socioscientific issues education. *Science Education, 89*(3), 357–377. doi:10.1002ce.20048

Section 4
Socioscientific Issues for Scientific Literacy in Preservice Teacher Training

Chapter 7
The Association Between Course Context and Preservice Teachers' Perceptions of SSI Instruction

Mark H. Newton
https://orcid.org/0000-0002-3616-319X
East Carolina University, USA

Melanie Kinskey
https://orcid.org/0000-0001-9789-1038
Sam Houston State University, USA

ABSTRACT

This chapter explores the association between elementary preservice teachers' (PSTs) perceptions of teaching using socioscientific issues (SSI) and the context in which the PSTs initially engaged with SSI. One course engaged with SSI while learning pedagogical strategies during an elementary science methods (decontextualized) course. The second course engaged with SSI via an environmental sciences course (contextualized). The contextualized course examined gray wolf management in Northern California as part of an ecology unit, which was followed by a series of debriefings regarding the implementation of the SSI. The findings indicate that while PSTs from both courses generally held more positive perceptions of SSI instruction after engaging with SSI, the students in the contextualized course perceived a greater ability to answer SSI-related questions and expressed more positive attitudes towards science content in the post-course data. Additionally, students in both courses perceived a greater need for teacher training on SSI implementation post-course.

DOI: 10.4018/978-1-7998-4558-4.ch007

INTRODUCTION

Much of the work associated with preservice teachers (PSTs) and socioscientific issues (SSI) neglects to consider the influence of the instructional context in which SSI is introduced. While Özden (2015) has previously explored elementary PSTs' views of SSI instruction, the aim of the study presented in this chapter was to extend this research and explore how the context in which the SSI framework was introduced to elementary PSTs may influence their views. Specifically, the authors examined the association between the context in which PSTs were introduced to the SSI framework and their perceptions of the SSI framework as it relates to student engagement, teacher effectiveness with implementation, and its existence in the elementary curriculum. The study highlighted in this chapter is unique in that it considers the contextual influences of science content and methods courses on PSTs perceptions of implementing SSI in the elementary classroom rather than the development of skills necessary to effectively engage with SSI. With these goals in mind, the following research questions guided this study:

1. To what extent, if any, were PSTs perceptions of teaching using SSI associated with the instructional context within which they were introduced to SSI instruction?
2. How, if at all, do elementary PSTs' perceptions of SSI instruction in elementary school change after being exposed to SSI as part of their teacher preparation program?

BACKGROUND

There is a concerning reluctance in elementary science teaching to engage children in rigorous science instruction, despite existing research demonstrating young children have capabilities to actively participate in challenging science lessons (Roth, 2014). Anecdotally, it has been observed that PSTs' reluctance has resulted in the reliance of prescriptive textbook-based science instruction and a lack of our students developing their own science lessons. Novice teachers often derive instructional decisions from their experiences in teacher preparation programs (Appleton & Kindt, 1999), resulting in a call to reform science teacher preparation, specifically in teachers' abilities to deliver science instruction focused on improving scientific literacy (NRC, 2012). In responses to these reforms, a socioscientific issues (SSI) approach has been suggested as one possible improvement to science teacher preparation programs (Zeidler, Herman, & Sadler, 2019). The pedagogical power of an SSI approach to science teaching has been well documented in the extant literature (see Sadler,

2009; Zeidler, 2014). Students have demonstrated gains in content knowledge, socioscientific reasoning (SSR), socioscientific perspective taking (SSPT), informal reasoning, and pro-environmental behaviors (Herman, Zeidler, & Newton, 2018; Sadler, 2011; Sadler, Barab, & Scott, 2007).

Socioscientific Issues Instruction and Training Preservice Teachers

Socioscientific issues are ill-structured, controversial issues that require students to engage in evidence-based reasoning (Sadler, 2011; Zeidler, 2014; Zeidler & Keefer, 2003). These issues involve the intentional use of science to provide opportunities for students to discuss, argue, and participate in dialogue concerning the moral and ethical components surrounding the societal issue (Sadler, 2011; Zeidler, 2014; Zeidler & Keefer, 2003). Teachers who are prepared to facilitate SSI in the classroom must have adequate knowledge and skills to guide students through this instruction. Specifically, teachers must: 1) have a thorough understanding of the science content that is connected to the issue and the ability to help students make those connections; 2) understand the moral and ethical considerations regarding the issue and the ability to facilitate moral and ethical development in students; 3) understand levels of argumentation and how to evaluate arguments as well as the ability to facilitate discussion and argumentation in the classroom; and 4) hold informed views of the nature of science (NOS) and have the ability to facilitate NOS in the context of SSI (Lederman, 2003; Simmons & Zeidler, 2003). Many elementary classroom teachers lack these specific SSI knowledge and skills and often express doubts in their ability to facilitate SSI in the elementary context (Kılınç, et al., 2013). PSTs in the present study had limited or no prior experiences engaging in SSI instruction as either students or teachers. This lack of prior experience with SSI provided an opportunity to explore the PSTs' perceptions of SSI instruction and its place in the elementary classroom. To expose PSTs to SSI, we provided them with vicarious experiences through modeling SSI lessons within our courses.

Vicarious experiences occur while observing others perform a task, and through these observations, observers gain knowledge and confidence in their abilities to perform the same task (Bandura, 1997). The authors modeled the knowledge and skills necessary to effectively facilitate SSI instruction as PSTs engaged in SSI experiences through two different contexts. The participants in this study were either enrolled in a science methods course, designed to provide preparation in the methods for science instruction, or an environmental science content course, designed to build science content knowledge of future elementary teachers. Understanding the context in which preparation for SSI is provided may influence how SSI is perceived, and therefore the goal for this study was to identify how, if at all, PSTs' perceptions of

SSI instruction in the elementary classroom may be influenced by the context in which they were introduced to the SSI framework.

Contexts of Socioscientific Issues with PSTs

Empirical literature concerning PSTs and SSI pays much attention to specific skills associated with SSI rather than the context in which SSI is delivered. While researchers do take the time to identify and describe the contexts in which PSTs are engaging with SSI, we were unable to find literature explicitly comparing how different contexts within science education potentially influence PSTs' development regarding SSI. To resolve this limitation, and highlight the important role context plays, we examined literature concerning PSTs and SSI with a focus on how the contexts were utilized for the purpose of the studies. We identified a majority research within science methods courses, few studies within field experiences, and a deficit of studies focusing on SSI within the context of a science content course.

The purpose of many science methods courses is to develop pedagogical content knowledge in future teachers to develop their understandings and abilities in transferring subject matter knowledge to effective science instruction (Abell, Appleton, & Hanuscin, 2010). We found this purpose also to be true of literature regarding SSI and PSTs development. Forbes and Davis (2008) utilized an elementary science methods course to help their elementary PSTs develop instructional understandings of SSI as they critiqued and adapted existing SSI-oriented science curricula. They found the PSTs' evaluation of the SSI-oriented lesson to be influenced more by what they learned about in their methods course, specifically the alignment of learning objectives, standards, and inquiry-based activities, than characteristics of SSI. While not the focus of this study, the findings demonstrated how the context in which SSI is introduced may sway PSTs' perceptions of what is important regarding SSI instruction. In a control-experimental study of 82 elementary PSTs enrolled in a science methods course, Yapıcıoğlu and Kaptan (2017) implemented SSI based instruction to 40 elementary preservice, while 42 elementary PSTs in the control group did not receive SSI based instruction. All participants completed the basic science and technology literacy scale instrument. Results showed the PSTs who engaged in SSI based instruction during the methods course had significant gains in their understandings of scientific and technological literacy. The qualitative data illustrated the PSTs in the experimental group had more accurate and positive perceptions of SSI-based instruction. This study illustrates the benefits of developing PSTs' scientific literacy skills by engaging them in SSI during methods courses.

Past research has also utilized methods courses to examine PSTs' self-efficacy regarding specific SSI or concerns regarding SSI instruction. For instance, Kılınç et al. (2014) examined how 441 secondary PSTs' content knowledge, risk perception, moral beliefs and religious beliefs concerning the contextualized SSI of genetically modified foods influenced their efficacy beliefs regarding teaching SSI. Results showed the PSTs had high confidence with teaching about genetically modified foods in general, but efficacy with teaching this SSI decreased when they considered religious backgrounds of their students' families, parent involvement with the instructional decisions regarding SSI, and the content knowledge necessary for teaching this science content in the context of SSI. In another study of beliefs regarding SSI during a science methods course, Borgerding and Dagistan (2018) explored 12 secondary preservice and one secondary in-service teachers' concerns about facilitating SSI instruction. Similar to Kılınç et al. (2014) they identified concerns about parental influences with SSI instruction, but also revealed concerns about student maturity levels and a lack of student understanding regarding SSI topics. Despite these concerns, however, the participants in this study did identify the benefits and importance of facilitating SSI.

Literature found within the context of field experiences focuses on middle or high school classrooms and utilizes the classroom environment to gain an understanding of PSTs' instructional practices concerning SSI. In their study of 10 secondary PSTs, Genel and Topcu (2016) analyzed lesson plans and video recordings of participants' SSI-focused science instruction to understand the abilities of these PSTs to organically integrate SSI into their teaching practice. Their findings highlight the challenge these participants faced in centralizing SSI rather than teaching science content in isolation. While Genel and Topcu's findings revealed struggles with SSI instruction by chance, Pitiporntapin, Yutakom, and Sadler (2016) intentionally focused on the struggles experienced during practicum experiences with SSI as they explored how 52 secondary PSTs used SSI in their science teaching, what struggles the participants identified, and what they felt they needed to improve their SSI-based instruction. Through observations and interviews, these researchers identified the struggle to utilize controversial issues in the classroom as well as a lack of confidence in participants' abilities to facilitate SSI, and more explicit professional development illustrating what effective SSI instruction looks like. Both of these studies enrich the knowledge base of SSI instruction in secondary classrooms because they were conducted within the context of the field, thus illustrating the important role context plays in developing PSTs' abilities for SSI-based teaching.

As mentioned earlier, empirical evidence for PSTs and SSI within science content courses is extremely limited. One relevant study, conducted by McDonald (2014), focused on the quality of five primary PSTs' written arguments after engaging with a global warming SSI during a general science content course. Findings revealed engaging in argumentation regarding an SSI during a science content course is an effective method for helping future teachers write high quality arguments, which is believed to transfer to their practice as a primary teacher. In another study that took place outside the context of a methods course, Yerdelen, Cansiz, Cansiz, and Akcay (2018) explored 54 preservice science, and 32 preservice social-science, teachers' attitudes toward SSI before and after they participated in an SSI course. Rather than having a pedagogical focus on SSI as a methods course would, the purpose of their SSI course was to expose the preservice teachers to the theoretical ideas of SSI with the goal of enhancing their informed decision-making skills and improving their attitudes toward SSI instruction. From their findings they argue that a background in science content is not a necessary prerequisite to improving attitudes toward SSI, but emphasis should be placed on developing critical thinking skills and exposure to socially relevant issues. As with the SSI scholarship discussed previously, these studies focused on skill development of PSTs, rather than the relationship between the learning context and their perceptions of SSI instruction.

Theoretical Underpinnings

Because perceptions are how individuals make meaning of the world, social psychologists have extensively researched how individuals develop perceptions (see Kruger & Dunning, 1999; Pickens, 2005). As a result of this common interest, much is known about this topic. A current explanation of perception development comes from research in management. Developing perceptions is a four-stage process: (1) stimulation, (2) registration, (3) organization and (4) interpretation (Pickens, 2005). However, when an individual is exposed to stimuli, that person makes meaning by interpreting the stimuli through their prior experiences (Lindsay & Norman, 1977). The individual will selectively choose stimuli based on their beliefs, attitudes, motivation, and personality. The individual selects stimuli that meet an immediate need, while stimuli that create anxiety are ignored (Assael, 1995; Sherif & Cantril, 1945). The selected stimuli are further organized and interpreted based on prior experiences. If the interpretation of the selected stimuli creates positive feedback, the perception reinforces the individual's reality, while negative feedback causes internal conflict (Pickens, 2005).

In an educational context, it is valuable to understand teachers' perceptions because those perceptions impact a multitude of decisions made in the classroom. An example of perceptions influencing decision making can be found in studies on pedagogical choices to select student experience. Montes and Rockley (2002) found that secondary level chemistry teachers tended to implement verification laboratory activities, which have an outcome to an existing question that is already known and requires students to follow a procedure that has already been developed; rather than inquiry-based activities. The reason for this decision is that the teachers perceived verification activities to be more convenient, despite acknowledging the benefits of inquiry instruction and shortcomings of verification activities on students. Montes and Rockley found that providing professional development for the teachers changed their perception and led to more inquiry-based instruction. Several other studies have demonstrated that teachers who held high perceptions of science teaching tended to incorporate more inquiry-based instruction (Choi & Ramsey, 2009; Hanley et al., 2002; Marshall et al. 2009).

Some researchers have studied preexisting perceptions associated with aspects of SSI. Topçu, Sadler, and Yilmaz-Tuzun (2010) provided 39 secondary PSTs with SSI-based scenarios to gain an understanding of their informal reasoning skills. They found that informal reasoning skills were not issue-dependent and emotional responses were highly influential in these PSTs' decision-making. Similarly, Lee, Chang, Choi, Kim, and Zeidler (2012) investigated 18 secondary PSTs' predispositions regarding character and values as they engaged in discourse related to three SSI scenarios. Results showed engaging in dialogic discourse did not cause the PSTs to change their views but found ways to apply their own values to the various perspectives presented. The foundational understanding each of these studies provides regarding PSTs' initial abilities to engage with SSI may be used to guide formal training in SSI development, specifically informal reasoning skills and reflective character and values processes, in both science content and pedagogy- focused undergraduate courses, as well as examine how PSTs' prior experiences influence how they contemplate SSI.

The current study investigates the general perceptions elementary PSTs hold of SSI instruction as it relates to student engagement, teacher effectiveness, and place in elementary curriculum. The SSI framework that guided our research provides opportunities for students to: conduct research on the various perspectives of stakeholders concerning a contentious, real-world issue and practice empathy toward the different perspectives; recognize the moral and ethical implications of decisions regarding the issue; and consider the varying perspectives, science content, and moral and ethical considerations when developing their own opinions and engaging in argumentation (Zeidler, 2014).

Previous research conducted by Ozden (2015), which informed the design of our present research, utilized the Socioscientific Issues in Science Course questionnaire to measure 113 elementary PSTs' views of SSI after they engaged in two weeks of SSI lessons while enrolled in a Science and Technology Teaching course. Additionally, eight PSTs were interviewed about their views of SSI instruction. Results showed that after their two-week engagement with SSI, PSTs accurately understood the general purpose of SSI instruction but lacked understanding of the moral and ethical components of such instruction. Additionally, Ozden found that prospective teacher participants identified the benefits of SSI instruction to enhance critical thinking skills of elementary aged children.

The fact that teachers' perceptions influence their pedagogical decisions, coupled with the notion that early career teachers tend to rely on their methods courses from their preparation program, as discussed previously, is what influenced the current study. It has been established in the extant literature (see above) that teachers identify curriculum standards, lack of content knowledge, and student maturity as factors impeding SSI implementation. Using the perceptions framework described earlier, it is possible to see that the perceived impediments of SSI implementation result from the teachers' prior experiences. To that end, the current study seeks to understand how initial exposure to SSI instruction influences how elementary PSTs perceive SSI instruction in elementary school.

Design and Methodology

A pre/post concurrent mixed methods design (Greene, 2007) was implemented to address the research questions, where PSTs responses to a Likert-type survey regarding their beliefs about SSI instruction related to the following three constructs: student engagement, teacher effectiveness with implementation, and its existence in the elementary curriculum. Qualitative data were collected simultaneously as an extension to the Likert-type survey and asked the PSTs to provide specific written examples to explain their responses in the survey regarding each specific construct. The survey was administered once before participants engaged with SSI, and again after they engaged with SSI. The time of the survey administration varied based on when each course instructor began SSI implementation, which is outlined in more detail below. PSTs initial beliefs and potential changes in those beliefs could be explored using concurrent data collection in the pre/post format allowed us to gauge their initial beliefs while also gaining a deeper understanding of the PST's perceptions through the qualitative responses. The Likert-type survey used for the quantitative data collections was a modified version of Özden's (2015) survey, statements were removed that did not address the research questions for the current study, while also adding the qualitative questions asking for explanations to the survey responses. The open-ended questions were added to the survey in lieu of conducting interviews to increase the sample size of data to include all PSTs enrolled in the two courses.

Context and Participants

A convenience sample of two groups of PSTs participated in the study. One group (Group 1, $n = 13$) of PSTs attended a large public urban university in the southeastern United States, while the second group (Group 2, $n = 34$) attended a small public university in the western United States. Matched pairs of data were collected and students' responses who did not complete both surveys were not analyzed.

Decontextualized SSI Course

The decontextualized SSI course, instructed by the second author, engaged with the SSI framework and various SSI-based scenarios while learning pedagogical strategies during their elementary science methods course. The students in the course were third year undergraduates and completing the second semester of a formal elementary teacher preparation program. The course met once a week for 2 hours 45 minutes per class. During their time in the methods course, PSTs were also completing an elementary field experience where they were responsible for working closely with a mentor teacher and plan and facilitate instruction one day per week.

Throughout the semester the instructor would model research-based instructional strategies for the purpose of exposing them to best practices in science teaching, thus SSI lessons were modeled as PSTs in the course assumed the role of elementary children as they worked through aspects of SSI (i.e. argumentation / evidence-based reasoning, perspective-taking, empathetic concerns, and moral and ethical considerations) for the purpose of gaining an understanding of how to engage their students in similar real-world scenarios. Prior to modeling SSI lessons, the instructor assigned a reading about SSI (Zeidler & Kahn, 2014) to provide background knowledge of the framework prior to engaging with it in the methods course. After completing the reading, SSI were tied to inquiry-based and interdisciplinary teaching strategies that followed the 5E lesson structure (engage, explore, explain, elaborate, evaluate) and covered topics such as the use of pesticides (as an upper elementary grade level example) and plastic in the ocean (as a lower elementary grade level example). Due to the nature of the course aiming to develop PSTs' pedagogical skills, the incorporation of SSI came later in the 14-week semester, after students had built foundational understanding of nature of science, levels of inquiry-based teaching, and the 5E instructional model. Table 1 provides a brief overview of the course layout.

Table 1. Summary of decontextualized SSI course

Week	Decontextualized SSI Course Topics
1	Course Overview / Syllabus Chat Science Notebooks Draw a scientist test Reflections on scientist images
2	Decontextualized nature of science: aspects of nature of science (subjective, collaborative, corrigible, communication, evidence-based) Reflections on NOS / connections to classroom
3	Decontextualized nature of science: explicit, reflective nature of science instruction; contextualized vs. decontextualized nature of science instruction Reflections on NOS / connections to classroom
4	Nature of science: image of scientists Science standards and reform Reflections / connections to classroom
5	Four levels of inquiry (confirmation, structured, guided, open) Long term investigations project: open inquiry Mealworm investigations Reflections on strategies / connections to classroom
6	5E instructional model (Engage, Explore) Assessing prior knowledge Learning Focused Strategies Levels of inquiry as explore
7	5E instructional model (Explain, Elaborate) Evidence-based reasoning Claim-Evidence-Reasoning organizers Student-led learning
8	5E instructional model (Evaluate) Authentic assessments in science Project-based learning (rubrics) Discussion (rubrics) Formative assessments in science Probes / Learning Focused Strategies Reflections on strategies / connections to classroom In class, students took the SSI survey for the first time. Homework was to read about SSI instruction.
9	Real-world connections in science Introduction to SSI Framework SSI and the 5E model (butterflies and pesticides – intermediate example) Engage: prior knowledge/ solicit initial opinions Explore: owl pellets
10	SSI and the 5E model Explain: food chains Elaborate: assign perspectives, complete readings from different views, claim-evidence-reasoning
11	SSI and the 5E model Evaluate: review authentic assessment / class debate Reflections on strategy / connections to the classroom
12	SSI and the 5E model (plastic in the ocean – primary example) Engage: prior knowledge and personal connections to disposable plastics / opinion about disposable plastics on the beach Explore: plastic in the ocean simulation Explain: What did you learn discussion / video of diver in ocean of litter
13	SSI and the 5E model Elaborate: different perspectives of plastics on beach/ what is right and wrong? Evaluate: what do you think? Illustration and paragraph. Share in pairs. Reflections on strategy/ connections to the classroom In class, students took the post-SSI survey.
14	Authentic science fair presentations

Contextualized SSI Course

The contextualized SSI course, instructed by the first author, was exposed to SSI within the context of an environmental science course for PSTs. The students in the course were in at least their third year of postsecondary education and all aspired to be elementary teachers. The course met for lab activities twice a week for two hours per class and one hour per week for lecture. Prior to this course, the students had limited exposure to elementary classrooms, primarily consisting of observing a single classroom. The internship portion of students' training occurs as part of a postbaccalaureate credentialing program. The primary focus of this course was to teach science content knowledge as well as effective pedagogical approaches for teaching environmental science to K-8 students. The SSI was embedded with an ecology unit that used wolf management in northern California as the SSI. The instructor used extant literature on SSI instruction to design the course (Herman, Sadler, Zeidler, & Newton, 2018; Herman, Zeidler, Newton, 2019; Newton, 2016; Zeidler & Newton, 2017). The issue of wolf management in Northern California was introduced early in the unit and provided relevance for the scientific knowledge and experiences that followed. The instructor intentionally scaffolded experiences that fostered the development of socioscientific reasoning (SSR):

1. Account for the complexity of SSI
2. Analyze the issue from multiple perspectives
3. Identify aspects of the issue that are subject to ongoing inquiry
4. Employing skepticism in the analysis of potentially bias information
5. Explore how science can contribute to the issue and the limitations of science (Sadler, Barab, & Scott, 2007).

The students were not provided with any information related to the SSI framework prior to the start of the unit nor did the instructor reference SSI during the unit to ensure that students' perceptions were not influenced. Furthermore, the instructor consulted with instructors from the prerequisite courses to ensure that SSI had not been discussed in those classes.

Student groups were assigned the perspective of a given stakeholder and used their experiences in class to develop a purposed management plan for California. These proposals were presented to a small group of peers in a debate-style presentation, which served as the culminating activity for the unit. Tables 2 summarizes the course design. After the unit was completed, the instructor led multiple class discussions deconstructing the pedagogical decisions made throughout the SSI unit, as well as how the experiences from the unit aligned with the Next Generation Science Standards (NGSS) (NGSS Lead States, 2013). The remainder of the course focused on various environmental science concepts and implemented other inquiry based pedagogical strategies unrelated to the current study.

Table 2. Summary of contextualized SSI content course design

Week	Contextualized SSI Course Content
1	Course Introduction and Information Decontextualized Nature of Science Activity Introduction of Wolf Management SSI News videos related to wolves reentering California Class discussion on students' initial thoughts on wolves reentering California Introduce Wolves in California Town Council Activity
2	Laboratory Activity – Primary Productivity and Energy Flow in an Ecosystem Assign groups stakeholder perspectives related to wolf management Jigsaw Cooperative Learning – readings covering wolf management in California from multiple perspectives Guest Speaker – local sheep rancher Assigned Reading – trophic cascade debate, course textbook chapter
3	Laboratory Activity – Habitat and Niche Laboratory Activity – Community Structure Guest Speaker – restorative ecologist with research experience related to wolf reintroduction in the Greater Yellowstone Ecosystem Assigned Reading – estimation techniques for wolf populations, course textbook chapter
4	Laboratory Activity – Population Dynamics Work time for Wolves in California Town Council Activity
5	Work time for Wolves in California Town Council Activity Wolves in California Town Council Activity and reflection Review for Exam 1
6	Exam 1 Lecture – Socioscientific Issues Framework Assigned Readings – teaching SSI in science, teaching SSI in Environmental Education

Data Collection and Analysis

Data were collected in a pre/post format, in which each instructor for the course administered the survey containing both the Likert-type and open-ended questions before the PSTs engaged with SSI in the course, and again after they engaged with SSI in the course. The quantitative data measured PSTs perceptions about SSI instruction by asking them to rank a range of agreement toward specific statements concerning SSI teaching The quantitative items are considered Likert-type, rather than Likert, because the authors make no attempt to create an individual scale based on the results, consequently, the responses can be treated as interval data and descriptive statistics can be reported (Clason & Dormody, 1994; Boone & Boone, 2012). The responses for the Likert-type items were converted quantitative data where *Strongly Agree* =1, *Agree* = 2, *Neither Agree or Disagree* = 3, *Disagree* = 4, and *Strong Disagree* = 5. A mean was calculated for each Likert-type item and organized by course because the unit of measure for each research question was the different courses. Therefore, a change in the pre/post mean that shifted toward 1.00 indicated a stronger agreement and a shift in mean toward 5.00 indicated a shift toward disagreement with the Likert-type item.

The purpose of the qualitative data provided insight beyond what the quantitative data showed, of the perceptions these PSTs had of SSI instruction, and so we analyzed the open-ended responses using a two-cycle coding approach (Saldana, 2009). The authors initially independently engaged with the data through an open-coding process in which generated analytic memos (Corbin & Strauss, 2008) of the responses were generated. After the initial stage the authors met to discuss the codes and reach 100% agreement. A second cycle of independent coding occurred, once again, where the codes were reorganized into pattern codes and themes. A final meeting occurred to discuss the final themes and reach consensus.

Results

Perceptions of Socioscientific Issues and Student Engagement

The quantitative data illustrates percentages of PST's responses related to how they perceived the appropriateness of implementing SSI to elementary children and are located in Table 3 below. As mentioned earlier, since the items were on a scale of 1 – 5, with 1 corresponding to "Strongly Agree" and 5 corresponding to "Strongly Disagree", lower means indicate the PSTs in that group more strongly agree with the statement than the group with the higher mean, while a higher mean indicates the group more strongly disagrees with the statement.

Overall, after engaging with SSI, both groups showed a stronger sense of disagreement with the statement that SSI is difficult for elementary children. The decontextualized group, however, had a larger shift toward disagreeing with that statement than the contextualized, group. Also noteworthy, is the shift in each group's perception of elementary student maturity and abilities to engage with SSI. After engaging with SSI, the mean of the decontextualized course shifted toward more strongly agreeing that elementary students were not mature enough for SSI, while the contextualized course shifted more toward disagreeing with that statement. After engaging with SSI both groups more strongly agreed with statements related to the ability of SSI instruction to increase students' interest in science and increase student engagement as evidenced in the unanimous agreement by both classes regarding increasing student interest and the unanimous agreement from the decontextualized course regarding student engagement.

Table 3. Elementary PSTs quantitative perceptions of elementary students ability to engage with SSI

Perceptions of Students' Engagement with SSI Curriculum							
Survey Statement		**Strongly Agree (%)**	**Agree (%)**	**Neither Agree nor Disagree (%)**	**Disagree (%)**	**Strongly Disagree (%)**	**Mean**
SSI is difficult for elementary students	Decontextualized Course						
	Pre-course	0.00	15.4	30.8	53.8	0.00	3.23
	Post-course	0.00	7.69	7.69	53.8	30.8	4.08
	Contextualized Course						
	Pre-course	0.00	5.88	32.4	52.9	8.82	3.59
	Post-course	0.00	5.88	17.6	61.8	14.7	3.85
Elementary students are not mature enough for SSI	Decontextualized Course						
	Pre-course	0.00	0.00	15.4	76.9	7.69	3.92
	Post-course	7.69	15.4	7.69	30.8	38.5	3.77
	Contextualized Course						
	Pre-course	0.00	8.82	14.7	64.7	11.8	3.79
	Post-course	0.00	0.00	8.82	67.6	23.5	4.15
Academically successful students would be more interested in SSI	Decontextualized Course						
	Pre-course	7.69	30.8	46.2	15.4	0.00	2.69
	Post-course	23.1	30.8	23.1	23.1	0.00	2.46
	Contextualized Course						
	Pre-course	11.8	58.8	26.5	2.94	0.00	2.21
	Post-course	35.3	44.1	17.6	2.94	0.00	1.88
Integrating SSI instruction would increase elementary students' interest in science	Decontextualized Course						
	Pre-course	46.2	38.5	15.4	0.00	0.00	1.69
	Post-course	30.8	69.2	0.00	0.00	0.00	1.69
	Contextualized Course						
	Pre-course	26.5	67.6	5.88	0.00	0.00	1.79
	Post-course	32.4	67.6	0.00	0.00	0.00	1.68
Elementary students can learn science better by engaging in SSI instruction	Decontextualized Course						
	Pre-course	30.8	38.5	23.1	7.69	0.00	2.08
	Post-course	46.2	53.8	0.00	0.00	0.00	1.54
	Contextualized Course						
	Pre-course	29.4	67.6	2.94	0.00	0.00	1.74
	Post-course	38.2	55.9	5.88	0.00	0.00	1.68

Table 4. Qualitative examples of elementary PSTs perceptions of elementary students engaging with SSI

	Pre-SSI response	Post-SSI response	Researchers' interpretation
Decontextualized course example	PST 39: I am very in the middle with this because I believe it depends on the grade the students are in. If they are in K-2, I do not think they are mature enough for it, but I think grades 3-5 are a little more mature for it.	PST 39: SSI is a very good way to bring in different things about science into the classroom that connect to real world things.	This PST initially considers the differences in lower and upper elementary students' abilities, but after engaging with SSI, generally believes SSI is a good way to help all students connect science to the real world.
Contextualized course example	PST 24: I think the SSI instruction sparks and interests in students because it allows them to relate what they are learning to situations in our world at the moment. I also think all students would be interested in SSI, not just those who are academically successful. Elementary students are not too young for SSI they just would need a different way of being taught then those who are in middle or high school.	PST 24: I think students would benefit from have real world examples and issues being used in the classroom. Instead of having students copy terms from a textbook, students will be able to work with issues that could teach them valuable life skills. Elementary students are not too young for SSI, the lessons will just have to be for grade level.	This PST initially believes SSI instruction would foster student interest and be appropriate for young children, but with modifications. Similar to the response from the decontextualized course, this PST also references the benefit SSI has to helping elementary children connect science to the real world.

Open-ended qualitative responses allowed the PSTs to provide insight into their quantitative responses. No qualitative evidence expanded the PST's perceptions regarding difficulty and maturity levels of SSI with elementary students; however, the responses provided support for both groups' positive perceptions regarding SSI instruction and student interest and improved science learning, exemplars are presented in Table 4. When considering student engagement with SSI, PST 39, from the decontextualized course, was initially focused on student ages and concern with students' developmental abilities to participate in SSI instruction. After SSI in the course, though, PST 39 made a more generalized statement of support for SSI instruction in elementary grades, without mention of age-based abilities. From the contextualized content course, PST 24 had displayed support for SSI instruction in elementary grades but recognized the need to differentiate instruction for younger children's abilities. In his/her post-SSI response, s/he continues supporting SSI instruction for elementary children and cites a concrete reason for moving away from textbook-based instruction to more real-world applicable science teaching with SSI.

Table 5. Elementary PSTs' quantitative perceptions of elementary teachers' preparation to teach SSI

Perceptions of Elementary Teachers' Preparation for SSI							
Survey Statement		**Strongly Agree (%)**	**Agree (%)**	**Neither Agree nor Disagree (%)**	**Disagree (%)**	**Strongly Disagree (%)**	**Mean**
Elementary teachers are not competent at integrating SSI	Decontextualized Course						
	Pre-course	0.00	15.4	38.5	30.8	15.4	3.46
	Post-course	0.00	7.69	53.8	15.4	23.1	3.54
	Contextualized Course						
	Pre-course	2.94	8.82	47.1	35.3	5.88	3.32
	Post-course	0.00	2.94	26.5	52.9	5.88	3.50
Elementary teachers can easily answer SSI-related questions	Decontextualized Course						
	Pre-course	0.00	7.69	61.5	30.8	0.00	3.23
	Post-course	0.00	38.5	23.1	38.5	0.00	3.00
	Contextualized Course						
	Pre-course	0.00	11.8	61.8	26.5	0.00	3.15
	Post-course	8.82	32.4	35.3	20.6	2.94	2.76
Elementary teachers should be trained in SSI instruction	Decontextualized Course						
	Pre-course	25.0	75.0	0.00	0.00	0.00	2.00
	Post-course	33.3	66.7	0.00	0.00	0.00	1.31
	Contextualized Course						
	Pre-course	32.4	61.8	2.94	2.94	0.00	1.76
	Post-course	35.3	64.7	0.00	0.00	0.00	1.53

Perceptions of Teacher Effectiveness with Socioscientific Issues Instruction

Table 5 provides data related to PSTs perceptions of elementary teachers' preparation for implementing SSI instruction. The mean scores demonstrate that both groups disagreed more with the notion that elementary teachers are not competent at implementing SSI instruction. Regarding elementary teachers' ability to easily answer SSI-related questions, the mean score for both groups indicated that PSTs agree more that elementary teachers' have the ability to easily answer SSI-related questions. Finally, both groups' mean scores indicated stronger agreement post-course that elementary teachers should be trained more in SSI instruction. The quantitative data indicates that context of the course did not influence PSTs' perceptions on teachers' competency, ability to answer SSI-related questions, or the need for teachers to be trained in SSI instruction.

The qualitative response examples in Table 6 show a difference in the way the PSTs from the decontextualized course and the PSTs in the contextualized course perceive elementary teachers' abilities to teach SSI. Prior to engaging with SSI, the PSTs in the decontextualized course perceived that most elementary teachers are not prepared to teach SSI, while the PSTs in the content course initially believes elementary teachers do have the preparation to teach SSI. After engaging with SSI, their responses align with the quantitative findings as both PSTs from the decontextualized content and contextualized methods courses recognize the gap in elementary science instruction and express the need for training for elementary teachers to adequately facilitate SSI.

Perceptions of Socioscientific Issues in Elementary Curricula

The quantitative data associated with the last three Likert-type items is located in Table 7. Concerning SSI being more suitable for science instruction in elementary school, the mean score for PSTs in the decontextualized course indicates a large shift toward disagreeing, while the mean score of PSTs in the contextualized course remained relatively unchanged. After completing their coursework, both groups' perceptions of the appropriateness of SSI in elementary school remained high. Both groups' mean score remained in the *Strongly Agree* range. Finally, both groups felt that SSI instruction would improve students' scientific literacy.

The qualitative responses provide insight into each groups' perception that SSI instruction not only belongs in elementary classrooms, but also that it is not restricted only to science. From their responses, an exemplar is presented in Table 8, we notice PST 45 from the decontextualized course initially believed SSI was more appropriate at "higher levels" and focused only on scientific literacy. S/he illustrated a shift in his/her post response when there was no mention of the "higher levels" and referenced SSI appropriateness in subjects other than science. This response provides an example of why there may have been a shift in the first statement for the decontextualized course regarding SSI inclusion in science. In contrast, the response from PST 20 in the content course highlights the consistency pre- to post-course of perceiving SSI's placement within elementary curricula, as s/he maintains their position of why SSI in elementary school is beneficial. This consistency was also evident in the minimal change of pre- to post course quantitative means for the contextualized course.

Table 6. Qualitative examples of PST's' perceptions of teacher effectiveness with socioscientific issues instruction

	Pre-SSI response	Post-SSI response	Researchers' interpretation
Decontextualized course example	PST 40: 1. Most elementary school teachers (at least in my experience) do not use/ know about SSI instruction, but some teachers may have researched or been taught about it and frequently use it in their classroom. 2. Teachers that are not well educated on SSI may not be able to answer questions about it, but those that are familiar with it may be able to. 3. SSI instruction will connect scientific practices/ ideas to real world and make the concepts more relatable and concrete for students. Therefore, it would be beneficial for teachers to be taught about this method so that they have that option in there "toolbox."	PST 40: Teachers who have been in the profession for a long time may not be competent to instruct with SSI because it is just now becoming more common, but it is training for it should be provided because this method is a great way to teach science authentically.	The PST is consistent in his/her perception that SSI instruction is not commonly taught in elementary classrooms and that teachers are not effective with teaching it. After engaging with SSI in the methods course, the PST focuses on benefits of training teachers for SSI instruction.
Contextualized course example	PST 9: Elementary teachers are very capable, just as any other teacher, of integrating SSI within their classroom. Teachers should already be prepared to teach students about topics such as air pollution, global warming, etc. because as teachers we also want to be able to create more of an interest within students minds and we'd also like to help them understand the rights and wrongs that might exist within our environment	PST 9: Teachers would benefit more by including and learning about SSI. I think that finding topics and issues that they are passionate about and including them witching their curriculum would help them teach better. This involvement would also help them in providing better quality lesson plans that they can accumulate as time passes by.	Prior to engaging with SSI, this PST was optimistic in his/her views of elementary teachers' subject matter knowledge regarding SSI and their ability to facilitate instruction concerning these issues. After participating in SSI during the course, this PST demonstrates a different understanding of elementary teachers and SSI, focusing more on how to improve SSI instruction in the elementary classroom.

Table 7. Preservice elementary teachers' perceptions of ssi instruction in elementary school

Perceptions of SSI Instruction in Elementary School							
Survey Statement		**Strongly Agree (%)**	**Agree (%)**	**Neither Agree nor Disagree (%)**	**Disagree (%)**	**Strongly Disagree (%)**	**Mean**
SSI is more suitable for science instruction than other academic disciplines	Decontextualized Course						
	Pre-course	23.1	30.8	30.8	15.4	0.00	2.38
	Post-course	0.00	23.1	53.8	23.1	0.00	3.00
	Contextualized Course						
	Pre-course	2.94	35.3	50.0	8.82	2.94	2.74
	Post-course	2.94	29.4	52.9	8.82	2.94	2.79
SSI instruction should be included in elementary school	Decontextualized Course						
	Pre-course	30.8	61.5	7.69	0.00	0.00	1.77
	Post-course	30.8	46.2	23.1	0.00	0.00	1.85
	Contextualized Course						
	Pre-course	20.6	73.5	5.88	0.00	0.00	1.85
	Post-course	32.4	64.7	0.00	0.00	0.00	1.67
Integrating SSI into science instruction will increase scientific literacy	Decontextualized Course						
	Pre-course	23.1	61.5	15.4	0.00	0.00	1.92
	Post-course	30.8	61.5	7.69	0.00	0.00	1.77
	Contextualized Course						
	Pre-course	11.8	73.5	14.7	0.00	0.00	2.01
	Post-course	26.5	67.6	2.94	0.00	0.00	1.76

SOLUTIONS AND RECOMMENDATIONS

The extant literature demonstrates a relationship between PSTs attitudes and experiences with science and how their science teaching may potentially impact the performance and attitudes of their students (Bursal & Paznokas, 2006; Denessen, Vos, Hasselman, & Louws, 2015; Tosun, 2000). One approach to support PSTs is by integrating SSI instruction into the teacher preparation curriculum. Previous studies have demonstrated that educating PSTs in SSI instruction has been successful in exposing PSTs to new methods of teaching science but has not been effective at creating a more sophisticated understanding of science and contentious issues related to science. For example, PSTs often conflate controversial scientific issues and scientific issues are controversial to teach (e.g., climate change, evolution) (Boderding

Table 8. Qualitative examples of PSTs' perceptions of elementary curricula and socioscientific issues

	Pre-SSI response	Post-SSI response	Researchers' interpretation
Decontextualized course example	PST 45: I think in higher levels it would be good to introduce these issues. I believe that it would make the students more literate in science. They will understand more ad will be curious to learn more.	PST 45: SSI can be included into different classes not just science.	Prior to engaging with SSI, this PST mentioned SSI would be appropriate for higher levels and scientific literacy but after engaging with SSI, his/her response recognizes the value of SSI instruction in promoting literacy across disciplines versus only scientific literacy.
Contextualized course example	PST 20: I really like the idea of introducing SSI instruction early on in elementary science because I think it will allow students to actively engage in science topics and appreciate how truly significant these types of courses and issues are in the real, present world.	PST 20: I think integrating the SSI instruction style early on in programs can help foster a better learning environment and allow students to start meaningful thinking at an earlier age.	This PST is consistent with his/her perception that SSI should be included in elementary curricula. There is a focus in the pre-SSI response on elementary science that evolves to an understanding of how SSI can be included in more general curricula, not only science.

& Dagistan, 2018). Furthermore, PSTs struggle to implement SSI instruction within a standards-based curriculum and cite concerns with their own content knowledge, the maturity level of their students, and a lack of student understanding as obstacles to effective SSI implementation (Forbes & Davis, 2008).

In response to the findings mentioned above, the current exploratory study sought to determine how, if at all, the context in which elementary PSTs were introduced to the SSI framework was associated with their perceptions of SSI. The findings from this study build on the findings of previous studies in that elementary PSTs perceive SSI instruction as an effective tool in the elementary classroom regardless of the instructional context in which PSTs were introduced to the SSI framework. Specifically, the quantitative data supports the association between elementary PSTs perceived SSI instruction as appropriate for elementary students and that SSI instruction should be part of the elementary curriculum, regardless of the context in which the SSI framework is introduced. The perception that SSI are highly engaging and effective at connecting science to the real world was also pervasive throughout the data in both contexts. Additionally, PSTs from both groups identified the possibility for SSI to develop critical thinking skills necessary to develop individuals capable of contributing to a democratic society and use evidence-based reasoning. This is captured in the following statement given by PST 38 from the decontextualized course: "Students will be more engaged and take into consider SSI besides it consists of real-life. It also informs students to make decisions in the future," as well as PST 32 from the contextualized course, "This is a great learning instruction because it helps students develop critical thinking skills as well as helps shape and strengthen student to become and develop democratic principles." These results are promising in that pervious research has demonstrated that early career teachers tend to rely on what was learned in methods courses as they consider the pedagogical strategies they employ in their classes (Forbes & Davis, 2008).

The perceptions indicated in this study also align with several international science education reforms for science education. For example, the United Nations Educational, Scientific, and Cultural Organization (UNESCO) has identified the need for future generations to develop the requisite skills to address the scientific and social problems they are destined to face (UNESCO, 2019). Returning to the quotes by PSTs 38 and PST 32, it is clear they recognized the opportunity SSI provides to foster the connection between science and the participation of citizens in making decisions at the local, regional, national, and global level. The findings of this study support the argument that exposure to the SSI framework is a critical component to science teacher training programs and supports the growing body of research advocating for the SSI framework to be included in teacher preparation programs (see Özden, 2015). It is reasonable to hope that if early career teachers rely heavily on courses from their teacher preparation programs then early career teachers with SSI exposure will promote the desired skill development identified internationally because the perceived skills identified by the students in this study are consistent with the findings throughout the SSI literature (Herman, Zeidler, & Newton, 2018; Sadler, 2011; Sadler, Barab, & Scott, 2007) and align with current calls for science curriculum reforms (NGSS Lead States, 2013; UNESCO, 2019).

Regarding findings associated with teacher effectiveness with SSI instruction, the quantitative findings were consistent among both groups indicating similar shifts in perceiving how well, or not well, elementary teachers are capable of facilitating SSI instruction and identifying they should be trained. The qualitative data, however, did show differences in how the context may have an influence on PSTs' perceptions. For instance, when responding to the open-ended question associated with elementary teacher effectiveness, PST40 in the decontextualized course initially stated elementary teachers are not educated enough to facilitate SSI, while PST 9, from the contextualized content course, did initially believe elementary teachers were capable of facilitating SSI. After learning more about SSI in their courses, both PSTs confirmed what Kilinc et al. (2003) found and identified the need of formal training in SSI for elementary teachers. When considering the context of the two courses, and the potential reason for the initial differences in perceptions, we noted that the PST in the decontextualized course were simultaneously enrolled in a field experience where they collaborated closely with their mentor teacher and taught lessons once a week. The PSTs in the content course, however, were not enrolled in a field experience and had limited, if any, exposure to the process of planning and teaching science lessons. The PSTs in the decontextualized course engagement with practicing elementary teachers may have helped them become more informed about deficiencies in teachers' abilities, while the PSTs in the content course's limited experience working alongside an elementary teacher may have contributed to a lack of awareness regarding teachers' self-identified inadequate science content knowledge associated with common SSI that literature has previously identified as a concern (see Kilinc, Demiral, & Kartal, 2017).

Another important finding from the current study indicates that exposing PSTs to the SSI framework within the context of a science content course is associated with an increased confidence in engaging with science and answering SSI-related questions, which was not evidenced in the decontextualized course data. For example, PST 20 provided the following quote on her post-course questionnaire, "…it made me feel more comfortable talking about science terms," and PST 21 offered, "I have learned for the first time in this class about SSI and have seen a great difference in how I view science." This is supported in the quantitative data as well, which indicated stronger agreement when asked about teachers' ability to answer SSI-related questions. This increased confidence of interacting with science is significant because previous research has identified science content knowledge as a factor in PSTs' perceptions of SSI (Forbes & Davis, 2008; Özden, 2015). Congruent with these findings is the notion that some students from the contextualized course included the caveat that teachers need to be cognizant of the developmental levels of their students when implementing SSI. Contextualizing the SSI framework within a science content course may help teachers understand the complexity of the science underlying the SSI and design developmentally appropriate SSI experiences for their students. This caveat is significant because developmental appropriateness was a deliberate focus of the decontextualized course, but not of the contextualized course. The results indicate that for some students in the contextualized course recognizing developmental appropriateness of science content occurred organically.

When considered the results and how perceptions are formed, these findings lend support to the claim that introducing PSTs to SSI instruction changes their perception of SSI instruction in elementary school. The qualitative and quantitative data show that regardless of context, PSTs perceived SSI in a more favorable manner post-course. The assumption here is that the favorable perceptions of SSI becomes part of the PSTs prior experiences and leads to implementation of SSI in their future classroom.

Furthermore, having PSTs engage with SSI as students first and observing effective implementation strategies creates prior experiences that are then considered as the PSTs develop perceptions about science as a discipline along with SSI instruction. As PSTs contemplate pedagogical strategies, their prior experiences with effective SSI implementation as students coupled with a robust theoretical understanding promote positive perceptions about the implementation of SSI.

In light of the current findings, perhaps it would be beneficial to consider an approach to SSI exposing PSTs to the SSI approach in a manner similar to the approach suggested for nature of science (NOS). For example, Clough (2011) discusses the value and appropriateness of discussing the NOS within both the context of a science lesson and also as a decontextualized learning objective. Clough and Herman (2016) also found that implementing a mixture of decontextualized and contextualized NOS experiences over the course of a teacher preparation program had positive and long-lasting impacts on teachers two to five years after they completed their teacher preparation programs. Like NOS, the SSI framework is a complex and nuanced way of thinking about science. Exposing PSTs to SSI in various contexts and over a period of time could potentially yield similar positive results as Clough and Herman found working with early career teachers and NOS.

FUTURE RESEARCH DIRECTIONS

Future studies related to PSTs' perceptions of SSI and the context in which they were exposed should consider implementing more sophisticated quantitative analysis to examine the significance of the changes in students' responses. In the current study, the quantitative data was used a method to confirm findings in the qualitative responses. Additionally, larger samples would improve the data collected in future studies. Finally, it would be beneficial to conduct interviews with students to help clarify their qualitative responses. For example, some students in this study answered the questions based on classrooms they had observed in the past, while others answered based on their perceptions of themselves in their future classrooms. Clarifying these responses could have provided a clearer interpretation of students' perceptions.

There are also several questions for future research on the topic of PSTs perception of SSI. Future research should examine the associations between the contextualized SSI course and students' perceptions more closely. For example, one of our noted findings illustrates PSTs in the science content course expressed more confidence in their abilities to facilitate SSI, and leads to a follow up question of, why do PSTs who are initially exposed to SSI in content courses feel better about science content? Another future study our current findings lead us to focuses on PSTs' feelings of engagement with SSI and their perceptions of SSI in the elementary classroom. A follow-up study to understanding how elementary PSTs feel about their own abilities to facilitate SSI would be to provide them opportunities in real classrooms to enact SSI-based science lessons.

CONCLUSION

The quantitative data revealed students from both groups shared similar changes to perceptions related to students' engagement with SSI. First, both groups indicated SSI is was not too difficult for students. Second, both groups' perceptions changed to indicate that SSI instruction would increase scientific literacy. Finally, both groups perceived that SSI was not best suited for science instruction. Additionally, the data showed similar changes related to teachers and SSI instruction. Both groups agreed more strongly that elementary teachers need to be trained in SSI instruction, that elementary teachers were competent at integrating SSI, and that elementary teachers could answer SSI-related questions. In total, the results show similarities that both groups agreed with the idea that SSI is an appropriate for elementary school and that elementary teachers are capable of effectively implementing SSI.

The quantitative data revealed similar changes in perception for both groups in the following areas, (1) disagreement that SSI is difficult for elementary students, (2) disagreement that elementary teachers are not competent at integrating SSI, (3) agreement that elementary teachers can easily answer SSI-related questions, (4) agreement that elementary teachers should be trained in SSI instruction, (5) disagreement that SSI is more suitable for science instruction and (6) agreement that integrating SSI instruction will increase scientific literacy.

Two differences in perceptions emerged in the quantitative data. First, the change in mean score for decontextualized course indicated that they did not perceive elementary students as mature enough to engage in SSI instruction, while the change in the contextualized course's mean scores indicated they felt elementary students were mature enough. It should be noted that despite the shift in mean scores, the post-course score still indicated that decontextualized students perceived elementary students as mature enough to engage in SSI instruction. The second difference occurred relative to PSTs' perceptions regarding the inclusion of SSI instruction in elementary school. The change in mean score for decontextualized indicated more disagreement that SSI should be included in elementary school, while the mean scores for the contextualized course indicated more agreement. Again, the change in perceptions was minor and still indicate a perceived agreement that SSI should be included in the elementary school curriculum.

REFERENCES

Abell, S. K., Appleton, K., & Hanuscin, D. L. (2010). *Designing and Teaching the Elementary Science Methods Course*. Routledge. doi:10.4324/9780203859131

Anderson, D., & Zhang, Z. (2003). Teacher perceptions of field trip planning and implementation. *Visitor Studies Today*, *6*(3), 6–11.

Appleton, K., & Kindt, I. (1999). Why teach primary science? Influences on beginning teachers' practices. *International Journal of Science Education*, *21*(2), 155–168. doi:10.1080/095006999290769

Assael, H. (1995). *Consumer behavior & marketing action* (5th ed.). PWS-Kent Publishing Company.

Australian Curriculum, Assessment, and Reporting Authority. (2015). Retrieved on October 14, 2019, from https://australiancurriculum.edu.au/media/5558/science_-_sequence_of_content.pdf

Bandura, A. (1997). *Self-efficacy: The exercise of control*. W. H. Freeman.

Boone, H. N., & Boone, D. A. (2012). Analyzing Likert Data. *Journal of Extension*, *50*(2), 2–5.

Borgerding, L., & Dagistan, M. (2018). Preservice science teachers' concerns and approaches for teaching socioscientific and controversial issues. *Journal of Science Teacher Education*, *29*(4), 283–306. doi:10.1080/1046560X.2018.1440860

Bursal, M., & Paznokas, L. (2006). Mathematics anxiety and preservice elementary teachers' confidence to teach mathematics and science. *School Science and Mathematics*, *106*(4), 173–180. doi:10.1111/j.1949-8594.2006.tb18073.x

Choi, S., & Ramsey, J. (2009). Constructing elementary teachers' beliefs, attitudes, and practical knowledge through an inquiry-based elementary science course. *School Science and Mathematics*, *109*(6), 313–324. doi:10.1111/j.1949-8594.2009.tb18101.x

Clarson, D. L., & Dormody, T. J. (1994). Analyzing data measured by individual Likert-type items. *Journal of Agricultural Education*, *35*(4), 31–35.

Clough, M. P. (2011). Teaching and assessing the nature of science. *Science Teacher (Normal, Ill.)*, *78*(6), 56–60.

Corbin, J., & Strauss, A. (2008). *Basics of qualitative research: Techniques and procedures for developing grounded theory* (3rd ed.). Sage Publications. doi:10.4135/9781452230153

Denessen, E., Vos, N., Hasselman, F., & Monika Louws, M. (2015). The relationship between primary school teacher and student attitudes towards science and technology. *Education Research International, 2015*, 534690. Advance online publication. doi:10.1155/2015/534690

Espeja, A. G., & Largarón, D. C. (2015). Socio-scientific issues (SSI) in initial training of primary school teachers: Pre-service teachers' conceptualization of SSI and appreciation of the value of teaching SSI. *Procedia: Social and Behavioral Sciences, 196*, 80–88. Advance online publication. doi:10.1016/j.sbspro.2015.07.015

Evagorou, M., Guven, D., & Mugaloglu, E. (2014). Preparing elementary and secondary pre-service teachers for everyday science. *Science Education International, 25*(1), 68–78.

Forbes, C., & Davis, E. (2008). Exploring preservice elementary teachers' critique and adaptation of science curriculum materials in respect to socioscientific issues. *Science & Education, 17*(8-9), 829–854. doi:10.100711191-007-9080-z

Genel, A., & Topcu, M. S. (2016). Turkish preservice science teachers' socioscientific issues-based teaching practices in middle school science classrooms. *Research in Science & Technological Education, 34*(1), 105–123. doi:10.1080/02635143.201 5.1124847

Haney, J. J., Lumpe, A. T., Czerniak, C. M., & Egan, V. (2002). From beliefs to actions: The beliefs and actions of teachers implementing change. *Journal of Science Teacher Education, 13*(3), 171–187. doi:10.1023/A:1016565016116

Herman, B. C., & Clough, M. P. (2017). The role of history and nature of science in climate change teaching and learing. In D. Shepardson, R. Roychoudhury, & A. Hirsch (Eds.), *Teaching and learning about climate change: A framework of educators* (pp. 31–44). Routledge.

Herman, B. C., Sadler, T. D., Zeidler, D. L., & Newton, M. (2018). A socioscientific issues approach to environmental education. In G. Reis & J. Scott (Eds.), *International perspectives on the theory and practice of environmental education: A reader* (pp. 145–162). Springer. doi:10.1007/978-3-319-67732-3_11

Herman, B. C., Zeidler, D. L., & Newton, M. (2018). Students' emotive reasoning though place-based environmental socioscientific issues. *Research in Science Education*. https://doi- doi:10.100711165-018-9764-1

Kilinc, A., Demiral, U., & Kartal, T. (2017). Resistance to dialogic discourse in SSI teaching: The effects of an argumentation-based workshop, teaching practicum, and induction on a preservice science teacher. *Journal of Research in Science Teaching*, *54*(6), 764–789. doi:10.1002/tea.21385

Kılınç, A., Kartal, T., Eroğlu, B., Demiral, Ü., Afacan, Ö., Polat, D., Guler, M. P., & Görgülü, Ö. (2013). Preservice science teachers' efficacy regarding a socioscientific issue: A belief system approach. *Research in Science Education*, *43*(6), 2455–2475. doi:10.100711165-013-9368-8

Kılınç, A., Kartal, T., Eroğlu, B., Demiral, Ü., Afacan, Ö., Polat, D., Guler, M. P., & Görgülü, Ö. (2014). Preservice science teachers' efficacy regarding a socioscientific issue: A belief system approach. *Research in Science Education*, *43*(6), 2455–2475. doi:10.100711165-013-9368-8

Lederman, N. G. (2003). Introduction. In D. Zeidler (Ed.), *The role of Moral Reasoning on Socioscientific Issues and Discourse in Science* (pp. 1–4). Kluwer Academic Publishers. doi:10.1007/1-4020-4996-X_1

Lee, H., Chang, H., Choi, K., Kim, S., & Zeidler, D. L. (2012). Developing character and values for global citizens: Analysis of pre-service science teachers' moral reasoning on socioscientific issues. *International Journal of Science Education*, *34*(6), 925–953. doi:10.1080/09500693.2011.625505

Lindsay, P., & Norman, D. A. (1977). *Human information processing: An introduction to psychology*. Harcourt Brace Jovanovich, Inc.

Marshall, J. C., Horton, R. M., Igo, B. L., & Switzer, D. M. (2009). K-12 science and mathematics teachers' beliefs about and use of inquiry in the classroom. *International Journal of Science and Mathematics Education*, *7*(3), 575–596. doi:10.100710763-007-9122-7

McDonald, C. V. (2014). Preservice primary teachers' written arguments in a socioscientific argumentation task. *The Electronic Journal of Science Education*, *18*(7), 1–20.

Montes, L. D., & Rockley, M. G. (2002). Teacher perceptions in the selection of experiments. *Journal of Chemical Education*, *79*(2), 244–247. doi:10.1021/ed079p244

National Research Council. (2012). *A Framework for K-12 Science Education: Practices, Crosscutting Concepts, and Core Ideas*. The National Academies Press., doi:10.17226/13165

Newton, M. H. (2016). *A longitudinal examination of a SSI-embedded experiential environmental education course and environmental behaviors* (Order No. 10244480). Available from Agricultural & Environmental Science Collection.

NGSS Lead States. (2013). *Next Generation Science Standards: For States, By States*. The National Academies Press.

Olson, J., Tippett, C., Milford, T., Ohana, C., & Clough, M. (2015). Science teacher preparation in a north American context. *Journal of Science Teacher Education, 26*(1), 7–28. doi:10.100710972-014-9417-9

Özden, M. (2015). Prospective elementary teachers' views about socioscientific issues: A concurrent parallel design study. *International Journal of Elementary Education, 7*(3), 333–354. Retrieved July 28, 2019, from

Pickens, J. (2005). Attitudes and Perceptions. In N. Borkowski (Ed.), *Organizational Behavior and Health Care* (pp. 43–76). Jones & Barlett Publishing.

Pitiporntapin, S., Yutakom, N., & Sadler, T. (2016). Thai pre-service science teachers' struggles in using socio-scientific issues (SSIs) during practicum. *Asia-Pacific Forum on Science Learning and Teaching, 17*(2), 1–21.

Remillard, J. T., & Bryans, M. B. (2004). Teachers' orientations toward mathematics curriculum materials: Implications for teacher learning. *Journal for Research in Mathematics Education, 35*(5), 352–358. doi:10.2307/30034820

Roth, K. J. (2014). Elementary science teaching. In N. G. Lederman & S. K. Abell (Eds.), *Handbook of research on science education* (Vol. 2, pp. 361–394). Routledge.

Sadler, T. D. (2009). Situated learning in science education: Socioscientific issues as contexts for practice. *Studies in Science Education, 45*(1), 1–42. doi:10.1080/03057260802681839

Sadler, T. D. (2011). Situating socioscientific issues in classrooms as a means of achieving goals of science education. In T. Sadler (Ed.), *Socio-scientific issues in the classroom: Teaching, learning and research* (pp. 1–9). Springer. doi:10.1007/978-94-007-1159-4_1

Sadler, T. D., Barab, S. A., & Scott, B. (2007). What do students gain by engaging in socioscientific inquiry? *Research in Science Education, 37*(4), 371–391. doi:10.100711165-006-9030-9

Saldaña, J. (2009). *The coding manual for qualitative researchers*. Sage Publications Ltd.

Sherif, M., & Cantril, H. (1945). The psychology of attitudes: I. *Psychological Review*, *52*(6), 295–319. doi:10.1037/h0062252 PMID:21012631

Simmons, M. L., & Zeidler, D. L. (2003). Beliefs in the nature of science and responses to socioscientific issues. In D. L. Zeidler (Ed.), *The Role of Moral Reasoning on Socioscientific Issues and Discourse in Science Education* (pp. 81–94). Kluwer Academic Publishers. doi:10.1007/1-4020-4996-X_5

Topçu, M. S., Sadler, T. D., & Yilmaz-Tuzun, O. (2010). Preservice science teachers' informal reasoning about socioscientific issues: The influence of issue of context. *International Journal of Science Education*, *32*(18), 2475–2495. doi:10.1080/09500690903524779

Tosun, T. (2000). The beliefs of preservice elementary teachers toward science and science teaching. *School Science and Mathematics*, *100*(7), 374–379. doi:10.1111/j.1949-8594.2000.tb18179.x

UNESCO. (2019). *Science for a sustainable future*. Retrieved from https://en.unesco.org/themes/science-sustainable-future

Yapıcıoğlu, A., & Kaptan, F. (2017). A mixed method research study on the effectiveness of socioscientific issue based instruction. *Education in Science*, *42*(192), 113–137.

Yerdelen, S., Cansiz, M., Cansiz, N., & Akcay, H. (2018). Promoting preservice teachers' attitudes toward socioscientific issues. *Journal of Education in Science Environment and Health*, *4*(1), 1–11.

Zeidler, D. L. (2014). Socioscientific issues as a curriculum emphasis: Theory, research and practice. In N. G. Lederman & S. K. Abell (Eds.), *Handbook of research in science education* (Vol. 2, pp. 697–726). Routledge.

Zeidler, D. L., Herman, B. C., & Sadler, T. D. (2019). New directions in socioscientific issues research. *Disciplinary and Interdisciplinary Science Education Research*, *1*(11), 1–9. doi:10.118643031-019-0008-7

Zeidler, D. L., & Keefer, M. (2003). The role of moral reasoning and the status of socio-scientific issues in science education. In D. L. Zeidler (Ed.), *The role of moral reasoning on socio- scientific issues and discourse in science education* (pp. 7–38). Kluwer Academic Publishers.

Zeidler, D. L., & Newton, M. H. (2017). Using a socioscientific issues framework for climate change education: An ecojustice approach. In D. Shepardson, A. Roychoudhury, & A. Hirsch (Eds.), *Teaching and learning about climate change: A framework for educators* (pp. 56–65). Routledge. doi:10.4324/9781315629841-5

ADDITIONAL READING

Hancock, T. S., Friedrichsen, P. J., Kinslow, A. T., & Sadler, T. D. (2019). Selecting socio- scientific issues for teaching. *Science & Education, 28*(6-7), 639–667. doi:10.100711191-019-00065-x

Herman, B. C., Sadler, T. D., Zeidler, D. L., & Newton, M. H. (2018). A Socioscientific Issues Approach to Environmental Education. In G. Reis & J. Scott (Eds.), *International Perspectives on the Theory and Practice of Environmental Education: A Reader. Environmental Discourses in Science Education* (Vol. 3). Springer. doi:10.1007/978-3-319-67732-3_11

Owens, D. C., Sadler, T. D., & Zeidler, D. L. (2017). Controversial issues in the science classroom. *Phi Delta Kappan, 99*(4), 45–49. doi:10.1177/0031721717745544

Sadler, T. D. (2009). Situated learning in science education: Socio-scientific issues as contexts for practice. *Studies in Science Education, 45*(1), 1–42. doi:10.1080/03057260802681839

Sadler, T. D., Barab, S. A., & Scott, B. (2007). What do students gain by engaging in socioscientific inquiry? *Research in Science Education, 37*(4), 371–391. doi:10.100711165-006-9030-9

Zeidler, D. L., Applebaum, S. M., & Sadler, T. D. (2011). Enacting a socio-scientific issues classroom: transformative transformation. In T. D. Sadler (Ed.), *Socio- scientific issues in the classroom*. Springer. doi:10.1007/978-94-007-1159-4_16

Zeidler, D.L., Herman, B.C. & Sadler, T.D. (2019). New directions in socioscientific issues research. *Disciplinary and Interdisciplinary Science Education Research,* (1), 11. doi:10.118643031-019-0008-7

Zeidler, D. L., Sadler, T. D., Simmons, M. L., & Howes, E. V. (2005). Beyond STS: A research-based framework for socioscientific issues education. *Science Education, 89*(3), 357–377. doi:10.1002ce.20048

KEY TERMS AND DEFINITIONS

Contextualized Science Content Course: A course focused on developing the science content knowledge by providing real-world context to students from various majors.

Decontextualized Science Methods Course: A pedagogically focused course commonly taken by college students pursuing a degree in education.

Elementary Curriculum: The standards, objectives, and resources kindergarten through grade five teachers are expected to implement in the classroom.

Elementary Science Teaching: The process of facilitating science content to children between the ages of five and ten.

Elementary Teacher Effectiveness: A kindergarten through grade five teacher's ability to accurately implement specific teaching strategies and transfer subject matter knowledge to students.

Perceptions: One's beliefs or attitudes toward a specific topic which are influenced by the individual's prior experiences.

Preservice Teacher: A college student pursuing a career as a teacher.

Student Engagement: Students' active participation in a lesson; explicitly expressing interest in the subject matter.

Chapter 8
Use of the SEE-SEP Model in Preservice Science Teacher Education:
The Case of Genetics Dilemmas

Dilek Karisan
https://orcid.org/0000-0002-1791-9633
Adnan Menderes University, Turkey

Umran Betul Cebesoy
https://orcid.org/0000-0001-7753-1203
Usak University, Turkey

ABSTRACT

This study aims to assess how preservice teachers' supporting reasons vary in relation to six subject areas, namely, sociology/culture (S), environment (E), economy (E), science (S), ethics/morality (E), and policy (P), in different socioscientific issues (SSI) and how preservice teachers' decisions regarding these issues interact with the aspects of value, knowledge, and personal experience. Exploratory sequential design was utilized in this study and 47 third-grade preservice science teachers. Data were collected via written reports in two different SSI scenarios, namely, gene therapy and preimplantation genetic diagnosis. Preservice teachers' written reports were analyzed using the SEE-SEP model. The results showed that preservice teachers' supporting reasons mainly stemmed from the subject areas of ethics/morality (41%) and science (32%). The results also revealed little use of justifications from the subject areas of policy and sociology/culture (11% and 10%, respectively), and subject area of economy (6%) was the least referred.

DOI: 10.4018/978-1-7998-4558-4.ch008

INTRODUCTION

To date, the science education committee has affirmed numerous "*Science Education Standards*" aimed at describing essential components of science education (Bybee, 2014). The ultimate goals of these standards are to enhance students' knowledge and understanding of scientific concepts, engage them in decision-making activities, and let them participate in cultural events. These ultimate goals are referred to as required skills for a scientifically literate person. Due to its complex nature, there is no clear-cut formula to raise scientifically literate individuals; however, educators have explicitly integrated relevant subjects (i.e. technology, society, environment, ethics) to enhance scientific literacy and tried to describe the characteristics of a scientifically literate person (Bybee, 1995). A scientifically literate person is someone who would have an understanding of Nature of Science (Halbrook & Rannikmae, 2007), Scientific Inquiry (Lederman, Antink & Bartos, 2014), who would be able to evaluate the quality of scientific information, engage in moral-ethical issues, and understand the complexity of socioscientific issues (hereafter SSI) (Zeidler, 2014).

One of the alternative ways used to enhance scientific literacy is known as the Science, Technology, and Society (STS) movement (Solomon & Aikenhead, 1994) which has been enhanced by the inclusion of the Environment component and converted to a more extensive term known as the Science, Technology, Society, and Environment (STSE) movement (Pedretti, 2003). Although the STSE movement addressed the connections and interdependency of the four components and supported achieving scientific literacy, it still has missing parts to be defined as it is a well-articulated approach to science education (Pedretti, 2003). Researchers have argued that STSE has such limitations as not including ethical issues nor does it aim to enhance moral or character development (Zeidler, Sadler, Simons & Howes, 2005) and that it lacks a theoretical framework (Sadler, 2004b). Thus, it is suggested that STSE be remodeled and improved by focusing on one important component -- each student's moral and ethical development -- to increase students' ethical, individual, and social responsibility (Zeidler et al., 2005). SSI covers the issues all that STSE includes as well as its missing parts such as considering the ethical dimensions of science and the moral reasoning of the student (Zeidler Walker, Ackett, & Simmons, 2002, p. 344).

The integration of SSI in formal or informal learning environments is emphasized as an important concept in numerous studies across the world (Chang, Wu, & Hsu, 2013; Eggert & Bögeholz, 2010; Sadler, 2004a; Powell, 2014). The role of SSI in the science classroom is highlighted as:

[Socioscientific issues] are usually controversial in nature but have the added element of requiring a degree of moral reasoning or the evaluation of ethical concerns in

the process of arriving at decisions regarding possible resolution of those issues. The intent is that such issues are personally meaningful and engaging to students, require the use of evidence-based reasoning, and provide a context for understanding scientific information. (Zeidler & Nichols, 2009, p.49).

Students are actively engaged in SSI-based activities because of the complex nature of these issues such as having real-life content and being ill-structured (Sadler, 2011). Exploration of SSI is widely recommended to educators as these issues are suitable not only for gaining the objectives of STSE (Ramsey, 1993), but also promoting scientific literacy (Abd-El Khalick, 2003; Zeidler, Herman, Sadler, 2019). Furthermore, SSI positively affects students' motivation toward science lessons as well as their career choices (Sadler, 2009). It also provides an opportunity for educators to implement character education in their classrooms and to practice education for citizenship. In those classrooms, students are asked to engage in democratic decision-making, fostered to take into consideration human values and nurtured emotional intelligence, recognized as the building blocks of character (Zeidler & Nichols, 2009).

The Purpose and the Research Questions

Although SSI helps researchers to examine the issues by considering multiple viewpoints (Sadler, Barab, & Scott, 2007), there is still limited study in SSI research that addresses its multidimensional structure (Es, & Ozturk, 2019). SSI, by its very nature, has contradictions and problematic issues that need to be solved. It is worth investigating how students deal with SSI, exploring their arguments and analyzing their reflection. There are many alternative analysis methods dealing with the effects of SSI-integrated research on students' epistemological thinking such as argumentation, reflective judgment, or informal reasoning. The SEE-SEP model is one of these alternative ways to explore students' argumentation and decision-making skills in numerous studies (i.e. Chang Rundgren, 2011; Christenson, Chang Rundgren & Rundgren, 2012; Christenson, Chang Rundgren & Zeidler, 2014; Es & Ozturk, 2019). These studies mainly explored upper secondary students' decisions (Christenson et al. 2012; Christenson et al. 2014) or elementary students' decisions in various SSI topics including genetically modified organisms and fishing bans (Es & Ozturk, 2019). However, little is known as to whether future teachers themselves can adopt these kinds of multiple perspectives in their classes. As a result of this gap, this study explored preservice science teachers' use of multiple perspectives in their decisions.

The SEE-SEP model in the present study illustrates how preservice teachers problematize the issues regarding gene therapy and the issue of using preimplantation genetic diagnosis. A substantial amount of research suggests the significance of embedding moral and ethical perspectives into SSI-based instruction (Sadler, 2004b; Zeidler et al., 2009). Genetic engineering issues were selected for this study because of their appropriateness to discover students' ethical and moral concerns (Sadler &Zeidler, 2004), and these issues are suitable as cross-disciplinary subjects (Christenson, 2015). Such issues help the learner to consider multiple perspectives; they contribute to higher-order thinking skills and enhance the quality of learners' decision-making on SSI (Wu and Tsai 2007). The SEE-SEP model, by its very nature, illustrates the multiple perspectives people have regarding SSI and make people aware of the complex features of SSI. Furthermore, people could develop a more holistic view of SSI based on the SEE-SEP model while arguing about SSI in daily life (Rundgren and Rundgren, 2010). The SEE-SEP model is thought to be useful for exploring preservice teachers' reasoning and for visualizing their mental abstractions on such issues. Being the science teachers of the future, it is beneficial for the participants of the present study to think about the irreversible consequences of the interactions between science, technology, society, and the environment, and to explore their reasoning, personal values, and beliefs about controversial issues. The SEE-SEP model is applied as an analytical framework in this study to analyze preservice teachers' supporting reasons in relation to six subject areas -- Science, Environment, Ethics/morality, Sociology, Economy, and Policy -- in the issues of gene therapy and preimplantation genetic diagnosis, and to explore how preservice teachers' decisions regarding these issues interact with the value, knowledge, and personal experience aspects.

The research questions guiding this study are:

1. How do preservice teachers' decisions interact with the value, knowledge, and personal experience aspects?
2. What are the different perspectives adopted by preservice science teachers in dealing with different SSI topics?

BACKGROUND

The Fruitfulness of Implementing SSI in Science Classrooms

Previous research stressed that incorporating SSI into the science classroom allows science educators to engage their students in real-life problems. Zeidler (2014) argued that SSI education ideally should include ill-structured problems that require individuals to engage in scientific argumentation and make evidence-based reasoning. In SSI-based classrooms, students are expected to express their understandings and make decisions in real-world situations that enhance their ability to make complex reasoning. Dealing with real-life problems in science classrooms and embedding SSI lessons are vitally important since science interacts with many other domains including Sociology, Environment, Economy, Ethics, and Policy; and all these domains have multifaceted reciprocity. Taking science into the center and enriching it with all those related concepts is a subject that has been dealt with in SSI research for decades (Sadler, 2004a, Zeidler et al., 2009). Engaging SSI not only enhances students' science learning, it also makes a significant contribution to their ethical sensitivity (Sadler et al., 2007). For example, learning about nuclear power in an SSI context enhances students' understanding of alternative energy in a scientific context and also helps them adopt a position in that context. Additionally, SSI can also give teachers the chance to bring global issues into the science classroom and enhance students' characters and values as global citizens (Lindahl, Folkesson, & Zeidler, 2019). In SSI-based learning environments, moral, social, and political issues shape individuals' decisions; they have to synthesize science concepts with its social practices (Sadler et al. 2007), and this requires an examination of the issue from multiple perspectives. By taking into consideration the fruitful nature of SSI context, the present study used gene therapy and preimplantation genetic diagnosis as the contexts to engage the preservice teachers' in SSI-based learning environments. The nature of these two issues allows students to evaluate the problem by considering multiple perspectives and connect them with out-of-school life experiences, and also prepares students for future citizenship (Aikenhead, 2006) to empower them for future making complicated decisions with moral implications. Morris (2014) also highlighted the importance of the genetic issue by analyzing science textbooks, focusing in detail on two SSI. She highlighted the popularity of genetic issues in science textbooks and its appropriateness for gaining interdisciplinary perspectives. Morris (2014) suggested that issues like genetic technology should be included in the curriculum to enhance SSI-based curricula.

The Role of Teacher Training

As well as the importance of including SSI into the curriculum, teachers, who are the implementer of the curriculum, are also crucial in SSI education. Preservice teacher education programs are vitally important to form the link between SSI and the aims of science teaching (Sadler, Amirshokoohi, Kazempour, & Allspaw, 2006). Teacher training is seen as one key factor in implementing SSI successfully in science classrooms (Christenson, 2015). Engaging students with SSI enhances their cognitive skills by allowing them to have multiple perspectives on ill-structured issues; it fosters critical thinking in them and engages them in problem-solving activities. As well as cognitive skills, preservice teachers' affective skills such as empathy, ethical, and moral characteristics are also developed at the end of SSI-based instruction. However, it is difficult to integrate SSI in the teaching process as it requires higher-order intellectual skills from teachers. As well as teachers' skills, various factors influence the teaching practices of teachers such as the school's physical structure, cultural values, epistemological orientations, and beliefs (Hancock & Gallard, 2004). It has been revealed that personal values or beliefs are also very often involved in an individual's reasoning about SSI (Albe, 2008; Lee, 2007). Previous research highlights that teachers' attitudes towards SSI, beliefs, personal experiences, etc., affect their instructions (Khishfe, 2014; Walker and Zeidler, 2007). Teachers who play a vital role in deciding how to teach SSI in the science classroom are the key element in SSI teaching. Thus, teacher education programs have crucial importance in the development of teachers' beliefs about teaching and learning (Forbes & Davis, 2008).

SSI-focused interventions are one of the alternatives to support student-centered approaches that enhance students' reasoning skills (Albe, 2008). Therefore, teaching models for SSI integration in teacher education programs are valuable (Sadler, 2011), and teaching how to implement SSI in science classrooms to preservice teachers is crucial. Preservice teachers should be given a chance to engage in SSI-based scenarios during their undergraduate education as they are learners of science now, but will be teachers of science for the next generation. To this end, in this research, preservice teachers, being future teachers, were engaged in two ill-structured genetic scenarios, namely, gene therapy and preimplantation genetic diagnosis. The investigators choose these particular scenarios for two major reasons. First, the appropriateness of the issue for selected participants (i.e. preservice teachers), and second, the popularity of the issues on social media as well as being morally divisive issues.

Exploring preservice teachers' views on these issues is of the utmost importance in several ways: first, these issues have a multidisciplinary nature that requires students to think and consider multiple perspectives; second, as well as students, teachers are also engaged in multifunctional tasks such as moderating socioscientific discussions, evaluating students' arguments, and implementing this kind of demanding activity in science classrooms (Christenson, 2015). Previous research has consensus on the difficulty of bringing SSI into science classrooms and managing SSI-based discussions for science teachers. Thus, it is important to engage in these discussions and gain some SSI-based discussion experience. Second, fostering SSI enhances students' character and values as global citizens (Lee et al, 2013). If an educator wants to have global citizens, s/he should know how to engage her/his students in SSI-based discussions. This is another reason why is it crucial to have teachers experienced in these kinds of discussions. Third, teachers' attitudes towards SSI affect how they teach. If the teacher does not feel comfortable to teach these issues, s/he might avoid bringing them into the classroom. Thus, the sooner the preservice teachers engage in these issues the better it is for them to gain experience and to realize how they perceive, negotiate, and resolve these issues.

SSI can be used in science classrooms in two ways; either as an instructional tool to improve some student skills (i.e. argumentation skills, critical thinking skills, reflective judgment skills, etc.) or it is put into the center to explore students' characters, values, and epistemological beliefs as well as how they make decisions. For example, it has been used as an instructional tool to explore students' reflective judgment skills (Zeidler, Sadler, Applebaum, Callahan, 2009; Karisan, Yilmaz-Tuzun, Zeidler, 2017), argumentation skills (Jimenez-Aleixandre et al., 2000; Karisan et al., 2017; Sadler and Donnelly, 2006), and character and values (Kim et al., 2020; Lee et. al., 2013). The common conclusion of the previous research is that SSI should include complex dilemmas that require students to consider multiple perspectives by using skepticism, thus enhancing learning and promoting scientific literacy, which are longstanding goals of science education. In addition to the aforementioned points of view regarding SSI, Chang Rundgren and Rundgren (2010) emphasized the multidisciplinary nature of SSI and presented a model entitled the "SEE-SEP," which allows the researchers to analyze SSI-based reasoning by its multidisciplinary nature. The model and its use in SSI research are explained below.

The SEE-SEP Model and Its Use

Previous research has concluded undeniably that SSI is multidisciplinary in nature. This multidisciplinary nature of SSI allows researchers to connect complex science-related issues to society. Chang, Rundgren, and Rundgren (2010) suggested an analysis framework for SSI that highlights its cross-disciplinary character by addressing the relationship between science and society. The model, namely, SEE-SEP, has integrated six divergent subject areas of SSI -- Sociology/Culture (S), Environment (E), Economy (E), Science (S), Ethics/ morality (E), and Policy (P) -- with three aspects, namely, value, personal experience, and knowledge. Students' reasoning in SSI when using these six subject areas is usually rooted in the value, knowledge, and personal experience aspects. The model allows the researchers to explore students' holistic views derived from divergent areas. The six components of the model are commonly referred to as the factors that influence students' reasoning in SSI (i.e. Bell & Lederman, 2003; Chang & Chiu, 2008; Halverson, Siegel & Freyermuth, 2009; Grace & Ratcliffe, 2002). The model, illustrated by Chang Rundgren and Rundgren (2010) can be seen in Figure 1.

Figure 1. The SEE-SEP Model of SSI (Chang, Rundgren, & Rundgren, 2010 p.11)

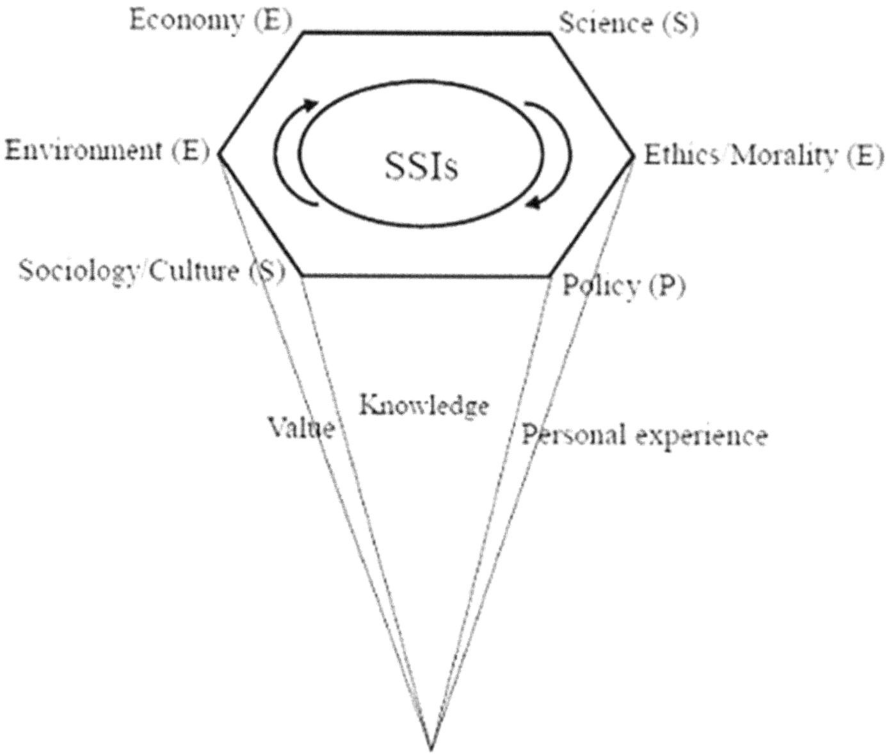

Chang Rundgren and Rundgren (2010) refer to people's affective domain by the term of value. They explained that each student makes decisions in SSI based on their own values. The aspect of personal experience was explained as students tending to use their own experiences while making decisions in real-life SSI. Moreover, the aspect of knowledge refers to students' explanations based on scientific knowledge. The subject area of sociology/culture refers to the effect of culture and societies on students' decisions. By the subject area of environment, Chang Rundgren and Rundgren (2010) expressed the importance of the explicit integration of the environment in students' decisions. Sometimes, students' decisions can be affected by the subject area of economy. Thus, using the subject area of economy, the researchers aimed to explain the effects of economic conditions on students' SSI-based decision-making (E.g. using DDT is allowed in poor countries whereas it is forbidden in rich countries as peoples' priorities are affected by economic wealth). One of the most important aspects, the science aspect, is described as follows: the scientific knowledge in different subjects including chemistry, biology, medicine, or technology affects the thinking process and so it also affects SSI-based decisions. The ethics/morality aspect is another subject area included in the SEE-SEP model. Religious views or personal ethics and moral viewpoints can emerge in students' decisions. The final subject area is policy, which refers to the trust of people who rely on authority (i.e. governments and laws). Some SSI-based issues such as constructing nuclear power plants or abortion are determined by law or by governments, and some students might think that governments have the right to make laws to constrain some regulations. Students' decisions can be based on one or more subject areas at the same time while interacting with the value, knowledge, and personal experience aspects. Furthermore, the subject areas and aspects referred to in students' decisions can change based on the SSI topic (Chang Rundgren and Rundgren, 2010).

Figure 2. The exploratory sequential design

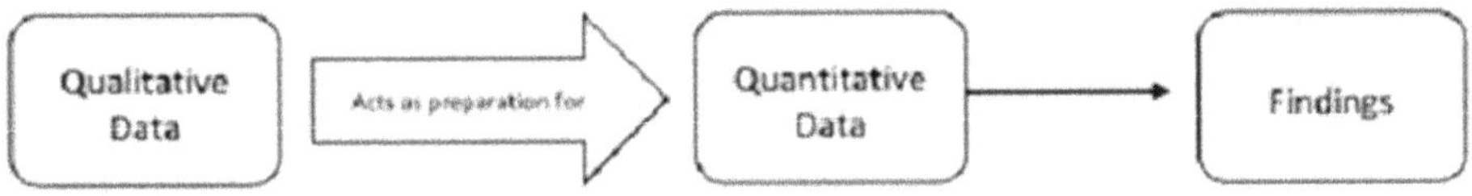

METHOD

Exploratory sequential mixed-method design, was utilized in this study (Figure 2). This design is characterized by an initial phase of qualitative data collection followed by quantitative data analysis (Creswell & Plano-Clark, 2011). In this study, researchers adopted the quantizing approach, which involves transforming qualitative data into numerical codes (Miles & Huberman, 1994; Teddlie &Tashakkori, 2006; Tashakkori & Teddlie, 1998). A qualitative approach was used to code Preservice Science Teachers' (PSTs) written reports in two scenarios based on the SEE-SEP model which served as an analytical framework for qualitative analysis (Chang Rundgren & Rundgren, 2010). Then, the qualitative data transformed into numerical codes using the quantizing approach (Miles & Huberman, 1994). This allowed the researchers to quantify the patterns in PSTs' supporting reasons and make comparisons between different scenarios. Then, to further elaborate and exemplify the patterns in PSTs' supporting reasons qualitative methods were applied.

Participants

Forty-seven third-grade PSTs from a four-year science teacher education degree program participated in the study. All PTs were enrolled in a small-sized public university located in the western part of Turkey. All PTs enrolled in Genetics and Biotechnology, a compulsory course offered in the sixth semester (third year of science teacher education program). Teacher education programs in Turkey are centralized and regulated by the Higher Education Council of Turkey (HEC). The Council determines the length of the programs and all the courses offered in all teacher education programs throughout Turkey. The science teacher education program consists of eight-semester-long instruction (two semesters a year). The graduates are entitled as science teachers and it is a necessary criterion for teaching science in primary school from the fifth to the eighth grade. In the science teacher education program, PTs complete a series of compulsory courses including Physics, Chemistry, Biology, and Mathematics, in addition to pedagogical and pedagogical content courses (HEC, 2007).

The Design and Data Collection

This study was conducted during the genetics and biotechnology course over the 2016-2017 spring semester. While the first author was the researcher in the study, the second author acted as the lecturer and researcher at the same time. The course consisted of two parts, namely, introductory genetics and biotechnology (1st part) and the discussion of scenarios (Table 1). Each week included a two-hour class.

Table 1. The design of the course

	The course content	The topics covered
Genetics and biotechnology course	1st part - introductory Genetics and Biotechnology (8 weeks)	An overview of modern genetics Classic (Mendelian Genetics) Human genetics Molecular genetics Biotechnology
	2nd part - Discussion of Scenarios (5 weeks)	Introduction to scenarios S-1: Gene Therapy S-2: Preimplantation Genetic Diagnosis

The first part lasted for seven weeks and PTs were taught the basic principles of classic and modern genetics followed by the concept of biotechnology. When the lecturer first introduced how genetics evolved as a scientific discipline, she presented classic genetics (Mendelian) as the precursor of modern genetics. In the following weeks, inheritance, Mendel's laws of inheritance, human inheritance, and pedigree diagrams were introduced. The lecturer also explained how molecular genetics was developed to uncover the unclear points in classic genetics with advances in technology. The discovery of DNA and its mechanism, the genetic code, its replication, and the central dogma process were included in this part. Lastly, the lecturer presented the concept of biotechnology and the new gene technologies including genetic engineering issues.

The second part of the course included an introduction to the dilemmas created by these new genetic technologies. Then, the lecturer explained the aim of the scenarios and the scenarios were sent to PTs via e-mail before the discussion of each scenario in the class. Each scenario was discussed over two weeks (100 minutes each week). PTs were required to write individual reports about what they would decide in each scenario and how before the discussion and to submit their reports individually. These written reports constituted the qualitative data for this study. PTs submitted 94 reports in total.

Both scenarios aimed to develop PTs' decision-making skills in SSI. Following recent revisions in 2013 and 2018 (Ministry of National Education [MoNE], 2013; 2018), the Turkish primary science curriculum emphasizes the importance of developing students' decision-making and informal reasoning skills in SSI. In line with this emphasis, some objectives in the eighth-grade curriculum aim for students to discuss dilemmas arising from genetics and the application of biotechnology (MoNE, 2018, p. 49). As a result, the teachers' role in handling discussions about dilemmas related to genetics and biotechnology becomes more critical. However, there is no direct course that aims to develop preservice teachers' decision-making skills in genetics-related dilemmas before graduation. Therefore, this study aims to close this gap by providing PTs with the means to join discussions in genetics-related SSI. The scenarios used in this study provide an appropriate context to develop PTs' decision-making skills by considering multiple perspectives. The scenarios and the questions used in the scenarios were prepared by reviewing previous research (Bell & Lederman, 2003; Sadler & Zeidler, 2004). In addition, the sources provided in the Personal Genetics Education Project (http://pged.org/lesson-plans/) were reviewed and scenarios were formulated. These scenarios were pilot tested with 42 PTs in the previous semester (2016 spring semester). As the researchers aimed to explore whether PTs could consider multiple perspectives in their decisions, a context was provided which requires PTs to make decisions and provide justifications for their decisions based on the scenarios. The scenarios and their content are presented below:

SSI Scenarios and Prompts

Scenario 1: Gene therapy

In this scenario, PTs were given some background information about gene therapy and the possible uses of gene therapy in some cases including Huntington Disease (which is a fatal genetic disease, symptoms beginning in the 30s) and increasing human intelligence. PTs were asked to make decisions about the following: In which cases should gene therapy be used? Treatment of Huntington Disease genes or increasing the intelligence of an unborn baby?

Participants were asked to consider possible stakeholders (family members, family and child rights, ethical, legal, social, and technological concerns).

Scenario 2: Preimplantation Genetic Diagnosis

In this scenario, some background information about preimplantation genetic diagnosis (i.e. how it is done and under what circumstances and possible uses of preimplantation genetic diagnosis in some cases) was introduced to PTs. Then, a hypothetical case was presented regarding a family having a sick child who desperately needs a bone marrow transplant. After the hypothetical case presentation, PTs were asked to discuss the use of preimplantation genetic diagnosis to create a bone marrow donor for a sick child. Participants were asked to consider their possible stakeholders (family members, family and child rights, ethical, legal, social, and technological concerns) and explain their views on having a second child who will be a perfect match for a bone marrow transplant for their sick child by using preimplantation genetic diagnosis.

Table 2. The definitions of codes and examples from PTs' written reports

Codes	Definition	Examples
Sociology/culture (So)	PTs provided supporting reasons in relation to the subject area of sociology/culture.	'For our society, family means rallying together for each family member to overcome difficulties.' (PT-40, S2)
Economy (Ec)	PTs provided supporting reasons in relation to the subject area of economy.	'With this kind of application [gene therapy], everybody can have smart children if they have enough money.' (PT-1, S-1)
Science (Sc)	PTs provided supporting reasons in relation to the subject area of science (i.e. biology, chemistry, technology, or medicine).	'In gene therapy, one gene in the gametes of parents is changed meaning only sick genes are replaced.' (PT-40, S1) 'The unborn baby might be sick but the probability of this is 50%.' (PT-31, S-2)
Ethics/morality (Et)	PTs provided supporting reasons in relation to the subject area of ethics/morality.	'Doing gene therapy is a matter of ethical codes. For instance, when gene therapy is done to a child, other people might find it unpleasant.' (PT-18, S-1) 'It is ethical to do this [preimplantation genetic diagnosis] for health reasons. It is not something that is arbitrarily done.' (PT-30, S-2)
Policy (Po)	PTs provided supporting reasons in relation to the subject area of policy and laws.	'It [gene therapy] could be used to treat sick people but there should be some kind of rules and legislation to regulate its application.' (PT-7, S-1) 'This method [preimplantation genetic diagnosis] should be used under some predetermined rules.' (PT-42, S2)

Table 3. The codes generated by the SEE-SEP model (adopted from Chang Rundgren and Rundgren, 2010)

	Aspects		
Subject areas	Knowledge (K)	Value (V)	Personal experience (P)
Sociology/culture (So)	SoK	SoV	SoP
Environment (En)	EnK	EnV	En
Economy (Ec)	EcK	EcV	EcP
Science (Sc)	ScK	ScV	ScP
Ethics/morality (Et)	EtK	EtV	EtP
Policy (Po)	PoK	PoV	PoP

Data Analyses

The SEE-SEP analytical framework, which allowed to explore PTs' use of multiple perspectives in their decisions, was used to code PTs' written documents. As PTs' decisions did not refer to any connection with the subject area of environment, only five subject areas (sociology/culture, economy, science, ethics/morality, and policy) were used in the analyses. While coding PTs' written justifications, unit of analysis was used to code PTs' written reports (Merriam, 1998; Trochim, 2020). The unit of analysis in a research represents the major entity that is being analyzed such as individuals, groups or artefacts including books or photos (Trochim, 2020). In this research, the coding units (unit of analysis) are based on PT's written responses. Each written sentence constituted a coding unit for the analyses. The frequency of appearance for each subject area and aspect was calculated. Here, 'PT' indicates the preservice science teacher while a number is also given to identify a specific PT (i.e. PT-3). Square brackets ([]) were used to complete the meaning. The different scenarios were coded as S-1 and S-2. Detailed descriptions of each subject area are presented in Table 2.

The six subject areas can be connected with one or more broad aspects of knowledge (K), value (V), and personal experience (P). As a result, the SEE-SEP model also contains the connection of these three aspects with the six subject areas generating 18 codes for analysis (see Table 3).

The PTs' written justifications may contain one or more subject areas and/or aspects, or the subject area and aspect interaction (i.e. EtK) at the same time. Thus, PTs' written justifications were coded using the 18 codes generated from the SEE-SEP model with respect to the aspects of knowledge, attitude, and personal experiences. An example of coding is shown below:

The family should not have a second child just for the reason of being a donor. It is not right. A family should have a second child only if they are economically capable of supporting a second child. (PT-2, S-2)

In this excerpt, PT-2 connected her supporting reason of economy with her beliefs. Thus, this excerpt was coded as EcV (economy connected to value).

If this method [gene therapy] became an easily applicable method, there could be wars between different countries. While doing good, bad things can also be done with this method. Thus, it is not ethical to use this method. (PT-5, S-1)

While PT-5 explains why it is not ethical to conduct gene therapy, she explains how good and bad things (showing the aspect of value) could happen as a result of gene therapy. Therefore, this excerpt was coded as EtV.

Trustworthiness of The Study

To ensure and validate the results of this study, several methods were adopted in the study. First, the researcher triangulation method was used to prevent misinterpretation of data. Both authors, having more than 10-year teaching experience in science education research, coded 10% of the PTs' written reports to ensure a consensus on the use of the SEE-SEP model during data analysis. After reaching a consensus on the use of the SEE-SEP model, all the written reports were coded by the second author. The first author coded half of the written reports, which were randomly selected. Then, inter-rater reliability showing the consistency of agreement among coders was calculated by dividing the total number of agreements and disagreements in the coded data by the number of agreements (Miles & Huberman, 1994). Inter-rater reliability was found to be 80.9% in the first round. Then, the authors met and discussed the codes used in coding the data. After the ambiguity in coding was resolved, both authors continued their coding. Inter-rater reliability was calculated as 90.4% showing that the data analysis between different coders was compatible and reliable.

To ensure the credibility and validity of the study, researchers also adopted prolonged involvement and rich description procedures in present study. The second author, who was both the lecturer and the researcher in the study, stayed in the research cited during the study to confirm the prolonged involvement of the researcher. (Creswell & Miller, 2000). In addition, two different scenarios were used to collect data that provided a rich description of the context (Maxwell, 2009).

FINDINGS

Preservice Science Teachers' Supporting Reasons in Relation to the SEE-SEP Model

First, it was analyzed how PTs' supporting reasons varied with respect to the six subject areas (sociology/culture, economy, environment, science, ethics/morality, and policy).

The number of supporting reasons representing each aspect (sociology/culture, economy, environment, science, ethics/morality, and policy) presented in this section was divided by the total number of PTs' supporting reasons. The frequency (%)

of each aspect was analyzed to show the distribution. The same method was used to analyze the distribution of the three aspects of knowledge, value, and personal experience.

The Subject Areas Expressed in Preservice Science Teachers' Supporting Reasons

Second, it was explored how PTs' supporting reasons varied in relation to the six subject areas in both scenarios (Figure 3). It is found that PTs mostly used ethics and morality (42%) in their decisions, followed by the subject area of science (32%). Sample excerpts referring to the subject areas of ethics/ morality and science are presented below:

I think it is ethical [referring to the use of preimplantation genetic diagnosis]. Saving a life is always ethical. You are bringing your child, who struggled with the disease from a very early age, back to life. (PT-5, S2, Et)

When necessary, people can donate their organs to someone they have never known before. There is no problem if this person [referring being donor for the brother] is his brother/sister. This affects treatment positively as siblings have high tissue compatibility. (PT-2, S2; Sc)

The excerpts above present two PTs' written justifications in which they approved the use of preimplantation genetic diagnosis. While the preservice science teacher in the first excerpt (PT-5) approves use of preimplantation genetic diagnosis for ethical reasons as she thinks it is for 'saving life,' the preservice teacher in the second excerpt (PT-2) approves the use of preimplantation genetic diagnosis as siblings have 'high tissue compatibility.'

While different PTs could have different reasons (as presented above) for supporting their decisions even though they made similar decisions on the use of preimplantation genetic diagnosis, one pre-service science teacher could use multiple reasons in the same scenario. For instance, in gene therapy scenario, PT-12 stated:

... If I had Huntington disease, there is a possibility to transmit the altered gene to my child. It is a 50% chance of getting sick. I would not want that. Thus, I want to try germ-line gene therapy. (ScK)

In a similar manner, I am positive about the use of gene therapy for increasing intelligence of human beings. If it was used and the child became a successful scientist who had creative ideas and inventions, for instance, then, it would be good for our country. As a result, there would become economic and social relief in our country. (So and Ec)

PT-12 explained why she supported the use of gene therapy in different cases provided in the scenario (Huntington disease and increasing human intelligence). Here, PT-12 used her knowledge about inheritance patterns with scientific concepts provided in the scenario (Huntington related gene (altered gene) in the germ cells will be inherited to the next generations and the chance of getting a sick child is 50% as Huntington disease has autosomal dominant pattern). Thus, these statements were coded as ScK. Moreover, the same pre-service science teacher made similar decision about the use of gene therapy by providing other concepts from sociology (being good for our country and our society) and economy (economic relief).

Figure 3. Average distribution of PTs' supporting reasons with respect to the six subject areas

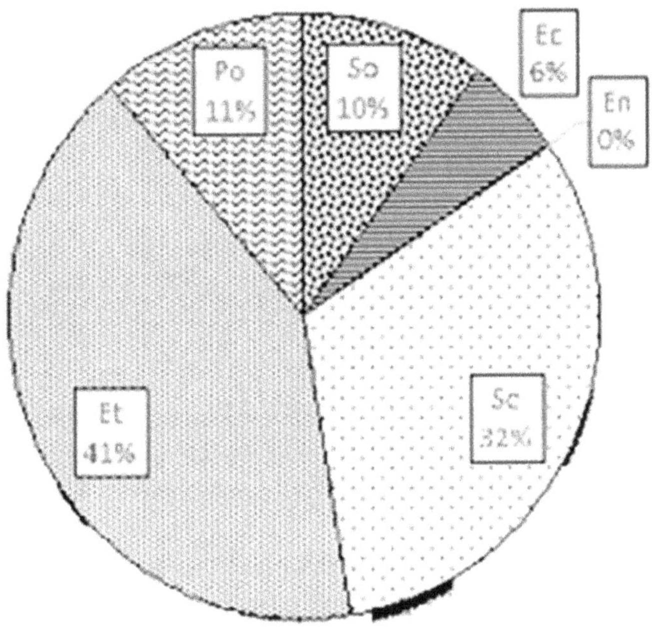

The subject areas of policy and sociology/culture were less mentioned in PTs' written reports (11% and 10% respectively). Furthermore, the subject area of economy was the least referred to (6%) in PTs' supporting reasons, and there were not any supporting reasons in the subject area of environment. PTs' written justifications with respect to these subject areas are also strongly related to the aspects of knowledge, value, and experiences. For instance, PT-9 expressed:

Figure 4. Average distribution of PTs' supporting reasons with respect to the aspects of knowledge, value, and personal experience

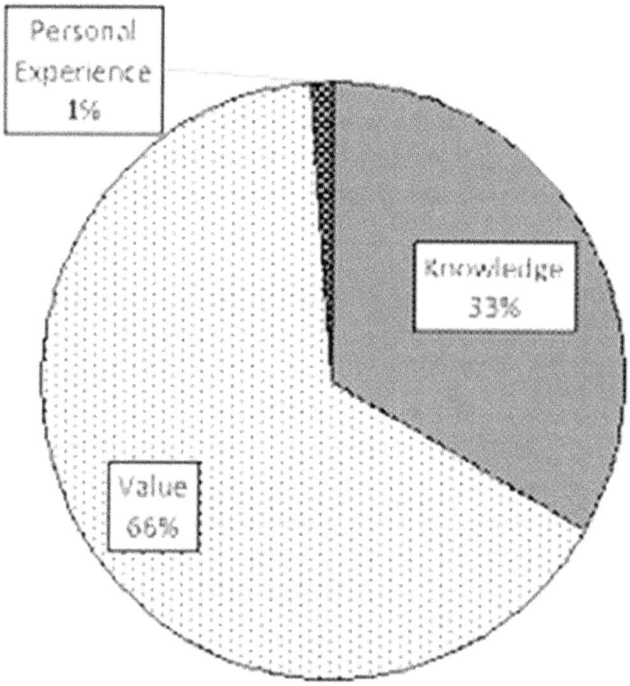

Financially well-off people can use this treatment method [referring gene therapy] easily while other people who are not aware or are not financially well-off will not use this treatment method. It can result in bad consequences. (S-1)

This statement is coded as EcV since this preservice teacher supports his claim of not using gene therapy with his economic concerns stemmed from his values (he believes it can result in bad consequences).

Another statement, expressed by PT-44, is:

It should be used [Preimplantation genetic diagnosis] but I think there should be some laws regulating the use of preimplantation genetic diagnosis since this method will help to create more healthy generations and the genes related with diseases will not be passed down to future generations. (S-2)

Figure 5. Distribution of PTs' supporting reasons in relation to the aspects of knowledge, value, and personal experience in the different scenarios

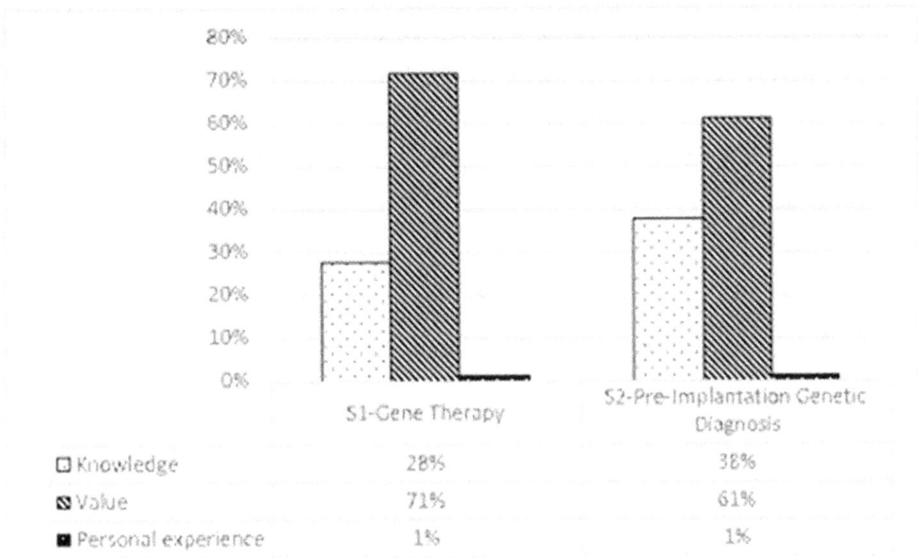

This statement is coded as PoK since PT-44 explained his reason why he believed that preimplantation genetic diagnosis should be used with the need for laws, referring to the subject area of policy in relation to the aspect of knowledge as he stated that the genes related with diseases will not be passed down to future generations (which is related with inheritance patterns).

The Aspects of Knowledge, Value, and Experiences in Preservice Science Teachers' Supporting Reasons

Figure 4 presents the distribution of PTs' supporting reasons in the two scenarios with respect to the aspects of knowledge, value, and personal experience. The results showed that PTs' supporting reasons mainly stemmed from their values (66%). PTs referred to knowledge in their supporting reasons to a lesser extent (33%), while personal experience was rarely noted in their written responses (1%).

How Pre-service Teachers' Supporting Reasons Differ in Different SSI Scenarios

After examining PTs' supporting reasons with respect to the aspects of knowledge, value, and personal experience in addition to the six subject areas, it was explored whether PTs' supporting reasons differ in relation to different scenarios. Even though both scenarios were based on genetics related SSI, PTs' supporting reasons with respect to the aspects of knowledge, value, and personal experience differed in different scenarios. Even though the percentages differ in the two scenarios (71% and 61% respectively), PTs generated more supporting reasons in relation to the aspect of value. The aspect of knowledge was referred to less in the written reports. The distribution for the aspect of knowledge was 28% in the gene therapy scenario and 38% in the pre-implantation genetic diagnosis scenario. Furthermore, only 1% of the supporting reasons stemmed from the subject area of personal experience (Figure 5).

Figure 6. Distribution of PTs' supporting reasons in relation to the six subject areas of the SEE-SEP model in the different scenarios

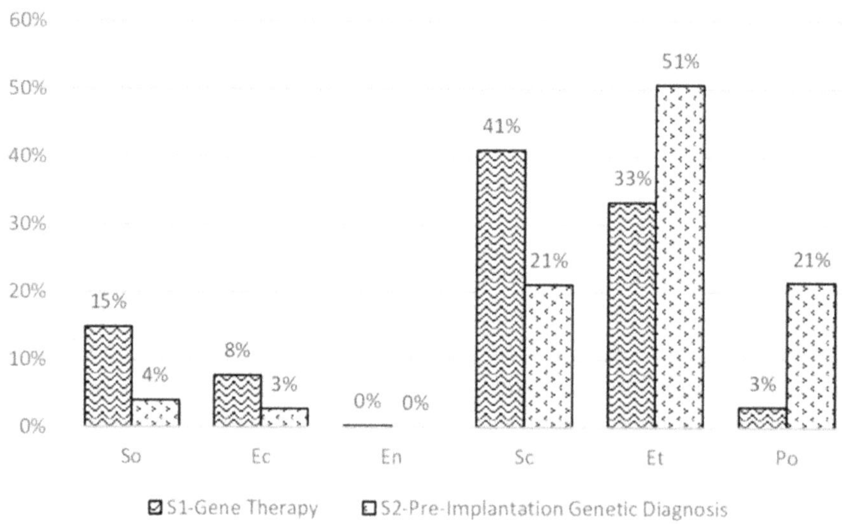

Then it was analyzed how PTs used different subject areas (sociology/culture, economy, environment/ecology, science, ethics/morality, and policy) in their decisions with respect to different scenarios (Figure 6). Regardless of the scenarios, the subject areas of science and ethics/morality were more dominant in their decisions. To be more specific, in the gene therapy scenario, the distribution for the subject area of science was 41%. This was followed by ethics/morality at 33% in PTs' decisions. While 15% of their reasons were related to sociology/culture, the subject areas of economy and policy were seldom presented in their decisions (8% and 3% respectively). In the pre-implantation genetic diagnosis scenario, participants mostly used the subject area of ethics/morality in their decisions (51%). Unlike the gene therapy scenario, the subject area of science was less used (21%) while the subject area of policy was equally found in their supporting reasons (21%). Supporting reasons from the subject area of sociology (4%) and economy (3%) were used less frequently in their decisions.

DISCUSSION

The aim of this study is to assess how PTs' supporting reasons varied in relation to six subject areas (Science, Environment, Ethics/morality, Sociology/culture, Economy, and Policy) in gene therapy and preimplantation genetic diagnosis issues and how PTs' decisions regarding these issues interact with the aspects of value, knowledge, and personal experience. To accomplish these tasks, the researchers designed an SSI-based course and presented two case scenarios involving genetics issues and analyzed PTs' written reports about the issues discussed in the class.

The results showed that PTs commonly used justifications from the subject area of ethics/morality when arguing on both scenarios. These findings are in line with previous research highlighting the importance of moral issues in an SSI context (e.g. Zeidler & Keefer, 2003). Sadler and Zeidler (2004) aimed at exploring students' ability to make informal reasoning on genetic engineering issues. Students' responses to genetic engineering scenarios were analyzed and researchers reported that moral implications affected participants' reasoning skills while moral emotions guided their tendency. In addition to the ethics/morality aspect, scientific knowledge was another common aspect of the participants' decision-making. The present findings demonstrate that PTs rely on the subject area of science when making decisions on genetic issues.

The second research question aimed at exploring how PTs' decisions regarding gene therapy and preimplantation genetic diagnosis issues interact with the aspects of value, knowledge, and personal experience. The results showed that PTs' personal values (66%) guided their decisions in both genetic scenarios followed by the aspect of knowledge at 33% with personal experiences emerging from the data as the last and least stated aspect (1%). The aspects of knowledge and value were the most dominant aspects in the participants' responses, whereas the aspect of personal experience played an insignificant role for a minority (1%) of the respondents. It can be concluded that the most dominant aspect PTs' decision-making was value when compared to the aspects of knowledge and personal experience. These findings are in line with several studies (Albe, 2008; Binder et. al., 2012; Chang & Chiu, 2008; Christenson 2015; Grace & Ratcliffe, 2002). Sources of information, personal experiences, and values are listed as possible factors that influence individuals' decisions (Cebesoy & Oztekin, 2016; Binder et al., 2012). In another study, Grace and Ratcliffe (2002) also found that students rely on their values rather than relying on scientific knowledge on the preservation of species topic. This supports the aforementioned claim. Moreover, Bell and Lederman (2003) also reported that students' decisions relied primarily on their personal values. Zeidler et al. (2002) support this argument by pointing out that people have a tendency to rely on their value judgments in controversial issues such as religion, belief, values, etc. Genetic issues in this study might have been one such controversial issue. Thus, even if the PTs attempted to evaluate the gene therapy and preimplantation genetic diagnosis issues using their scientific knowledge, their personal values influenced what actions they chose.

The aspect of knowledge was the second most common code found among PTs' supporting reasons (33%). Numerous studies support the idea that content knowledge is helpful for enhanced decision-making (Cebesoy & Oztekin, 2016; 2018; Means & Voss 1996; Sadler & Donnelly, 2006). Rundgren and Rundgren (2010) argue that content knowledge is helpful in solving complex real-life problems via SSI-based implementations. The positive relationship between content knowledge and the quality of informal reasoning was also addressed by Sadler and Zeidler (2005) and Zeidler et al. (2005). Moreover, Lewis and Leach (2006) reported that the quality of discussions in genetic issues was highly influenced by students' content knowledge. It has been acknowledged that scientific knowledge is important in decision making or informal reasoning in SSI (Sadler, 2004a) as they contribute to the quality of SSI (Sadler &Donnelly, 2006) and to the development of the ability to make well-grounded, informed decisions.

However, students fail to relate scientific data to their claims (Sadler, Chambers, & Zeidler, 2004). Although the aspect of knowledge was the second most common code in this study, 33% is not a big slice of the cake. Indeed, 67% of PTs' responses did not rely on scientific evidence or content knowledge. Jho et. al., (2014) reported that there are non-scientific factors (i.e personal experience, emotion, trust, value) that play a greater role than scientific knowledge. Most of the participants in this study rarely applied scientific knowledge in arguing genetic issues. There is an undeniable number of previous studies that support the idea that content knowledge is a prerequisite for making informed decisions (Means &Voss, 1996). For instance, Lewis and Leach (2006) stated that students must have specific scientific knowledge and understanding of key elements of the topic being discussed to have relevant discussions on genetic issues. Although the aforementioned research emphasizes a positive relationship between science content knowledge and the ability to make well-constructed decisions, Jho et. al. (2014) challenge this view in their studies. In the study by Jho et al. (2014), the attitudes of students were found to be effective in decision-making while scientific knowledge was reported as being ineffective.

Even if Jho et al. (2014) found a contradictory result, there is a consensus on the role of content knowledge in SSI-based decision-making. Exploring 80 upper secondary students' use of scientific knowledge to support their arguments, Christenson et al. (2012) found that the students rarely gave responses using the aspect of knowledge. The present study supports the argument that if students understand the issue and have specific content knowledge, they make coherent arguments about the issue. The fact that the aspect of scientific knowledge was used to a lesser extent might be because the PTs lack sufficient content knowledge because the issues being discussed are only covered in the scope of this study. The curriculum does not include a mandatory course for genetic engineering dilemmas or biotechnology.

Contrary to the previous literature that supports the idea of individuals using personal experiences to solve an ill-structured problem (Zeidler & Schafer, 1984), the aspect of personal experience is the least stated code found among PTs' supporting reasons in the present study. Researchers (Sadler et al. 2004a; Zeidler et al., 2002) have noted that personal experiences play a central role in decision-making. However, the findings of the present study suggest that relying on personal experiences is dependent on the issue being discussed. For instance, in the study by Patronis et al. (1999), researchers used a local issue about the construction of a road. Students probably had significant personal experiences concerning the area in which the road was to be built. Their personal experiences let them construct their arguments based on their previous experiences. However, the findings of the current showed that only 1% of the explanations made by PTs were connected with their personal experiences in the gene therapy and preimplantation genetic diagnosis issues. The reason for PTs not referring to their own experiences when making decisions about these issues might be their lack of experience in relation to those issues. In this study, the participants were still undergraduate students and all of them were single with no children. None of them had faced the scenarios in person or in their families. It is obvious that the two scenarios had no bearing on their personal life for this reason. Therefore, in the scope of this study, having 1% in their decisions is reasonable for the aspect of personal experience.

Regardless of scenarios, the subject areas of science and ethics/morality and the aspect of value were more dominant in PTs' decisions in both scenarios. The results showed that PTs used value to a greater extent (66%) than scientific knowledge (33%) for both SSI topics. Although the dominant aspect (value) did not change between the two scenarios, the percentages of the aspects differed among the SSI topics (see Figure 5 and Figure 6). For instance, the percentage for the aspect of value was 71% for gene therapy and 61% for preimplantation genetic diagnosis. Furthermore, the percentages for the aspect of knowledge were 28% and 38% respectively in the scenarios. This result is somewhat similar to work by Jho et al. (2014) claiming that apart from content knowledge, contextual knowledge also determines the quality of decision-making. Kolstø (2001) also highlighted a similar finding stating that as well as content knowledge, contextual knowledge might contribute to decision-making. It is also influenced by the specificity of the context under discussion and personal experience (Lewis &Leach, 2006). This can be explained by Christenson et al.'s study (2012) as the level of scientific knowledge in SSI topics can change from one issue to another. Some issues require individuals to rely on scientific knowledge (e.g. molecular biology, biotechnology), some on environment (global climate change, alternative energy sources), while others require relying on sociology/culture (consumption behaviors, etc.). It is obvious that the nature of the issue affects the way individuals formulate reasoning and use it to support their views. The SEE-SEP model helps teachers to decide which topics should be integrated into science lessons by analyzing how students contextualize and respond to controversial issues. Furthermore, the model helps researchers to get an overview of their students' multidisciplinary viewpoints when dealing with SSI (Chang Rundgren & Rundgren, 2010).

FUTURE RESEARCH DIRECTIONS

The SEE-SEP model illustrated is an analytical framework used in the analysis of PTs' written arguments about two genetic dilemmas in this chapter. The model helps the educators to observe the participants' multiple perspectives about the given issues. However, the present study and the previous literature that used this model to explore which subject areas (science, environment, ethics/morality, sociology, economy, and policy) emerge in the participants' reasoning are found to be context-dependent. While some issues are suitable for the subject area of policy to emerge (constructing a nuclear power plant or building a highway in a district), others are more suitable for environment (climate change, renewable energy sources) or ethics/morality (gene therapy, cloning). In the present study, the researchers could not refine any finding with respect to the subject area of environment due to the irrelevant nature of the issues being discussed (genetic dilemmas) for the environment. Considering this, future researchers should use divergent scenarios if they want to explore all six subject areas.

Previous research and also the present study noted that most of the participants rely on their explanations on personal values rather than knowledge. Although personal values and experiences are commonly seen in SSI-based reasoning, scientific knowledge is also a prerequisite for having advance reasoning. This common finding could alert future researchers to plan what to do to enhance scientific knowledge. If future researchers aim to achieve the goals of science education standards and raise scientifically literate people, they should recognize that scientific knowledge is rarely seen in students' SSI-based discussions but value and personal experiences are. It is also recommended that future research should focus on how to improve scientific knowledge, which is inherent in SSI. Future researchers could also investigate the quantity of scientific knowledge that is presented as being sufficient or not, as well as its quality of it (based on scientific fact or pseudoscience).

Another recommendation for future research is about the use of the SEE-SEP model. The present study used mixed methods to analyze PTs' written reports. The focus of the study was to explore the frequencies of the six subject areas and their relation to three aspects. To this end, the researchers adopted a quantizing approach that involves transforming qualitative data into numerical codes to quantify the patterns in PTs' supporting reasons and make comparisons between different scenarios. The frequency of appearance for each subject area and aspect was calculated. The presentation of the results in this way could be a limitation for exploration. Future researchers could focus on investigating the interaction between the six subject areas (science, environment, ethics, sociology, economy, and policy) with the three aspects (value, knowledge, and personal experiences). Focusing on the interactions between these dimensions could yield deeper insight into how participants reason.

CONCLUSION

The overarching goal of science education is to enhance scientific literacy. There is no direct formula to do this; however, educators have listed numerous ways (i.e. engaging in inquiry-based learning activities, bringing real-life issues into the classroom, forming interdisciplinary learning environments, etc.) to achieve this. SSI-based instruction is one of these alternative paths that contribute to scientific literacy. Although the inclusion of SSI in science lessons is nothing new, it is still difficult for educators or teachers to integrate it and even when integration is done, it is still difficult for implementers to evaluate it. The SEE-SEP model is a clear and understandable method to evaluate SSI-based discussions. It helps the investigator to explore how participants reason in SSI. The model also lets the researcher explore the multidimensional nature of SSI by revealing the interaction of science, environment, ethics, sociology, economy, and policy with the three aspects of knowledge, value, and personal experiences.

REFERENCES

Abd-El-Khalick, F. (2003). Socioscientific issues in pre-college science classrooms. In D. L. Zeidler (Ed.), *The role of moral reasoning on socioscientific issues and discourse in science education* (pp. 41–61). Springer. doi:10.1007/1-4020-4996-X_3

Aikenhead, G. S. (2006). *Science education for everyday life: Evidence-based practice*. Teachers College Press.

Albe, V. (2008). Students' positions and considerations of scientific evidence about a controversial socioscientific issue. *Science and Education, 17*(8-9), 805–827. doi:10.100711191-007-9086-6

Bell, R. L., & Lederman, N. G. (2003). Understandings of the nature of science and decision making on science and technology based issues. *Science Education, 87*(3), 352–377. doi:10.1002ce.10063

Binder, A. R., Cacciatore, M. A., Scheufele, D. A., Shaw, B. R., & Corley, E. A. (2012). Measuring risk/benefit perceptions of emerging technologies and their potential impact on communication of public opinion toward science. *Public Understanding of Science (Bristol, England), 21*(7), 830–847. doi:10.1177/0963662510390159 PMID:23832561

Bybee, R. W. (1995). Achieving scientific literacy. *Science Teacher (Normal, Ill.), 62*(7), 28.

Bybee, R. W. (2014). NGSS and the next generation of science teachers. *Journal of Science Teacher Education, 25*(2), 211–221. doi:10.100710972-014-9381-4

Cebesoy, U. B., & Oztekin, C. (2016). Relationships among Turkish pre-service science teachers' genetics literacy levels and their attitudes towards issues in genetics literacy. *Journal of Baltic Science Education, 15*(2), 159–172.

Cebesoy, U. B., & Oztekin, C. (2018). Genetics literacy: Insights from science teachers' knowledge, attitude, and teaching perceptions. *International Journal of Science and Mathematics Education, 16*(7), 1247–1268. doi:10.100710763-017-9840-4

Chang, H. Y., Wu, H. K., & Hsu, Y. S. (2013). Integrating a mobile augmented reality activity to contextualize student learning of a socioscientific issue. *British Journal of Educational Technology, 44*(3), E95–E99. doi:10.1111/j.1467-8535.2012.01379.x

Chang, S. N., & Chiu, M. H. (2008). Lakatos' Scientific Research Programmes as a Framework for Analysing Informal Argumentation about Socioscientific Issues. *International Journal of Science Education, 30*(13), 1753–1773. doi:10.1080/09500690701534582

Chang Rundgren, S. N., & Rundgren, C. J. (2010). SEE-SEP: From a separate to a holistic view of socioscientific issues. *Asia-Pacific Forum on Science Learning & Teaching*, *11*(1), 1–24.

Christenson, N. (2015). *Socioscientific argumentation: Aspects of content and structure* (Doctoral dissertation, Karlstads University). http://www.diva-portal.org/smash/record.jsf?pid=diva2%3A806023&dswid=-8412

Christenson, N., Chang Rundgren, S. N., & Höglund, H. O. (2012). Using the SEE-SEP model to analyze upper secondary students' use of supporting reasons in arguing socioscientific issues. *Journal of Science Education and Technology*, *21*(3), 342–352. doi:10.100710956-011-9328-x

Christenson, N., Chang Rundgren, S. N., & Zeidler, D. L. (2014). The relationship of discipline background to upper secondary students' argumentation on socioscientific issues. *Research in Science Education*, *44*(4), 581–601. doi:10.100711165-013-9394-6

Creswell, J. W., & Miller, D. L. (2000). Determining validity in qualitative inquiry. *Theory into Practice*, *39*(3), 124–130. doi:10.120715430421tip3903_2

Creswell, J. W., & Plano Clark, V. L. (2011). *Designing and conducting mixed methods research* (2nd ed.). Sage Publications.

Czerniak, C. M., & Lumpe, A. T. (1996). Relationship between teacher beliefs and science education reform. *Journal of Science Teacher Education*, *7*(4), 247–266. doi:10.1007/BF00058659

Eggert, S., & Bögeholz, S. (2010). Students' use of decision-making strategies with regard to socioscientific issues: An application of the Rasch partial credit model. *Science Education*, *94*(2), 230–258.

Es, H., & Ozturk, N. (2019). An Activity for transferring the multidimensional structure of SSI to middle school science courses: I discover myself in the decision-making process with SEE-STEP! *Research in Science Education*, 1–22. doi:10.100711165-019-09865-1

Forbes, C. T., & Davis, E. A. (2008). Exploring preservice elementary teachers' critique and adaptation of science curriculum materials in respect to socioscientific issues. *Science & Education*, *17*(8-9), 8–9, 829–854. doi:10.100711191-007-9080-z

Grace, M. M., & Ratcliffe, M. (2002). The science and values that young people draw upon to make decisions about biological conservation issues. *International Journal of Science Education*, *24*(11), 1157–1169. doi:10.1080/09500690210134848

Halverson, K. L., Siegel, M. A., & Freyermuth, S. K. (2009). Lenses for framing decisions: Undergraduates' decision making about stem cell research. *International Journal of Science Education, 31*(9), 1249–1268. doi:10.1080/09500690802178123

Hancock, E. S., & Gallard, A. J. (2004). Preservice science teachers' beliefs about teaching and learning: The influence of K-12 field experiences. *Journal of Science Teacher Education, 15*(4), 281–291. doi:10.1023/B:JSTE.0000048331.17407.f5

Higher Education Council (HEC). (2007). *Öğretmen Yetiştirme ve Eğitim Fakülteleri (1982-2007)* [Teacher training and faculties of education (1982-2007)]. https://www.yok.gov.tr/Documents/Yayinlar/Yayinlarimiz/ogretmen-yetistirme-ve-egitim-fakulteleri.pdf

Holbrook, J., & Rannikmae, M. (2007). The nature of science education for enhancing scientific literacy. *International Journal of Science Education, 29*(11), 1347–1362. doi:10.1080/09500690601007549

Jho, H., Yoon, H. G., & Kim, M. (2014). The relationship of science knowledge, attitude and decision making on socio-scientific issues: The case study of students' debates on a nuclear power plant in Korea. *Science & Education, 23*(5), 1131–1151. doi:10.100711191-013-9652-z

Jiménez-Aleixandre, M. P., Bugallo Rodríguez, A., & Duschl, R. A. (2000). "Doing the lesson" or "doing science": Argument in high school genetics. *Science Education, 84*(6), 757–792. doi:10.1002/1098-237X(200011)84:6<757::AID-SCE5>3.0.CO;2-F

Karisan, D., Tüzün, Ö. Y., & Zeidler, D. L. (2017). Quality of preservice teachers argumentation in socioscientific issues context. *Journal of Human Sciences, 14*(4), 3504–3520. doi:10.14687/jhs.v14i4.4949

Khishfe, R. (2014). Explicit nature of science and argumentation instruction in the context of socioscientific issues: An effect on student learning and transfer. *International Journal of Science Education, 36*(6), 974–1016. doi:10.1080/09500693.2013.832004

Kim, G., Ko, Y., & Lee, H. (2020). The effects of community-based socioscientific issues program (SSI-COMM) on promoting students' sense of place and character as citizens. *International Journal of Science and Mathematics Education, 18*(3), 399–418. doi:10.100710763-019-09976-1

Kolstø, S. D. (2001). Scientific literacy for citizenship: Tools for dealing with the science dimension of controversial socioscientific issues. *Science Education, 85*(3), 291–310. doi:10.1002ce.1011

Lederman, N. G., Antink, A., & Bartos, S. (2014). Nature of science, scientific inquiry, and socio-scientific issues arising from genetics: A pathway to developing a scientifically literate citizenry. *Science & Education, 23*(2), 285–302. doi:10.100711191-012-9503-3

Lee, H., Yoo, J., Choi, K., Kim, S. W., Krajcik, J., Herman, B. C., & Zeidler, D. L. (2013). Socioscientific issues as a vehicle for promoting character and values for global citizens. *International Journal of Science Education, 35*(12), 2079–2113. doi:10.1080/09500693.2012.749546

Lee, Y. C. (2007). Developing decision-making skills for socio-scientific issues. *Journal of Biological Education, 41*(4), 170–177. doi:10.1080/00219266.2007.96 56093

Lewis, J., & Leach, J. (2006). Discussion of socio-scientific issue: The role of science knowledge. *International Journal of Science Education, 28*(11), 1267–1287. doi:10.1080/09500690500439348

Lindahl, M. G., Folkesson, A. M., & Zeidler, D. L. (2019). Students' recognition of educational demands in the context of a socioscientific issues curriculum. *Journal of Research in Science Teaching, 56*(9), 1155–1182. doi:10.1002/tea.21548

Maxwell, J. A. (2009). Designing a qualitative study. In L. Bickman & D. J. Rog (Eds.), *The Handbook of Applied Social Research Methods* (pp. 214–253). Sage. doi:10.4135/9781483348858.n7

Means, M. L., & Voss, J. F. (1996). Who reasons well? Two studies of informal reasoning among children of different grade, ability, and knowledge levels. *Cognition and Instruction, 14*(2), 139–178. doi:10.12071532690xci1402_1

Merriam, S. B. (1998). *Qualitative research and case study applications in education*. Jossey-Bass.

Miles, M. B., & Huberman, A. (1994). *Qualitative data analysis: An expanded sourcebook*. Sage.

Ministry of National Education (MoNE). (2013). *Fen Bilimleri dersi öğretim program (İlkokul ve ortaokul 3, 4, 5, 6, 7 ve 8. Sınıflar)* [Science teaching program (Primary and middle 3rd, 4th, 5th, 6th, 7th and 8th grades]. Board of Education. Retrieved from http://mufredat.meb.gov.tr/ProgramDetay.aspx?PID=325

Morris, H. (2014). Socioscientific issues and multidisciplinarity in school science textbooks. *International Journal of Science Education, 36*(7), 1137–1158. doi:10.1080/09500693.2013.848493

Olafson, L. J., & Schraw, G. (2002). Some final thoughts on the epistemological melting pot. *Issues in Education, 8*(2), 233–247.

Patronis, T., Potari, D., & Spiliotopoulou, V. (1999). Students' argumentation in decision-making on a socio-scientific issue: Implications for teaching. *International Journal of Science Education, 21*(7), 745–754. doi:10.1080/095006999290408

Pedretti, E. (2003). Teaching science, technology, society and environment (STSE) education. In D. L. Zeidler (Ed.), *The role of moral reasoning on socioscientific issues and discourse in science education* (pp. 219–239). Springer. doi:10.1007/1-4020-4996-X_12

Powell, W. A. (2014). *The effects of emotive reasoning on secondary school students' decision-making in the context of socioscientific issues*. https://scholarcommons.usf.edu/etd/5385/

Ramsey, J. (1993). The science education reform movement: Implications for social responsibility. *Science Education, 77*(2), 235–258. doi:10.1002ce.3730770210

Sadler, T. D. (2004a). Informal reasoning regarding socioscientific issues: A critical review of research. *Journal of Research in Science Teaching, 41*(5), 513–536. doi:10.1002/tea.20009

Sadler, T. D. (2004b). Moral and ethical dimensions of socioscientific decision-making as integral components of scientific literacy. *Science Educator, 13*(1), 39–48.

Sadler, T. D. (2009). Situated learning in science education: Socio-scientific issues as contexts for practice. *Studies in Science Education, 45*(1), 1–42. doi:10.1080/03057260802681839

Sadler, T. D. (Ed.). (2011). *Socio-scientific issues in the classroom: Teaching, learning and research*. Springer. doi:10.1007/978-94-007-1159-4

Sadler, T. D., Amirshokoohi, A., Kazempour, M., & Allspaw, K. M. (2006). Socioscience and ethics in science classrooms: Teacher perspectives and strategies. *Journal of Research in Science Teaching, 43*(4), 353–376. doi:10.1002/tea.20142

Sadler, T. D., Barab, S. A., & Scott, B. (2007). What do students gain by engaging in socioscientific inquiry? *Research in Science Education, 37*(4), 371–391. doi:10.100711165-006-9030-9

Sadler, T. D., Chambers, F. W., & Zeidler, D. L. (2004). Student conceptualizations of the nature of science in response to a socioscientific issue. *International Journal of Science Education, 26*(4), 387–409. doi:10.1080/0950069032000119456

Sadler, T. D., & Donnelly, L. A. (2006). Socioscientific argumentation: The effects of content knowledge and morality. *International Journal of Science Education, 28*(12), 1463–1488. doi:10.1080/09500690600708717

Sadler, T. D., & Zeidler, D. L. (2004). The morality of socioscientific issues: Construal and resolution of genetic engineering dilemmas. *Science Education, 88*(1), 4–27. doi:10.1002ce.10101

Sadler, T. D., & Zeidler, D. L. (2005). The significance of content knowledge for informal reasoning regarding socioscientific issues: Applying genetics knowledge to genetic engineering issues. *Science Education, 89*(1), 71–93. doi:10.1002ce.20023

Solomon, J., & Aikenhead, G. (1994). *STS Education: International Perspectives on Reform. Ways of Knowing Science Series.* Teachers College Press.

Tashakkori, A., & Teddlie, C. (1998). *Mixed methodology: Combining qualitative and quantitative approaches.* Sage.

Teddlie, C., & Tashakkori, A. (2006). A general typology of research designs featuring mixed methods. *Research in the Schools, 13*(1), 12–28.

Topcu, M. S., Sadler, T. D., & Yilmaz-Tuzun, O. (2010). Preservice science teachers' informal reasoning about socioscientific issues: The influence of issue context. *International Journal of Science Education, 32*(18), 2475–2495. doi:10.1080/09500690903524779

Trochim, W. (2020). *Research methods data base.* Retrieved from https://conjointly.com/kb/

Walker, K. A., & Zeidler, D. L. (2007). Promoting discourse about socioscientific issues through scaffolded inquiry. *International Journal of Science Education, 29*(11), 1387–1410. doi:10.1080/09500690601068095

Wu, Y. T., & Tsai, C. C. (2007). High school students' informal reasoning on a socio-scientific issue: Qualitative and quantitative analyses. *International Journal of Science Education, 29*(9), 1163–1187. doi:10.1080/09500690601083375

Zeidler, D. L. (2014). Socioscientific issues as a curriculum emphasis: Theory, research, and practice. In N. G. Lederman & S. K. Abell (Eds.), *Handbook of Research on Science Education, 2* (pp. 711–740). Routledge.

Zeidler, D. L., Herman, B. C., & Sadler, T. D. (2019). New directions in socioscientific issues research. *Disciplinary and Interdisciplinary Science Education Research, 1*(1), 1–9. doi:10.118643031-019-0008-7

Zeidler, D. L., & Keefer, M. (2003). The role of moral reasoning and the status of socioscientific issues in science education. In D. L. Zeidler (Ed.), *The role of moral reasoning on socioscientific issues and discourse in science education* (pp. 7–38). Springer. doi:10.1007/1-4020-4996-X_2

Zeidler, D. L., & Nichols, B. H. (2009). Socioscientific issues: Theory and practice. *Journal of Elementary Science Education, 21*(2), 49–58. doi:10.1007/BF03173684

Zeidler, D. L., Sadler, T. D., Applebaum, S., & Callahan, B. E. (2009). Advancing reflective judgment through socio-scientific issues. *Journal of Research in Science Teaching, 46*(1), 74–101. doi:10.1002/tea.20281

Zeidler, D. L., Sadler, T. D., Simmons, M. L., & Howes, E. V. (2005). Beyond STS: A Research-Based Framework for Socioscientific Issues Education. *Science Education, 89*(3), 357–377. doi:10.1002ce.20048

Zeidler, D. L., & Schafer, L. E. (1984). Identifying mediating factors of moral reasoning in science education. *Journal of Research in Science Teaching, 21*(1), 1–15. doi:10.1002/tea.3660210102

Zeidler, D. L., Walker, K. A., Ackett, W. A., & Simmons, M. L. (2002). Tangled up in views: Beliefs in The Nature of Science and Responses to Socioscientific Dilemmas. *Science Education, 86*(3), 343–367. doi:10.1002ce.10025

Section 5

Integrating Modeling and STEM in Socioscientific Issues for Scientific Literacy

Chapter 9
Intersections of Integrated STEM and Socio-Scientific Issues

Gillian Roehrig
https://orcid.org/0000-0002-6943-7820
University of Minnesota, USA

Khomson Keratithamkul
https://orcid.org/0000-0001-9933-3128
University of Minnesota, USA

Benny Mart R. Hiwatig
https://orcid.org/0000-0003-0225-8630
University of Minnesota, USA

ABSTRACT

Current policy documents across the world call for changes in K-12 science teaching to use integrated STEM strategies to provide a more authentic learning environment for students. Though the importance of integrated STEM education is established through national and international policy documents, there remains disagreement on focus, models, and effective approaches for integrated STEM instruction. A primary focus of STEM policies is addressing STEM workforce issues. However, other important foci for global STEM initiatives are more equitable education, poverty reduction, and increased STEM literacy and awareness. This chapter critiques current implementations of STEM as focused only on technical aspects of engineering design which will not meet any of the goals of integrated STEM. Rather, the authors propose that integration of SSI into STEM lessons will promote the social thinking necessary in engineering design and enhance work toward achieving not only STEM workforce preparation, but also increased STEM literacy and equity within STEM.

DOI: 10.4018/978-1-7998-4558-4.ch009

INTRODUCTION

Current policy documents in the United States call for changes in K-12 science classrooms to employ integrated STEM strategies in order to provide a more authentic learning environment for students (National Academy of Sciences [NAS], 2014). Indeed, countries from across the world have established STEM policies and initiatives designed to improve the quality of STEM education (Freeman, Marginson, & Tytler, 2014). Though the importance of integrated STEM education is established through national and international policy documents, there remains disagreement on focus, models, and effective approaches for integrated STEM instruction (Bybee, 2010). However, a clearly stated, primary focus of STEM policies within the United States and many other Western countries is addressing STEM workforce issues. While less visible in the current STEM rhetoric, other equally important foci for global STEM initiatives are more equitable education, poverty reduction, and increased STEM literacy and awareness (Freeman et al., 2014; NAS, 2014).

While the argument that the United States is not producing adequate numbers of STEM graduates to meet the growing number of positions within these fields (Vilorio, 2014) dominates our national STEM rhetoric, STEM literacy is an equally important goal of current reforms (Honey, Pearson, & Schweingruber, 2014). Policy documents maintain that for students to become informed citizens who can make sense of our information rich, technologically advanced global environment, STEM literacy is necessary, even for those not working in a STEM field (Honey et al., 2014). Policy documents in the United States also call for STEM education access for *all* students (National Research Council [NRC], 2011; President's Council of Advisors in Science and Technology [PCAST], 2010). Yet, females and students of color remain under-represented in most STEM fields. Furthermore, a wide gap in STEM achievement persists between white students and under-represented minorities when compared to population demographics and the respective representation of these groups in the STEM fields (National Center for Science and Engineering Statistics, 2017). Indeed, reimagining STEM education to increase STEM participation for females and students of color would both help to meet workforce needs and ensure increased STEM literacy and participation for *all* students.

Teaching STEM from a workforce rationale is seen as problematic (e.g. Zeidler, 2016; Zeidler, Herman, Clough, Olson, Kahn, & Newton, 2016). Many proponents of teaching through socio-scientific issues (SSI) argue that goals such as increased STEM literacy cannot be achieved without the inclusion of humanistic and sociocultural values essential to acting as a democratic citizen (e.g. Kahn, 2015; Zeidler, 2014). Sadler, Barab and Scott (2007) argue that "the thoughtful negotiation of SSI is fundamental to modern notions of scientific literacy" (p. 372). Similarly, Kahn (2015) argues that by considering the humanistic approaches and moral development central to SSI, STEM can be reframed as "a vehicle for inclusivity, an incubator of innovation and inquisitiveness for all students, each of whom has a stake in STEM, perhaps as a future vocation, most certainly as a future global citizen" (p.151).

Problem Statement

In this chapter, we problematize an over-simplified focus of STEM on workforce needs and misrepresentation of the nature of engineering design in education settings as compared to the professional work of engineers. Through a deeper analysis and understanding of the work of engineers, we will argue that engineering design needs to move beyond the technical implementation of scientific and engineering practices to include explicit consideration of the societal impacts relevant to an engineering design problem. Implementation of engineering design lessons following the guidelines of standards document such as the Next Generation Science Standards (NGSS Lead States, 2013) reduces the design process to solving technical problems and minimizes or even removes relevant societal connections and issues necessary to providing possible solutions to a real-word problem. By incorporating the socio-scientific issues framework into integrated STEM lessons, the work of engineering becomes more authentically represented and also provides a stronger pedagogical framework to address the goals of meeting STEM literacy, equity within STEM, and meeting STEM workforce needs.

OVERVIEW OF K-12 STEM EDUCATION

Features of Quality Integrated STEM Education

In spite of these global policy calls for STEM education, there is no clear theoretical or practical definition of integrated STEM to implement the overarching goals of addressing STEM workforce needs, equity in STEM, and improved STEM literacy. A variety of approaches to STEM education have been suggested, for example, Bybee (2010) outlined a continuum of models ranging from STEM disciplines as completely separate and unconnected to full integration of all four STEM disciplines. However, there is a developing consensus around features indicative of quality integrated STEM. First, integrated STEM instruction uses real-world contexts to engage students in authentic and meaningful learning (Kelley & Knowles, 2016; Moore, Stohlmann, Wang, Tank, & Roehrig, 2014; Sanders, 2009). Second, student-centered pedagogies, including inquiry-based learning and design thinking, are central to quality integrated STEM (Kelley & Knowles, 2016; Moore, Stohlmann, et al., 2014). Third, quality STEM integration supports the development of 21st century skills such as creativity, collaboration, communication, and critical thinking (Honey et al., 2014; Moore, Glancy, Tank, Kersten, & Smith, 2014). Finally, connections between STEM disciplines should be made explicit to students (English, 2016; Herschbach, 2011; Honey et al., 2014; Kelley & Knowles, 2016; Moore, Stohlmann, et al., 2014). There remains a debate about how many of the STEM disciplines need to be evident in a lesson in order to consider it as integrated STEM. However, given the prominence of engineering within STEM policy documents (NRC, 2011, 2012; NGSS Lead States, 2013), many STEM approaches are structured around an engineering context or problem and use the engineering design process to come up with solutions (Hmelo, Holton, & Kolodner, 2000; Mehalik, Doppelt, & Schunn, 2008; Moore, Stohlmann, et al., 2014; Sadler, Coyle, & Schwartz, 2000). Thus, it is important to consider in detail how NGSS (NGSS Lead States, 2013) and the associated *Framework for K-12 Science Education* (NRC, 2012) describe the teaching of science within an engineering context and the work of engineering.

Engineering in K-12 Education

Design is widely considered to be the central activity of engineering (Dym, 1999) and is also the focus of K-12 STEM education (Moore, Tank, Glancy, & Kersten, 2015). Engineering is defined in K-12 settings as the processes of defining problems, generating and evaluating solutions, testing and optimizing solutions, and communicating solutions (Wilson-Lopez & Minichiello, 2017). Specifically, the *Frameworks for K-12 Science Education* (NRC, 2012) defines engineering design as:

a systematic process for solving engineering problems, is based on scientific knowledge and models of the material world. Each proposed solution results from a process of balancing competing criteria of desired functions, technological feasibility, cost, safety, esthetics, and compliance with legal requirements. There is usually no single best solution but rather a range of solutions. Which one is the optimal choice depends on the criteria used for making evaluations (p.52).

The *Frameworks* further specify that "engineering practices incorporate specialized knowledge about criteria and constraints, modeling and analysis, and optimization and trade-offs" (NRC, 2012, p.204). Students are called upon to understand the criteria and constraints of a problem (Watkins, Spencer, & Hammer 2014), yet within the K-12 setting, these constraints are limited to issues such as time, access to materials, and budget. Such definitions of problem scoping ignore the social, political, and ethical aspects of engineering design. For example, consider the construction of a new light rail line to connect suburban areas to downtown businesses. In addition to the technical aspects of the project, engineers have to evaluate the impact of proposed routes on communities and the environment. SSI researchers note that the NGSS and *Frameworks* (NRC, 2012) marginalize moral and ethical considerations related to engineering design (Kahn, 2015), and subsequently, K-12 curriculum largely ignores our responsibility as citizens to make decisions that mitigate local and global problems (Olson, 2013). It is also important to recognize that the representations of engineering within NGSS, as well as undergraduate engineering programs, are not representative of the work of professional engineers (Bairaktarova & Pilotte, 2020; Cunningham & Carlsen, 2014; Zoltowski, & Cardella, 2012).

School Engineering vs. Engineering

A hallmark of integrated STEM instruction is contextualizing instruction in real-world problems, typically by engaging students in design-based tasks (Kelley & Knowles, 2016; Moore, Stohlmann, et al., 2014; NAS, 2014; Sanders, 2009). Not only does solving real-world problems engage learners to apply, integrate, and expand their knowledge of the STEM disciplines (Monson & Besser, 2015), but real-world problems that are meaningful to students can also evoke interest and motivation in learning (Buxton, 2010; Monson & Besser, 2015). Real-world problems should be complex enough to foster multiple solutions (Lachapelle & Cunningham, 2014) and focused on generating solutions for a client, mirroring the work of professional engineers (Brophy, Klein, Portsmore, & Rogers, 2008; Cunningham & Carlsen, 2014). In the absence of a real-world context, design tasks can degenerate into simply making crafts; whereas, engineering requires the evaluation of a solution against criteria and constraints, as well as the consideration of the societal and environmental impacts of any proposed solution.

However, while real-world problems "can bring STEM fields alive for students and deepen their learning, they may also pose particular challenges for them" (NAS, 2014, p.82). Practitioners and researchers are concerned that students may be cognitively overwhelmed by the complexity of real-world problems, and subsequently, they create activities that are "engineering-like" rather than engaging in authentic engineering practices. For example, in considering K-12 science instruction, Ford and Forman (2006) state that "if students are merely engaged in trivial aspects of science (e.g., holding a test tube), then it is unlikely that they are "taking away" what is deemed important in that discipline" (p.11). Authentic engagement in the practices of science and engineering requires more than the simplistic notion that engineers solve problems to address a client's needs through consideration of only technical criteria and constraints while avoiding issues such as ethics and social justice.

It is important to understand the work of engineers that engineering design activities in K-12 classrooms are attempting to emulate. While engineers have made significant improvements for human life, many engineering projects have had adverse effects on the natural and social environment. Engineers must be aware of the cultural and environmental impacts of their work and the impact engineering design can make to societies and quality of life (De Vere, Johnson, & Thong, 2009). The role of an engineer is more than that of technical service provider and thus, engineers require competence beyond technical skills (Grandin & Hirleman, 2009). Engineering education researchers have criticized a singular focus on the technical skills:

Engineers handle an ever-growing body of engineering knowledge; many programs are crammed with technical information and leave little room for students to develop professional practices that enable them to become skillful communicators, ethical decision makers, team leaders, creative thinkers, and problem solvers. However, professional practices are essential, even critical, since engineers regularly interact with people in local, national, and international communities and create technical solutions that address complex social and environmental issues (Sauer, 2012).

The engineering and engineering education fields are experiencing a shift from "technology-centered design" to "human-centered design" (Krippendorff, 2006). In technology-centered design, engineers make design decisions which are imposed on the intended users (Krippendorff, 2006), whereas human-centered design starts with the relevant stakeholders (Zoltowski & Cardella, 2012). Zoltowski and Cardella (2012) emphasize that human-centered design involves and values stakeholders throughout the design process rather than checking for "user-friendliness" at the end of the process.

While engineering education is shifting toward human-centered design approaches that better reflect the work of professional engineers (Bairaktarova & Pilotte, 2020), engineering is still a market-driven enterprise; even when engaging in human-centered design the bottom line is profit. Human-centered design is promoted because it has been shown to "increase productivity, improve quality, reduce errors, reduce training and support costs, improve people's acceptance of new products, enhance companies' reputations, increase user satisfaction and reduce development costs" (Zoltowski & Cardella, 2012, p.30). Understanding the work of professional engineers expands engineering practices, such a problem-scoping, beyond technical aspects of the design process to include social, health, and environmental impacts. However, if STEM education is to go beyond providing education that better mirrors the reality of the STEM workforce and address STEM literacy, students need opportunities to develop moral and critical perspectives and to appreciate the social relevance and connections to their personal lives and community. We argue that by drawing on the existing framework of socio-scientific issues, integrated STEM can better address the development of both STEM literacy and equity within STEM, as well as informing engineering workforce development.

Overview of Socio-Scientific Issues

Socio-scientific issues (SSIs) are described as controversial, complex, social problems with conceptual and/or procedural links to science (Sadler, 2009; Zeidler, Walker, Ackett, & Simmons, 2002) that are affected by different social factors such as politics, economics, and ethics, in addition to scientific and technological knowledge. By engaging in an SSI, students develop conceptual understanding of science and are expected to apply that understanding in making and assessing personal, societal, and global decisions. SSIs not only allow for connections and application of science content to socially relevant science issues within the science curriculum, they also evoke values, ethics, and perspectives students need to consider as they evaluate scientific information (Sadler & Zeidler, 2005). Consequently, SSIs must include social and moral dilemmas that engage students in using critical thinking skills to analyze the data they are presented with. In addition, good SSIs are characterized as authentic, contemporary, controversial, data-supported, relevant and timely, pertinent to content standards, and illustrative of the nature and process of science (Levinson, 2006; Sadler, Amirshokoohi, Kazempour, & Allspaw, 2006; Stolz, Witteck, Marks, & Eilks, 2013). Current issues that possess these characteristics include topics such as global warming, cloning, genetic engineering, stem cell research, and captive breeding.

Similar to problem-based learning, SSIs are open-ended problems that are not expected to have single, correct answers. In problem-based learning, students engage in scientific and technical thinking to propose possible solutions to the problem or scenario provided (Hung, Jonassen, & Liu, 2008). On the contrary, students are posed with real-world scenarios through SSI that promote not only scientific and technical problem solving but also reflective thinking around ethical, moral, social and environmental responsibilities. Thus, SSI-based instruction employs authentic assessments and open-ended questions that require learners to investigate and explore the controversial issue through scientific reasoning integrated with various perspectives and social aspects (e.g., moral, ethical, racial, and cultural) to establish a position based on their gathered information or evidence (Klosterman & Sadler, 2010).

Potential of SSI to Enhance K-12 STEM Education

The following sections illustrate how SSI can be used to enhance K-12 STEM education to look beyond the technical aspects of engineering design to better promote the goals of K-12 STEM to promote STEM literacy, equity in STEM, and improve STEM workforce preparation. The inclusion of SSI moves STEM education from a simplistic, technical approach to an approach that explicitly considers the role of ethics and morality in engineering design.

SSI and STEM Literacy

Reforms in science education have long called for the development of a scientifically literate citizenry (DeBoer, 2000) with current reforms expanding to a broader notion of STEM literacy. Bybee (2013) clearly states that the overarching goal of STEM education is to further develop a STEM literate society. His definition of STEM literacy includes: (i) an individual's "knowledge, attitudes, and skills to identify questions and problems in life situations, explain the natural and designed world, and draw evidence-based conclusions about STEM-related issues" and (ii) an individual's "willingness to engage in STEM-related issues and with the ideas of science, technology, engineering and mathematics as a constructive, concerned, and reflective citizen." (p.101). While STEM literacy includes an understanding of the concepts and processes of science, technology, engineering, and mathematics, it must go beyond a traditional discipline-bound view and "consider the use and application of that knowledge, not just the acquisition of knowledge as a primary purpose of STEM education" (Bybee, 2013, p. 64). Aspects of STEM literacy related to the application of knowledge are well-aligned with problem-based learning approaches (Hung et al., 2008) and traditional approaches to STEM that employ an engineering design challenge that calls for the application of scientific, mathematical, and engineering knowledge to a real-world problem or design challenge (Moore, Glancy, et al., 2014). There are similar parallels to the Science-Technology-Society (STS) approach to teaching. According to NSTA (1982), STS focuses on developing scientifically literate individuals who,

understand how science, technology, and society influence one another and who are able to use their knowledge in everyday decision-making. The scientifically literate person has a substantial knowledge of the facts, concepts, and process skills which enable the individual to learn logically. (p.1)

However, STEM literacy is more than knowledge and application of STEM concepts. It also incorporates the social, political, moral, and ethical aspects related to the application of STEM to real-world problems. Zeidler, Walker, Ackett, and Simmons (2002) argue that the SSI framework goes "above and beyond" traditional implementation of science, technology, and society (STS) education, situating SSI as a "broader term that subsumes all that STS has to offer, while also considering the ethical dimensions of science, the moral reasoning of the child, and the emotional development of the student" (p. 344). Thus, SSI involves connecting STEM content to a social issue, processing information regarding the issue, considering moral and ethical values, and developing a position on the issue (Sadler, Chambers, & Zeidler, 2004).

SSI, which is designed as a pedagogical strategy to promote students' awareness of the interdependence between science and society (Zeidler, Sadler, Simmons, & Howe, 2005), aligns well with the overarching goal of STEM education to develop STEM literacy as "a constructive, concerned, and reflective citizen (Bybee, 2013, p.101)." SSI enhances the ways in which STEM considers the normative components of science critical to developing a STEM literate citizenry. While STEM does consider normative components such as prescribing courses of action and choosing to create selected products, in the STEM professions, even with human-centered design, these actions are focused on profit motives rather than moral reasoning. The development of a STEM literate citizenry cannot occur if we, as STEM educators, promote the "abdication of any sense of responsibility during the practice of science, or in the decision making process about the everyday use of science in the public sector (Zeidler, 2014, p.13)."

SSI and Equity

Given the socio-cultural and political history of STEM-related disciplines that favor White men from higher socio-economic status, other groups such as women and people of color remain under-represented in STEM (Mead & Metraux, 1957; Ong, 2005; Vakil & Ayers, 2019). For example, Foor and Walden (2015) looked at gender norms and identities within the engineering disciplines. They found that among female engineering students, there is a perceived hierarchy between the engineering fields. Students associated "soft" engineering disciplines, such as industrial engineering, with women due to their non-technical nature. Women are seen as being more adept at the social and business aspects of engineering than the technical aspects. Unfortunately, engineering reinforces these gender stereotypes when the singular focus is on technical education (Bairaktarova & Pilotte, 2020; Sauer, 2012). However, professional engineers require competence beyond technical skills (Grandin and Hirleman, 2009) and increasing the focus on social and cultural aspects of engineering will both better prepare all engineers for the reality of the work and help underrepresented students in STEM to identify with possible careers in engineering and other STEM fields. SSI-based instruction can refocus the norms and identities within the STEM fields to make visible and to value the non-scientific thinking that is essential to solving real-world problems and the work of engineering.

Research shows that engaging students in authentic engineering design experiences improves interest in both science and engineering (Guzey, Moore, & Harwell, 2014; Guzey, Moore, & Morse, 2016; Lachapelle & Cunningham, 2014). However, research also suggests girls and students of color are more motivated by projects focused on societal issues such as health, the environment, and social justice as opposed to more gendered engineering projects such as designing cars and rockets (Leammukda & Roehrig, 2020). Diekman, Brown, Johnston, and Clark (2010) argue that women's communal goal orientation can negatively intersect with beliefs that STEM careers do not involve helping people. Thus, if integrated STEM is to be successful in achieving one of its stated goals of increasing equity within STEM, it is critical that we engage students in the societal, as well as technical, aspects of engineering. Socio-scientific issues promote discourse around topics of interest that can open up communities of practice where students' identities can be expressed and where they can use previously-appropriated discourses (Sadler, 2009). Indeed, through the inclusion of ethics and morality in STEM lessons and degree programs, we would both diversify those who develop STEM identities and prepare a generation of engineers who engage in moral reasoning within the design process.

One way to promote learning among traditionally marginalized students in STEM is for instructors to draw upon their students' lived experiences and funds of knowledge or what Moll, Amanti, Neff, and Gonzalez (1992) called "knowledge and skills found in local households" (p. 132). When science lessons are contextualized in such experiences and knowledge, learning becomes meaningful to students (Upadhyay, 2005). It is important that both science and research surrounding science education be responsive to students' lived experiences (Barton, 2001; Gay, 2013) to promote meaningful learning. Connected through SSIs, culturally responsive teaching can promote thinking from other points of view, beyond the dominant or Western perspective. In addition to responsiveness, Carlone, Haun-Frank, and Webb (2011) push for equitable science classrooms as places where students get to "expand the roles they play … negotiate their participation, and share authority" (p. 480) which cannot be achieved by focusing only on technical aspects of engineering problems. With equity in mind, instructors need to pay more attention to both culture and normative identities (Carlone et al., 2011) which cannot be achieved with traditional, technical-focused STEM instruction unless students are challenged to investigate and explore real-world problems through various perspectives and social aspects, such as moral, ethical, racial, and cultural, in order to establish a position based on their evidence as modelled in SSI instruction (Klosterman & Sadler, 2010).

SSI and Workforce Development

In order to prepare students for the STEM workforce, it is critical that they are exposed to authentic experiences that parallel the work of engineers. These authentic experiences should engage student in the solving of real-world problems through scientific and engineering practices (NGSS Lead States, 2013; NRC, 2012), as well as critical thinking based on ethical and societal aspects of these real-world problems (De Vere, Johnson, & Thong, 2009; Grandin and Hirleman, 2009).

Knowledge and Skills for the STEM Workforce Development

The problems that we face in our ever-changing, increasingly global society are multidisciplinary, require the integration of multiple science, technology, engineering, and mathematics (STEM) concepts to solve them, and call for ethical and moral reasoning in developing possible solutions. These problems are complex and ill-structured and as such, multiple solutions are possible. Similarly, SSI problems "do not have single correct answers, cannot be meaningfully addressed through memorized or well-rehearsed responses, and are not subject to relatively simple algorithms" (Sadler, 2009, p. 11). Solutions to such problems require a range of critical thinking and reasoning processes (Levinson, 2006; NRC, 2012; Sadler et al., 2006; Stolz et al., 2013).

The practice of engineering requires the application of science, mathematics, and engineering knowledge. K-12 STEM education should emphasize this interdisciplinary nature by providing students with opportunities to apply developmentally appropriate mathematics or science in the context of solving engineering problems (Arık & Topçu, 2020; Moore, Glancy, et al., 2014). According to NGSS (2013), the design brief should clearly state all of the criteria, constraints, or limits of the problem (Arık & Topçu, 2020). Arık and Topçu (2020) provide a concrete example: "a civil engineer does not just think about a bridge; he thinks about one that can carry tonnes of goods, or one that is earthquake-resistant, or one at a certain height for a clear passage of ships" (p.13). Unfortunately, this example, like most engineering design tasks found in K-12 classrooms, focuses only on the technical aspects of engineering design and the application of STEM content. A professional engineer also has to weigh environmental and social considerations of the design such as the impact of the bridge design and location on the river ecosystem, and the impact on local neighborhoods due to increased traffic volume.

The 21st Century engineering profession now requires competence beyond technical skills (Grandin & Hirleman, 2009) and, both the American Society of Civil Engineers (ASCE) and the Accreditation Board for Engineering and Technology (ABET) all support the inclusion of the social sciences and humanities into engineering programs. At the undergraduate level, engineering education has traditionally focused on the physical sciences, design, and making students technical experts (Prados, 1998). However, researchers now recommend that engineers also be educated to understand, balance, and apply what they know amidst the complexities of the natural world, the built environment, and the needs of society (Russell, Hanna, Bank, & Shapira, 2007). Grimson (2012) critiques this technical focus as follows:

Engineers handle an ever-growing body of engineering knowledge; many programs are crammed with technical information and leave little room for students to develop professional practices that enable them to become skillful communicators, ethical decision makers, team leaders, creative thinkers, and problem solvers. However, professional practices are essential, even critical, since engineers regularly interact with people in local, national, and international communities and create technical solutions that address complex social and environmental issues (p.1).

These critiques have led to some limited but growing emphasis on ethics and empathy in engineering education. Martin and Schinzinger (1989) define engineering ethics as the study of moral issues and decisions that confront engineers. However, in practice, challenges such as the narrow division of labor which tends to cause moral tunnel vision, preoccupation with legalities, and the human tendency to place responsibility on an external authority can cause engineers to abdicate moral responsibility. Simply put, engineers either do not recognize the "moral challenges embedded in their decisions or they see their own role as neutral and pass the task on to managers or politicians to engage with the moral debate and make decisions" (Buser & Koch, 2012, p.258). However, any moral decision making by managers or policy makers would take place after a product is already developed. Roeser (2012) argues that technological design is not value neutral, the way a technology is designed determines its possibilities and as such has consequences for human well-being. Moral and ethical issues are part and parcel of the production of knowledge.

While engineering educators have acknowledged the need for students to develop both deep technical expertise and broader social competencies, the focus has been on communication, teamwork, and interdisciplinary collaborations. Explicit consideration of empathy, however, is far less common (Strobel, Hess, Pan, & Wachter Morris, 2013; Walter, Miller, & Sochacka, 2017). Ignoring the role of empathetic thought in engineering education is problematic. For example, Cech (2014) found an inverse relationship between empathic and analytic thought processes and that undergraduate engineering students' public welfare commitments concerns eroded throughout their degree program. Buser and Koch (2012) argue that engineers may develop a practical morality which draws on a pragmatic ethic of benefit that focus on only whether the project is a success and the client is satisfied. Empathy in engineering is most commonly treated as "a means to an end, or "tool to take off the shelf," when there is likely to be some personal or professional benefit from doing so (Walter, Milller, & Sochacka, 2017, p.130). Arguing to expand notions of empathic design beyond a focus on the user-designer interface, Walter and colleagues (2017) draw on Segal's (2011) model of social empathy to develop a framework for empathy in engineering that considers empathy as "not only a skill and practice orientation but also a way of being, that is, a lived worldview and values system in action that embraces the inherent humanism and social-embeddedness of engineering practice (p. 133)." Roeser (2012) argues that engineering education programs have an obligation to develop the moral expertise of prospective engineers, as engineers have a moral responsibility to reduce the risks of a technological product by exploring and developing different design solutions.

Arguments for the inclusion of explicit consideration of ethics, empathy, and moral reasoning in engineering education, are strongly aligned with the philosophical underpinnings in SSI instruction. SSI provides a useful framework for developing integrated STEM instruction that more appropriately represents STEM workforce skills that include non-technical skills, such as moral reasoning. The SSI pedagogical framework affords students the opportunity to consider ethical and moral aspects of solutions to real-world problems, as the complex structure of socio-scientific issues involves a multiplicity of social perspectives and the need to understand potential tensions between public and scientific perspectives and reasoning (Goodnight, 2005), with multiple logical solutions being possible (Levinson, 2006).

INTEGRATING STEM AND SSI IN K-12 CLASSROOMS

Engineering in K-12 Education: Understanding the Status and Improving the Prospects (NRC, 2009) emphasizes engineering design and calls for K-12 engineering to incorporate developmentally appropriate mathematics, science, and technology skills and promote engineering habits of mind. Specifically, this report highlights that students should understand and perceive engineering design processes as highly iterative and open-ended, whereby a problem may have many possible solutions. While researchers agree that engineering design does not represent everything one needs to know about engineering (Moore, Glancy, et al., 2014; Moore, Tank et al., 2015; NRC, 2009), the NGSS focuses solely on engineering design as a means for students to understand the work of engineers by stating: "It is important to point out that the NGSS do not put forward a full set of standards for engineering education, but rather include only practices and ideas about engineering design" (NGSS Lead States, 2013, Appendix I, p. 3).

There is no one engineering design process. However, common elements of this iterative process include: problem identification and scoping, ideation, design, testing, and redesign (Dym, 1999; Moore, Glancy, et al., 2014). The *Framework for Quality K-12 Engineering Education* (Moore, Glancy, et al, 2014) depicts engineering design processes as having three phases: Problem and Background, Plan and Implement, and Test and Evaluate (Table 1). This framework is aligned with NGSS, state standards in non-NGSS states, and parallel standards related to engineering design processes for undergraduate engineering programs (ABET, 2019). As previously noted, the focus is on the technical aspects of engineering design, thus providing a limited view of engineering for workforce preparation.

Table 1. Phases of engineering design processes

Phase	Description
Problem and background	Engineering design begins with the identification of an engineering problem. This phase begins with problem scoping, the process of developing and understanding of the relevant criteria and constraints (Watkins et al., 2014). It also includes researching about the problem and within the K-12 setting participating learning the necessary background knowledge.
Plan and Implement	During this stage, students are engaged in brainstorming, developing multiple solution possibilities, and evaluating the benefits and drawbacks of possible solutions to create a prototype.
Test and Evaluate	In this phase, the prototype is iteratively tested to identify strengths and weaknesses of the solution. Data collected through tests and experiments are used as feedback in redesign. This iterative testing provides data to improve the solution or product until it meets the specified design criteria.

There are interesting parallels between the phases of engineering design and frameworks for the implementation of SSI. Hancock, Friedrichsen, Kinslow, and Sadler (2019) presented a framework for SSI teaching and learning which has three sequential phases (see Table 2). The first phase entails the introduction of a focal issue to students. In the second phase, students explore pertinent science concepts and practices within the context of the issue. Finally, the third phase involves a culminating experience wherein students can synthesize their ideas, perspectives, and research pertinent to the issue they were investigating.

Table 2. Phases of the framework for SSI implementation

Phase	Description
Encounter focal issue	SSI-based instruction centers on a compelling social issue with strong connections to science. This issue should be presented at the beginning of instruction to serve as a true context for learning (Presley et al., 2013).
Develop	Students explore fundamental science concepts and practices related to the issue that serves as a context for learning. Through engagement with relevant science concepts and practices, students develop both increased awareness and understanding of the focal issue and socio-scientific reasoning (Sadler, Barab, & Scott, 2007).
Synthesize	Students synthesize their ideas, perspectives, and research related to the focal issue.

Integrating concepts from the Framework for SSI Instruction (Hancock et al., 2019) and K-12 engineering design processes (Moore, Glancy, et al., 2014), has the potential to better represent the professional work of engineers which is important for workforce development goals. In addition, the integration of SSI into instruction using the engineering design process will promote the development of engineers who embrace the moral and ethical values in their work and support engineers in fulfilling their ethical obligation to "hold paramount the health, safety and welfare of the public" (National Society of Professional Engineers, 2015). Additionally, by engaging students in SSI problems, there is increased potential for the development of STEM literacy (Sadler, 2011) and interest in STEM for under-represented students (Diekman, Brown, Johnston, & Clark, 2010; Authors, 2020). Kahn (2015) presents an example of how integration of SSI into a STEM lesson emphasizes moral discourse and evidence-based reasoning that draws on both STEM concepts and ethical and moral considerations.

Engineers engage in design processes within the context of real-world constraints and are expected to consider issues of equity, social justice, and the environment when proposing solutions or considering the impacts and consequences of proposed solutions. The work of engineers is a merging of the first phases from SSI and Engineering Design (Tables 1 and 2), i.e., engineers work within a context that influences the ways in which they draw on STEM concepts and practices and requires that engineers use moral reasoning as integral in design decisions (Roeser, 2012). By merging these initial phases, we ask students not only to understand the technical criteria and constraints of a problem but also to consider the problem within the context of a potentially difficult moral and ethical dilemma thereby expanding the limited representations of problem scoping represented in K-12 STEM education and undergraduate engineering programs (Roeser, 2012; Strobel, Hess, Pan, & Wachter Morris, 2013; Walter, Miller, & Sochacka, 2017). Teachers should seize such opportunity to guide students in sense-making, understanding the authentic nature of real-world problems, and approaching these problems with a critical perspective – drawing on knowledge from the STEM disciplines to substantiate their claims while recognizing the inter-connectedness of STEM, society, and moral values. In this light, STEM education can promote STEM careers *and* social responsibility by showing students that what they are learning is relevant to their lives, by exposing them to social justice issues situated in the STEM community, by strengthening their belief systems and teaching them how to defend and promote their ideals using evidence, and by empowering them to participate in critical discourse about issues affecting their communities.

ACKNOWLEDGMENT

This study was made possible by National Science Foundation grant DRL-1813342. The findings, conclusions, and opinions herein represent the views of the authors and do not necessarily represent the view of personnel affiliated with the National Science Foundation.

REFERENCES

ABET. (2019). *Criteria for accrediting engineering programs*. Retrieved from https://www.abet.org/accreditation/accreditation-criteria/accreditation-policy-and-procedure-manual-appm-2019-2020/

Arık, M., & Topçu, M.S. (2020). Implementation of engineering design process in the K-12 science classrooms: Trends and issues. *Research in Science Education*. doi:10.100711165-019-09912-x

Bairaktarova, D., & Pilotte, M. (2020). Person or thing oriented: A comparative study of individual differences of first-year engineering students and practitioners. *Journal of Engineering Education, 109*(2), 1–13. doi:10.1002/jee.20309

Barton, A. C. (2001). Science education in urban settings: Seeking new ways of praxis through critical ethnography. *Journal of Research in Science Teaching, 38*(8), 899–917. doi:10.1002/tea.1038

Berkowitz, M. W., & Simmons, P. (2003). Integrating science education and character education. In D. L. Zeidler (Ed.), *The role of moral reasoning on socioscientific issues and discourse in science education* (pp. 117–138). Kluwer. doi:10.1007/1-4020-4996-X_7

Brophy, S., Klein, S., Portsmore, M., & Rogers, C. (2008). Advancing engineering education in P-12 classrooms. *Journal of Engineering Education, 97*(3), 369–387. doi:10.1002/j.2168-9830.2008.tb00985.x

Buser, M., & Koch, C. (2012). Eyes wide shut? Loyalty, and practical morality in Engineering Education. In S. Christensen, C. Mitcham, L. Bocong, & Y. An (Eds.), *Engineering, Development, and Philosophy: American, Chinese, and European Perspectives* (pp. 253–267). Springer. doi:10.1007/978-94-007-5282-5_15

Buxton, C. A. (2010). Social problem solving through science: An approach to critical, place-based, science teaching and learning. *Equity & Excellence in Education, 43*(1), 120–135. doi:10.1080/10665680903408932

Bybee, R. W. (2010). Advancing STEM education: A 2020 vision. *Technology and Engineering Teacher*, *70*(1), 30–35.

Bybee, R. W. (2013). *The case for STEM education: Challenges and opportunities*. NSTA Press.

Carlone, H. B., Haun-Frank, J., & Webb, A. (2011). Assessing equity beyond knowledge- and skills-based outcomes: A comparative ethnography of two fourth-grade reform-based science classrooms. *Journal of Research in Science Teaching*, *48*(5), 459–485. doi:10.1002/tea.20413

Cunningham, C. M., & Carlsen, W. S. (2014). Teaching engineering practices. *Journal of Science Teacher Education*, *25*(2), 197–210. doi:10.100710972-014-9380-5

De Vere, I., Johnson, K. B., & Thong, C. (2009). *Educating the responsible engineer: Socially responsible design and sustainability in the curriculum*. International conference on engineering and product design education, University of Brighton, UK.

DeBoer, G. (2000). Scientific literacy: Another look at its historical and contemporary meanings and its relationship to science education reform. *Journal of Research in Science Teaching*, *37*(6), 582–601. doi:10.1002/1098-2736(200008)37:6<582::AID-TEA5>3.0.CO;2-L

Diekman, A. B., Brown, E. R., Johnston, A. M., & Clark, E. K. (2010). Seeking congruity between goals and roles: A new look at why women opt out of science, technology, engineering, and mathematics careers. *Psychological Science*, *21*(8), 1051–1057. doi:10.1177/0956797610377342 PMID:20631322

Dym, C. L. (1999). Learning engineering: Design, learning, and experiences. *Journal of Engineering Education*, *88*(2), 145–148. doi:10.1002/j.2168-9830.1999.tb00425.x

English, L. D. (2016). STEM education K-12: Perspectives on integration. *International Journal of STEM Education*, *3*(1), 1–8. doi:10.118640594-016-0036-1

Foor, C. E., & Walden, S. E. (2009). "Imaginary engineering" or "re-imagined engineering": Negotiating gendered identities in the borderland of a college of engineering. *NWSA Journal*, *21*(2), 41–64.

Ford, M. J., & Forman, E. A. (2006). Refining disciplinary learning in classroom contexts. *Review of Research in Education*, *30*(1), 1–32. doi:10.3102/0091732X030001001

Freeman, B., Marginson, S., & Tytler, R. (2014). *The age of STEM: Educational policy and practice across the world in science, technology, engineering and mathematics*. Routledge. doi:10.4324/9781315767512

Gay, G. (2013). Teaching to and through cultural diversity. *Curriculum Inquiry, 43*(1), 48–70. doi:10.1111/curi.12002

Goodnight, G. T. (2005). Science and technology controversy: A rationale for inquiry. *Argumentation and Advocacy, 42*(1), 26–29. doi:10.1080/00028533.2005.11821636

Grandin, J. M., & Hirleman, E. D. (2009). Educating engineers as global citizens: A call for action: A report of the national summit meeting on the globalization of engineering education. *Online Journal of Global Engineering Education, 4*(1), 10–25.

Grimson, J. (2010). Re-engineering the curriculum for the 21st century. *European Journal of Engineering Education, 27*(1), 31–37. doi:10.1080/03043790110100803

Guzey, S. S., Moore, T., & Harwell, M. (2014). Development of an instrument to measure students' attitudes toward STEM. *School Science and Mathematics, 114*(6), 271–279. doi:10.1111sm.12077

Guzey, S. S., Moore, T., & Morse, G. (2016). Student interest in engineering design-based science. *School Science and Mathematics, 116*(8), 411–419. doi:10.1111sm.12198

Herschbach, D. R. (2011). The STEM initiative: Constraints and challenges. *Journal of STEM Teacher Education, 48*(1), 96–112. doi:10.30707/JSTE48.1Herschbach

Hmelo, C. E., Holton, D., & Kolodner, J. L. (2000). Designing to learn about complex systems. *Journal of the Learning Sciences, 9*(3), 247–298. doi:10.1207/S15327809JLS0903_2

Honey, M., Pearson, G., & Schweingruber, H. A. (Eds.). (2014). *STEM integration in K-12 education: Status, prospectus, and an agenda for research.* The National Academies Press.

Hung, W., Jonassen, D. H., & Liu, R. (2008). Problem-based learning. In J. M. Spector, J. G. van Merrienboer, M. D. Merrill, & M. Driscoll (Eds.), *Handbook of Research on Educational Communications and Technology* (3rd ed., pp. 485–506). Erlbaum.

Kahn. (2015). Another "M" for STEM? Moral Considerations for Advancing STEM Literacy. *K-12 STEM Education, 1*(4), 149-156.

Kelley, T. R., & Knowles, J. G. (2016). A conceptual framework for integrated STEM education. *International Journal of STEM Education, 3*(11), 1–11. doi:10.118640594-016-0046-z

Klosterman, M., & Sadler, T. (2010). Multi-level assessment of scientific content knowledge gains associated with socioscientific issues-based instruction. *International Journal of Science Education*, *32*(8), 1017–1043. doi:10.1080/09500690902894512

Krippendorff, K. (2006). *The semantic turn: A new foundation for design*. CRC Press Taylor & Francis Group.

Lachapelle, C., & Cunningham, C. (2014). Engineering in elementary schools. In S. Purzer, J. Strobel, & M. Cardella (Eds.), *Engineering in pre-college settings: Synthesizing research, policy, and practices* (pp. 61–88). Purdue University Press. doi:10.2307/j.ctt6wq7bh.8

Leammukda, F. D., & Roehrig, G. H. (2020). *Community-Based Conceptual Framework for STEM Integration*. Paper presented at the annual meeting of the Association for Science Teacher Education, San Antonio, TX.

Levinson, R. (2006). Towards a theoretical framework for teaching controversial socio-scientific issues. *International Journal of Science Education*, *28*(10), 1201–1224. doi:10.1080/09500690600560753

Martin, M., & Schinzinger, R. (1989). *Ethics in Engineering*. McGraw-Hill.

Mead, M., & Métraux, R. (1957). Image of the scientist among high-school students: A pilot study. *Science*, *126*(3270), 384–390. doi:10.1126cience.126.3270.384 PMID:17774477

Mehalik, M., Doppelt, Y., & Schunn, C. D. (2008). Middle school science through design-based learning versus scripted inquiry: Better overall science concept learning and equity gap reduction. *Journal of Engineering Education*, *97*(1), 71–85. doi:10.1002/j.2168-9830.2008.tb00955.x

Moll, L., Amanti, C., Neff, D., & Gonzalez, N. (1992). Funds of knowledge for teaching: Using a qualitative approach to connect homes and classrooms. *Theory into Practice*, *31*(2), 132–141. doi:10.1080/00405849209543534

Monson, D., & Besser, D. (2015). Smashing milk cartons: Third-grade students solve a real-world problem using the engineering design process, collaborative group work, and integrated STEM education. *Science and Children*, *52*(9), 38–43. doi:10.2505/4c15_052_09_38

Moore, T. J., Glancy, A. W., Tank, K. M., Kersten, J. A., Smith, K. A., & Stohlmann, M. S. (2014). A framework for quality K-12 engineering education: Research and development. *Journal of Pre-College Engineering Education Research*, *4*(1), 1–13. doi:10.7771/2157-9288.1069

Moore, T. J., Stohlmann, M. S., Wang, H.-H., Tank, K. M., & Roehrig, G. H. (2014). Implementation and integration of engineering in K-12 STEM education. In J. Strobel, S. Purzer, & M. Cardella (Eds.), *Engineering in Pre-College Settings: Research into Practice*. Sense Publishers.

Moore, T. J., Tank, K. M., Glancy, A. W., & Kersten, J. A. (2015). NGSS and the landscape of engineering in K-12 state science standards. *Journal of Research in Science Teaching*, *52*(3), 296–318. doi:10.1002/tea.21199

National Academy of Sciences. (2004). *The engineer of 2020: Visions of engineering in the new century*. National Academies Press.

National Governors Association Center for Best Practices, Council of Chief State School Officers. (2010). *Common core state standards*. National Governors Association Center for Best Practices, Council of Chief State School Officers.

National Research Council. (2011). *Successful K-12 STEM education: Identifying effective approaches in science, technology, engineering, and mathematics*. National Academies Press.

National Research Council. (2012). *A framework for K-12 science education: Practices, crosscutting concepts, and core ideas*. National Academies Press.

National Science Teachers Association (NSTA). (1982). Science/technology/society (STS). Washington, DC: NSTA.

National Society of Professional Engineers. (2015). *Code of ethics for engineers*. NSPE Press.

NGSS Lead States. (2013). *Next generation science standards: For states, by states*. National Academies Press.

Olson, J. K. (2013). The purposes of schooling and the nature of technology: The end of education? In M. P. Clough, J. K. Olson, & D. S. Niederhauser (Eds.), *The nature of technology: Implications for learning and teaching*. Sense Publishers. doi:10.1007/978-94-6209-269-3_13

Ong, M. (2005). Body projects of young women of color in physics: Intersections of gender, race, and science. *Social Problems*, *52*(4), 593–617. doi:10.1525p.2005.52.4.593

Prados, J. W. (1998). *Engineering education in the United States: Past, present, and future*. Paper presented at the International Conference on Engineering Education.

President's Council of Advisors on Science and Technology. (2010). *Report to the president: Prepare and inspire: K-12 education in science, technology, engineering, and mathematics (STEM) for America's future*. Washington, DC: Executive Office of the President. Retrieved from https://obamawhitehouse.archives.gov/sites/default/files/microsites/ostp/pcast-stem-ed-final.pdf

Presley, M. L., Sickel, A. J., Muslu, N., Merle-Johnson, D., Witzig, S. B., Izci, K., & Sadler, T. D. (2013). A framework for socio-scientific issues based education. *Science Educator, 22*(1), 26–32.

Roeser, S. (2012). Emotional Engineers: Toward Morally Responsible Design. *Science and Engineering Ethics, 18*, 103–115.

Russell, J. S., Hanna, A., Bank, L. C., & Shapira, A. (2007). Education in construction engineering and management built on tradition: Blueprint for tomorrow. *Journal of Construction Engineering and Management, 133*(9), 661–668.

Sadler, P. M., Coyle, H. P., & Schwartz, M. (2000). Engineering competitions in the middle school classroom: Key elements in developing effective design challenges. *Journal of the Learning Sciences, 9*(3), 299–327. doi:10.1207/S15327809JLS0903_3

Sadler, T. D. (2009). Socioscientific issues in science education: Labels, reasoning, and transfer. *Cultural Studies of Science Education, 4*(3), 697–703. doi:10.100711422-008-9133-x

Sadler, T. D. (2011). Situating socio-scientific issues in classrooms as a means of achieving goals of science education. In T. D. Sadler (Ed.), *Socio-scientific issues in the classroom: Teaching, learning, and research* (pp. 1–9). Springer. doi:10.1007/978-94-007-1159-4_1

Sadler, T. D., Amirshokoohi, A., Kazempour, M., & Allspaw, K. M. (2006). Socioscience and ethics in science classrooms: Teacher perspectives and strategies. *Journal of Research in Science Teaching, 43*(4), 353–376. doi:10.1002/tea.20142

Sadler, T. D., Barab, S., & Scott, B. (2007). What do students gain by engaging in socioscientific inquiry? *Research in Science Education, 37*(4), 371–391. doi:10.100711165-006-9030-9

Sadler, T. D., Chambers, F. W., & Zeidler, D. L. (2004). Student conceptualisations of the nature of science in response to a socioscientific issue. *International Journal of Science Education, 26*(4), 387–409. doi:10.1080/0950069032000119456

Sadler, T. D., & Zeidler, D. L. (2005). The significance of content knowledge for informal reasoning regarding socioscientific issues: Applying genetics knowledge to genetic engineering issues. *Science Education*, *89*(1), 71–93. doi:10.1002ce.20023

Sanders, M. (2009). STEM, STEM education, STEM mania. *Technology Teacher*, *68*(4), 20–26.

Sauer, G. (2012). *Redesigning engineering curricula for the 21st century*. Iowa State University.

Segal, E. A. (2011). Social empathy: A model built on empathy, contextual understanding, and social responsibility that promotes social justice. *Journal of Social Service Research*, *37*(3), 266–277. doi:10.1080/01488376.2011.564040

Stolz, M., Witteck, T., Marks, R., & Eilks, I. (2013). Reflecting socio-scientific issues for science education coming from the case of curriculum development on doping in chemistry education. *Eurasia Journal of Mathematics, Science and Technology Education*, *9*(4), 361–370.

Strobel, J., Hess, J., Pan, R., & Wachter Morris, C. A. (2013). Empathy and care within engineering: Qualitative perspectives from engineering faculty and practicing engineers. *Engineering Studies*, *5*(2), 137–159. doi:10.1080/19378629.2013.814136

Upadhyay, B. R. (2005). Using students' lived experiences in an urban science classroom: An elementary school teacher's thinking. *Science Education*, *90*(1), 94–110. doi:10.1002ce.20095

Vakil, S., & Ayers, R. (2019). The racial politics of STEM education in the USA: Interrogations and explorations. *Race, Ethnicity and Education*, *22*(4), 449–458. doi:10.1080/13613324.2019.1592831

Vilorio, D. (2014). STEM 101: Intro to tomorrow's jobs. *Occupational Outlook Quarterly*, 2–12. https://www.bls.gov/careeroutlook/

Walther, J., Miller, S. E., & Sochacka, N. W. (2017). A model of empathy in engineering as a core skill, practice orientation, and professional way of being. *Journal of Engineering Education*, *106*(1), 23–148. doi:10.1002/jee.20159

Watkins, J., Spencer, K., & Hammer, D. (2014) Examining young students' problem scoping in engineering design. *Journal of Pre-College Engineering Education Research*, *4*(1), Article 5.

Wilson-Lopez, A., & Minichiello, A. (2017). Disciplinary literacy in engineering. *Journal of Adolescent & Adult Literacy*, *61*(1), 7–14. doi:10.1002/jaal.658

Zeidler, D. L. (2014). STEM education: A deficit framework for the twenty first century? A sociocultural socio-scientific response. *Cultural Studies of Science Education, 11*(1), 11–26. doi:10.100711422-014-9578-z

Zeidler, D. L., Herman, B. C., Clough, M. P., Olson, J. K., Kahn, S., & Newton, M. (2016). Humanitas emptor: Reconsidering recent trends and policy in science teacher education. *Journal of Science Teacher Education, 27*(5), 465–476. doi:10.100710972-016-9481-4

Zeidler, D. L., Sadler, T. D., Simmons, M. L., & Howes, E. V. (2005). Beyond STS: A research-based framework for socioscientific issues education. *Science Education, 89*(3), 357–377. doi:10.1002ce.20048

Zeidler, D. L., Walker, K. A., Ackett, W. A., & Simmons, M. L. (2002). Tangled up in views: Beliefs in the nature of science and responses to socioscientific dilemmas. *Science Education, 86*(3), 343–367. doi:10.1002ce.10025

Zoltowski, C. B., Oakes, W. C., & Cardella, M. E. (2012). Students' ways of experiencing human-centered design. *Journal of Engineering Education, 101*(1), 28–59. doi:10.1002/j.2168-9830.2012.tb00040.x

Chapter 10
Socio-Scientific Issues and Model-Based Learning

Benzegül Durak
https://orcid.org/0000-0002-1162-4658
Duzce University, Turkey

Mustafa Sami Topçu
https://orcid.org/0000-0001-5068-8796
Yildiz Technical University, Turkey

ABSTRACT

This chapter aims to provide a literature analysis on socio-scientific issues and model-based learning. The position of socio-scientific issues in the process of raising science literate students is indisputable. On the other hand, modeling gives students opportunities to construct their own models and use them through the learning process to formulate hypothesis, make investigations, explain scientific phenomena, and communicate and justify their ideas. Therefore, embedding modeling practice to SSI-based instruction through a framework is an innovative tool for scientific literacy in science education.

SCIENTIFIC MODELS AND MODELING

For over fifty years, modeling—and how students learn with models—has been researched in terms of philosophy of science, epistemology, and classroom practices (Harrison & Treagust, 1998; 2000). The concept of a model can be defined in many ways, but for science education, it is best explained by evaluating it in terms of various categories. A model can take one or more physical forms when published,

DOI: 10.4018/978-1-7998-4558-4.ch010

even though it always exists differently in the mind of its inventor or any subsequent user. These forms can be represented in a variety of media, as stated by Gilbert and Justi (2016):

> *...for example, in the form of a gesture (e.g., of the relative position of objects), in a material form (e.g., a ball-and-stick representation of a crystal structure), in a visual form (e.g., as a diagram of a metabolic pathway), in a verbal form (e.g., an analogy for the structure of an atom based on that of the solar system), in a symbolic form (e.g., as a chemical equation), and in a virtual form (e.g., as a computer simulation). The range of entities that can be represented is wide: objects (e.g., of a virus), systems (e.g., of a blood circulation system), processes (e.g., of the liberation of energy from foodstuffs), events (e.g., of the attack of a white blood cell on a virus), ideas (e.g., of a vector of a force), and arrays of data about any of these entities (p. viii).*

Specifically, scientific models may be a representation of a real object/process (like a model of the human skeletal system) or an abstract object/process (like an atom model) (Gilbert, 2004; Gobert & Buckley, 2000; Harrison & Treagust, 1998; Treagust, Chittleborough, & Mamiala, 2002). Therefore, scientific models can be used for different purposes in science lessons, such as to represent concrete or abstract objects/processes, to provide definitions or simplifications of complex scientific events, and to form the basis of both scientific explanations and predictions of phenomena (Gilbert, 2004; Gobert & Buckley, 2000; Harrison & Treagust, 2000).

Modeling, on the other hand, refers to a fundamental scientific practice that enables students to conceptualize, investigate, and explain natural phenomena, as well as convince others about their results (Gilbert, 2004; Harrison & Treagust, 2000). Hence, modeling is a process in which students create representations of their learning (Clement, 2000).

MODEL-BASED LEARNING

Model-based learning (MbL) is a learning and teaching approach in science education that supports learning via students' models (Louca & Zacharia, 2012). There are many studies about model-based learning from different perspectives (Louca & Zacharia, 2012). However, the authors take a constructivist perspective; they believe that learning requires individuals' creation of their own schema to assimilate new concepts (Yager, 1991). Thus, MbL approaches from constructivist learning theory are mentioned in the present chapter. The constructivist perspective argues that learning occurs through the interaction between experiences and ideas (Savin-Baden & Major, 2013). These can be externalized by the construction and transferring of models.

Each individual produces his/her own mental models by linking new information to the previous information that he/she uses to make sense of experiences. For this reason, learning is the process of adjusting mental models to adapt to new experiences (Schwarz et al., 2009). From this perspective, MbL facilitates the constructivist learning theory, because it can be used to “make sense of abstract, difficult, and non-observable science concepts to accommodate the explainer, the audience, the content, and the context” (Treagust & Harrison, 1999, p. 32).

Of the many studies on model-based learning, only a few are on how model-based learning is used effectively in the classroom (Clement, 2000; Louca & Zacharia, 2012). Clement (2000) developed a basic theoretical framework for MbL using the following concepts: preconceptions, natural reasoning skills, intermediate models, learning processes, target models, and expert consensus models.

Figure 1. A basic theoretical framework for model-based learning
Source: Clement, 2000

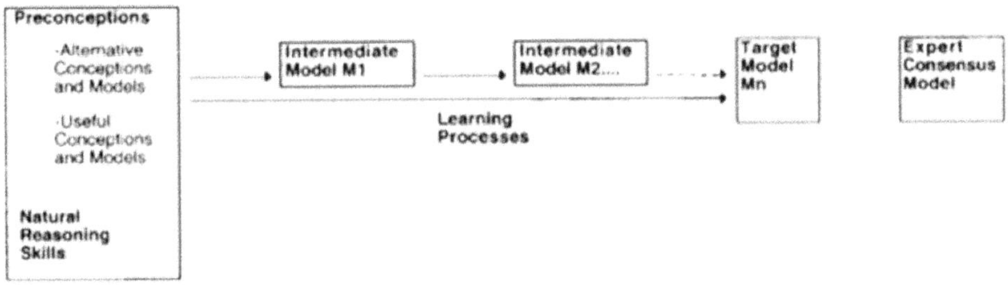

This framework is based on using learners’ preconceptions and natural reasoning skills to help them construct an intermediate model. Then, throughout the learning process, students construct new intermediate models until they reach the target model or desired knowledge level (Clement, 2000).

This framework gives students an opportunity to relate models to the learning process, however, the role of the teachers is challenging. Science teachers still need to be supported in order to use modeling effectively in lessons (Schwarz et al., 2009). Teachers have to participate in the design and development of MbL activities, coordinate implementation of these activities, and also guide and scaffold students in implementation (Fretz et al., 2002; Louca & Zacharia, 2012). However, teachers have limited knowledge about the use of models and modeling in science education. For instance, some teachers tend to use models just to illustrate physical systems that not directly observed (Justi & Gilbert, 2002; Van Driel & Verloop, 2002). Also, Justi and Gilbert (2002) found that teachers generally showed an awareness of the value of models in learning *of* science but not learning *about* science. In other words, when they used modeling in lessons, teachers were concerned with the content of the models more than the nature of models and the modeling process itself as a means for students to construct knowledge (Justi & Gilbert, 2002). Because many science teachers need support to effectively engage their students in scientific modeling, Schwarz et al. (2009) generated a learning progression for scientific modeling. The aim of the learning progression is to make modeling accessible and meaningful for learners.

Figure 2. Interaction of the elements of the practice and metamodeling knowledge
Source: Schwarz et al., 2009

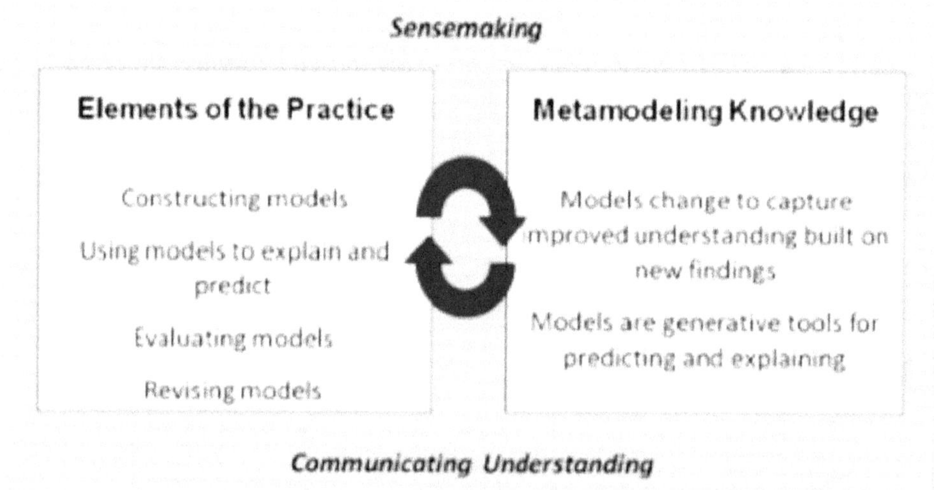

According to Schwarz et al. (2009), modeling is an interaction of two components: the elements of the practice and metamodeling knowledge (Figure 2). The elements of the practice are as follows:

- Students construct an initial model according to their pre-knowledge to illustrate, explain, or predict phenomena.
- Students use this model to illustrate, explain, or predict phenomena.
- Students evaluate the model on whether it is sufficient to represent and explain patterns in phenomena or predict new phenomena.
- Students revise the model to increase its explanatory or predictive power, using additional evidence or aspects of a phenomenon (Schwarz et al., 2009).

The second component, metamodeling knowledge, guides students to take part in the practices. If students are aware that models change to represent improved understanding based on new findings, and that models are producing tools for prediction and explanation, then the students will benefit more from the modeling process (Schwarz et al., 2009).

WHY ARE MODELING AND MODEL-BASED LEARNING IMPORTANT?

Models and modeling, when working toward scientific literacy, are of great importance according to researchers and education policy makers (Gobert & Buckley, 2000; National Research Council [NRC], 2012). According to the NRC (2012) framework:

More sophisticated types of models should increasingly be used across the grades, both in instruction and curriculum materials, as students' progress through their science education. The quality of a student-developed model will be highly dependent on prior knowledge and skill and also on the student's understanding of the system being modeled, so students should be expected to refine their models as their understanding develops. Curricula will need to stress the role of models explicitly and provide students with modeling tools (e.g., Model-It, agent-based modeling such as NetLogo, spreadsheet models), so that students come to value this core practice and develop a level of facility in constructing and applying appropriate models (p. 59).

The National Research Council proposes that teachers not only use models and modeling directly, but also participate in the many other science education programs and research efforts that prioritize students' engagement in modeling practices (Clement, 2000; Forbes, Zangori, & Schwarz, 2015; Louca & Zacharia, 2012; Ministry of National Education [MoNE], 2018; NRC, 2012; Schwarz et al., 2009). Using models and modeling as scientific practices in lessons gives students opportunities to construct their own models and use them to formulate hypotheses, make investigations, explain scientific phenomena, and communicate and justify their ideas. During the learning process, students develop their models, evaluate the models in the light of any new information, and revise their models according to their developing understanding (Namdar & Shen, 2015). This is very beneficial for science education because, as mentioned, integrating scientific practices into lessons is important to developing scientific literacy. Furthermore, modeling supports other skills, such as informal reasoning and decision making (Demir & Namdar, 2019). Through modeling, students find data to support their arguments—this provides a basis to develop informal reasoning and decision-making skills (Demir & Namdar, 2019). These are critical skills for scientifically literate citizens.

Like modeling, model-based learning (MbL) is an innovative and effective approach that serves science education in many ways. Shen, Lei, Chang and Namdar (2014) note three main advantageous features. First, MbL engages students to actively participate in the learning process as they develop, test, and revise their models (similarly to what scientists do in the field). It also encourages the use of different representations and alternative models (i.e., physical models, graphs, or mathematical formulas), so that different learning styles can be supported. Third, MbL increases peer-learning opportunities in the classroom—students develop models together, present their models to the class, and evaluate each other's models. Shen et al., (2014) refer studies that indicates the group of students who participate in co-building modeling activity showed more interaction in group and also performed better in content test (Birchfield & Megowan-Romanowicz, 2009, as cited in Shen et al., 2014, p.535) and collaboration in modeling and challenged by peers improve students' understandings (Simpson et al., 2006, as cited in Shen et al., 2014, p.535).

Modeling is one of the more complex scientific practices; it offers an opportunity for students to participate in other connected practices such as analyzing data and constructing explanations (Duschl et al., 2016; Louca & Zacharia, 2012; Sadler, Friedrichsen, Zangori, 2019). The model-based learning approach can also be used for assessment (Namdar & Shen, 2015). Modeling-oriented assessment is a dynamic process including three interrelated dimensions: assessment of modeling products, assessment of modeling practices, and assessment of meta-modeling knowledge, according to the Schwarz et al. (2009) modeling schema (Namdar & Shen, 2015).

INTEGRATING MBL AND SOCIO-SCIENTIFIC ISSUES

Using SSI in teaching and learning have been successful in supporting students to understand science concepts (Klosterman & Sadler, 2010; Sadler, Barab, & Scott, 2007), increase students' interest in learning science (Albe, 2008; Harris & Ratcliffe, 2005), develop students' reasoning skills (Sadler, 2009) and also improve students' argumentation skills (Dawson & Venville, 2013). However, as Zangori, Peel, Kinslow, Friedrichsen and Sadler (2017) stated that "with the exception of the practice of argumentation, the SSI approach has generally not been associated with promoting student engagement in scientific practices" (p.4). However, as stated in the previous section, education programs (e.g., MoNE, 2018; NRC, 2012; NGSS Lead States, 2013) gave priority to science and engineering practices such as modeling.

Zangori et al. (2017) embedded an SSI into a model-based learning unit to promote students' understanding of climate change and engagement in model-based reasoning.

Models were developed to answer the question, "Where does carbon come from and where does it go?" and then used in a debate about the climate change issue (Zangori et al., 2017, p. 7). Students used the Schwarz et al. (2009) SSI modeling cycle and created models at three time-points during the unit. They developed their initial models to reflect their pre-knowledge of and experiences with this issue. They used their models to make sense of their understanding about the issue at that time. As the learning process continued, students' prior knowledge was challenged and so students evaluated their models. Next, they revised the models to reflect their new understanding. In order to help this process, they answered the following questions:

- *How do you think carbon moves? How do you know?*
- *On a scale of 1 to 5, how well does your model explain carbon movement?*
- *Why did you give your model this rating? In what ways should your model be revised?*
- *Discuss your understanding of how the carbon cycle works when all variables are balanced and how changing certain variables might affect other parts of the carbon cycle and the overall system on earth* (Zangori et al., 2017, p. 8).

Learning takes place through the dialogue between the student and the model because the student uses, evaluates, and revises the model continuously in the light of new information (Zangori et al., 2017). This successful integration of MbL and SSI led to the development of model-oriented issue-based teaching.

A PROFESSIONAL DESIGN CYCLE TO SUPPORT MODEL-ORIENTED ISSUE-BASED (MOIB) TEACHING

Model-oriented issue-based (MOIB) teaching integrates SSI and model-based reasoning and teaching (Zangori, Foulk, Sadler, & Peel, 2018). Current education programs emphasize using science and engineering *practices* to support students' understanding of how scientific knowledge is developed (MoNE, 2013, 2018; NRC, 2012; NGSS Lead States, 2013). Moreover, science and engineering practices are crucial for the development of students' scientific reasoning skills (Peel, Zangori, Friedrichsen, Hayes, & Sadler, 2019; Zangori et al., 2018). However, as Sadler (2004) pointed out, SSI instruction has not been adequately associated with these practices, focusing more on the interaction between science and society (Peel, Sadler, Andrew, Zangori, & Friedrichsen, 2017; Sadler, 2004; Zangori et al., 2017). Interestingly, recent research claims that students' skills in scientific practices such as developing, processing, and concluding ideas in fact increases their ability to comprehend the interactions between science and society (Peel et al., 2017; Peel,

Zangori, et al., 2019; Zangori et al., 2017, 2018). Therefore, Zangori et al. (2017) theorized that science and engineering practices are essential to SSI instruction. After Zangori et al.'s (2017; 2018) and Peel et al.'s (2017) and Peel, Zangori et al. (2019) conclusions, Peel, Nguyen, Friedrichsen, Sadler, and Zangori (2019) presented a framework for MOIB teaching.

Figure 3. Model-oriented issue-based (MOIB) teaching model
Source: MOIB Model | RI[2], 2019 and Peel, Nguyen, et al., 2019

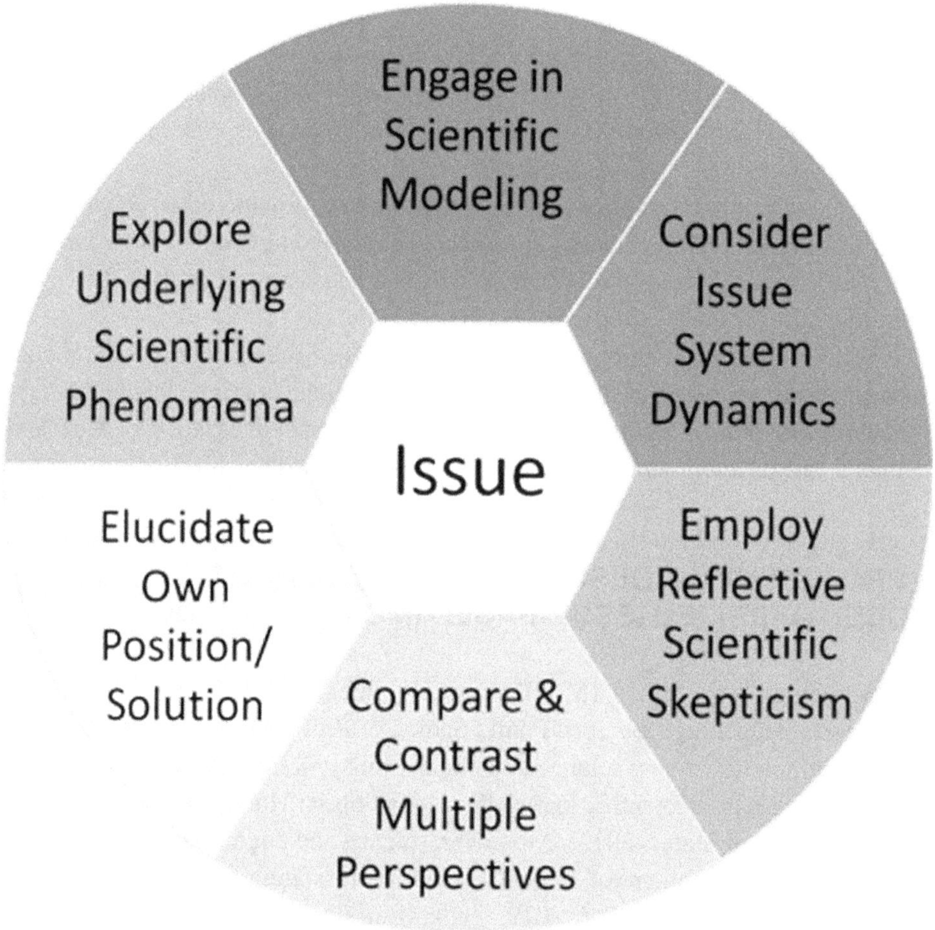

According to the MOIB teaching model, students engage in modeling related to a focal SSI to make sense of the science phenomenon involved in understanding and explaining the issue. Students examine the science phenomenon and go through the modeling steps to develop, use, evaluate, and revise their models in order to comprehend the science behind the focal SSI. This process allows students to apply their science knowledge to the issue. They can investigate the complexities, perspectives, and factors of the issue. The integration of model-based reasoning and SSI-based instruction allows students to develop meaningful and science-relevant perspectives toward real-world issues (Peel, Zangori, et al., 2019; Zangori et al., 2018).

A FRAMEWORK FOR TEACHING FOR SOCIO-SCIENTIFIC ISSUE AND MODEL-BASED LEARNING (SIMBL)

Mirroring the learning process itself, the tools and methods used in this process change. In the past few years frameworks for SSI-based instruction were developed (Friedrichsen, Sadler, Graham, & Brown, 2016; Presley et al., 2013; Sadler, 2011; Sadler, Foulk, & Friedrichsen, 2017). The most current framework for SSI teaching and learning (Sadler et al., 2017) can be summed up in three stages: 1) encountering the focal issue at the beginning; 2) developing science ideas, practices, and socio-scientific reasoning in the context of the focal issue; and 3) synthesizing ideas, practices, and reasoning through a culminating activity. In the past few years, Sadler et al. (2019) have focused on understanding the challenges teachers face when using SSI and on developing evolving frameworks for SSI-based instruction (Friedrichsen et al., 2016; Presley et al., 2013; Sadler, 2011; Sadler et al., 2017). Then, Sadler et al. (2019) stated that working with this sequence led them to three conclusions:

First, applying primary focus to one scientific practice, in our case scientific modeling, may be more supportive of teacher learning and classroom practices than highlighting all eight NGSS practices simultaneously. Second, socio-scientific reasoning needs to be further unpacked in order to be useful for most teachers. Finally, for many teachers who are naturally inclined to modify and adapt teaching and learning materials, a prescriptive sequence may not afford enough flexibility (p. 22).

Based on these conclusions, Sadler et al. (2019) provided a different way to integrate modeling into SSI-based instruction. After working with teachers in professional development programs (see Sadler et al., 2019 and Zangori et al., 2018), researchers determined a need for a tool created for teachers who design SSI instruction. This style of teaching demands the use of practices such as argumentation and modeling. When integrating MbL and SSI, these epistemic practices that teachers struggle with are potentially the most productive. Therefore, Sadler et al. (2019) designed a framework that does not cover all scientific practices but instead focuses on modeling.

This framework consists of six interrelated features, like the MOIB model, that define the nature of SIMBL (Figure 4). Similarly, SIMBL emphasizes the centrality of the SSI in the approach and the need for continuous connections with the issue throughout the curriculum unit (Sadler et al., 2019). The six features are: 1) explore underlying scientific phenomena, 2) engage in scientific modeling, 3) consider issue system dynamics, 4) employ information and media literacy strategies, 5) compare and contrast multiple perspectives, and 6) elucidate own position/solution. The ordering of the features given in this framework may vary, depending on the specific needs of a class or the characteristics of the SSI.

In the initial SSI teaching and learning frameworks (Sadler et al., 2017), presenting the focal issue at the beginning and then connecting the issue to science ideas and societal concerns were the first step. Then, in the context of the focal issue, science ideas and practices were interconnected with socio-scientific reasoning process. The last step was synthesizing the ideas, practices, and reasoning through an issue-focused culminating activity (Sadler et al., 2017). However, the purpose of the SIMBL is primarily to encourage teachers to include each of the essential elements in their SSI teaching, so Sadler et al. (2019) do not recommend a specific order for these features in SIMBL. Teachers have the ability to sort the key features of the unit and adjust the time allocated to each feature. As explained by Sadler et al. (2009), it gives teachers a flexible tool in SSI teaching. As a result, the only visible change between the two frameworks (Figure 3 and 4) is replacing "Employ Reflective Scientific Skepticism" to "Employ Information and Media Literacy Strategies." Thus, though the SIMBL is different from other SSI teaching and learning frameworks in both use and aim, there is no need to compare and contrast MOIB and SIMBL. Instead, the MOIB cycle can be viewed as the one version of the SIMBL model.

Figure 4. The socio-scientific issue and model-based learning (SIMBL) framework
Source: Sadler et al., 2019, p. 17.

The following sections provide a description of each key SIMBL feature and examples from two SSI units implemented by teachers working with Sadler et al. (2019). The Vanishing Prairie unit was designed for high school students; it focuses on climate change and ideas about carbon cycling and ecological interactions. The MONARCH (Modeling Natural Systems, Restoration, and Conservation of Habitat) unit was designed for elementary school students, and it focused on investigating the reduction of the population of monarch butterflies as a result of habitat loss (Sadler et al., 2019).

Explore Underlying Scientific Phenomena

This feature may be the most familiar to science teachers because it emphasizes the science content related to the selected SSI. In teaching science content, teachers should provide students opportunities to explore the scientific phenomenon related to the focal issue. The idea of engaging students' investigations by linking science concepts to a phenomenon comes from reform-oriented science teaching (Sadler et al., 2019). For this investigatory process, it is recommended that teachers use anchor phenomena that are "relevant to students' everyday experiences; observable; complex; have associated data, text and images; and have a stakeholder community or audience interested in the findings" (NGSS Lead States, 2013; Sadler et al., 2019, p. 18). The field trip was served as an anchor phenomenon for both Vanishing Prairie and MONARCH units which are example projects of Sadler et al. (2019). In the Vanishing Prairie unit students visit the prairie and made activities includes collecting data about soil, identifying native plants, estimating biodiversity, etc. In the MONARCH unit students take field trip to explore monarch habitat.

Engage in Scientific Modeling

Throughout the history of science education, models' role in lessons evolved from learning via textbooks or teachers to learning through models constructed by students. Although the SIMBL framework includes many forms of modeling, both through online platforms and mathematical modeling, it is ideal to include a modeling packet -including different forms of modeling- in which students develop their own models in response to a question or problem targeting scientific phenomena. After the first drawing session, students may rate their models and explain why they gave that score to their drawing. In lower grades, students are asked to use colored pencil to label their first model with what should be changed according to their new knowledge. Then, they draw a second model. This process is repeated at the end of the unit by asking the students to evaluate and rate their second models and draw a final model. Evaluating and revising is a very important step in using models. Students realize that their ideas are dynamic learning tools, not static entities (Sadler et al., 2019).

Consider Issue System Dynamics

Learning about system dynamics is a required skill for students; however, there is not sufficient support for its instruction. Therefore, in SSI-based learning, when students think about a system, they also think about both the scientific elements and the social elements of the issue. Social elements may be political, economic, ethical, or religious considerations. Within the SIMBL framework, discussion about complex systems should be made explicit so that students can take into account the parts of each system that interact, impact, and/or affect other systems. For example, in the Vanishing Prairie unit, students are asked to consider how their strategies to minimize climate change (i.e., reduce atmospheric carbon emissions) interact with economic and political effects. In the MONARCH unit, elementary school students are encouraged to think about the biological needs of butterflies, but also their own desire for a play area (Sadler et al., 2019).

Employ Information and Media Literacy Strategies

Socio-scientific issues are contemporary issues and are therefore open to change with developments in ongoing research, new ideas, and new perspectives. The information surrounding the SSI is dynamic, unlike other science subjects, which are relatively static and easily represented in textbooks. Because information about SSI is constantly changing or increasing, more dynamic sources of information are needed. Accordingly, the use of media tools is necessary for students to access relevant information, but this skill is not taught in traditional science education. Students need to be able to interpret the media and make decisions about the quality of information.

When teaching SSI, Sadler et al. (2019) make suggestions about media literacy and encourage students to think about the following four factors: "1) meanings and messages of an information report, 2) authors and their intent in preparing information reports, 3) the intended audiences, and 4) potential biases that may influence the generation of the information or the ways in which it is reported" (p. 20). They provide students with a list of questions labeled *Know Your Resources* and students are encouraged to question every report or website they interact with. Some questions from the list are:

- Who is the author or organization disseminating this information?
- What is the purpose of the publication?
- What expertise and/or relevant experiences does the author have?
- What biases could affect the presentation of the information? (Sadler et al., 2019, p. 20).

Compare and Contrast Multiple Perspectives

This feature refers to an important skill: the ability to think beyond one's own perspective. It is also a characteristic of socio-scientific reasoning and a crucial step of developing empathy (Kahn & Zeidler, 2017; Sadler et al., 2019). Socio-scientific issues are ill-structured and complex problems and open to multiple solutions (Sadler, 2004). This provides students the opportunity to exercise multiple-perspective-taking (Sadler et al., 2019). In order to introduce multiple perspectives to students, some activities could be designed specifically to fit the unit. For example, Sadler et al. (2019) presented a multiple-perspective-taking activity in the MONARCH unit. In that unit, the case was turning a school soccer field into a butterfly garden; elementary school students were asked to think about how student athletes and conservationists may react to different outcomes (Sadler et al., 2019).

Elucidate Own Position/Solution

In this key feature, students are asked to defend their position and/or offer a solution to the focal issue. Generally, this step is introduced at the beginning of the unit to allow teachers to reveal students' initial positions and re-visit them at the end of the unit. Usually in SIMBL units, a culminating activity is designed that enables students to develop a written argument that defends their positions and/or offers a solution. When students write their cases, they are also expected to synthesize the science embedded in the issue. Sadler et al. (2019) emphasized that this culminating activity should be handled after the students' exercise in taking multiple perspectives. Clarifying their perspective is important; it serves a scientific literacy aim by giving the students opportunities to create solutions, and to defend their suggestions for complex problems, and to learn science in a context similar to what they will face as citizens (Sadler, 2011; Sadler et al., 2019).

CONCLUSION

In science education, scientific literacy is expressed as a goal by many researchers (Bybee, 1997; Laugksch, 2000; Ogunkola, 2013; Roberts & Bybee, 2014). According to Grant and Lapp (2011), to achieve scientific literacy, teachers should implement four main strategies: 1) identify interesting science topics related to the real world and integrate them into lessons; 2) familiarize students with how science is produced; 3) make it a habit for students to use science practices and to link their prior knowledge to new situations; and 4) familiarize students with using multiple sources and evaluating their quality. The use of the Socio-Scientific Issue and Model Based Learning (SIMBL) framework (Sadler et al., 2019) in lessons would meet the standards by Grant and Lapp (2011) and highly support the promotion of scientifically literate citizens. Unlike previous SSI frameworks (Friedrichsen et al., 2016; Presley et al., 2013; Sadler, 2011; Sadler et al., 2017), this framework (Sadler et al., 2019) is flexible in terms of the sequences and time allocation. Based on classroom community dynamics and focal issue characteristics, the lessons may be started from any feature and the allotted time is adapted accordingly.

In the end, the main difference of the SIMBL framework (Sadler et al., 2019) and MOIB cycle (Peel, Nguyen, et al., 2019) from previous SSI teaching and learning frameworks (Friedrichsen et al., 2016; Presley et al., 2013; Sadler, 2011; Sadler et al., 2017) is that these most recent frameworks are designed to guide teachers by adding essential features to existing frameworks. This means that SIMBL and MOIB would not replace the SSI-based teaching and learning framework; they are intended exclusively for modeling practice within the scope of that framework (Sadler et al., 2019). As such, these two frameworks (SIMBL and MOIB) have enabled a key scientific practice, modeling, to take place effectively within the framework of socio-scientific issues, the importance of which is indisputable for the goal of scientific literacy.

REFERENCES

Albe, V. (2008). When scientific knowledge, daily life experience, epistemological and social considerations intersect: Students' argumentation in group discussion on a socio-scientific issue. *Research in Science Education*, *38*(1), 67–90. doi:10.100711165-007-9040-2

Bybee, R. W. (1997). *Achieving Scientific Literacy: From Purposes to Practices.* Heinemann.

Clement, J. (2000). Model based learning as a key research area for science education. *International Journal of Science Education*, *22*(9), 1041–1053. doi:10.1080/095006900416901

Dawson, V., & Venville, G. (2013). Introducing high school biology students to argumentation about socioscientific issues. *Canadian Journal of Science. Mathematics and Technology Education*, *13*(4), 356–372. doi:10.1080/14926156.2013.845322

Demir, A., & Namdar, B. (2019). The effect of modeling activities on grade 5 students' informal reasoning about a real-life issue. *Research in Science Education.* Advance online publication. doi:10.100711165-019-09896-8

Duschl, R. A., Bismack, A. S., Greeno, J., Gitomer, D. H., Bismack, A. S., Greeno, J., & Gitomer, D. H. (2016). Coordinating preK-16 STEM education research and practices for advancing and refining reform agendas. In R. A. Duschl & A. S. Bismack (Eds.), *Reconceptualizing STEM education: The central role of practices* (pp. 1–32). Routledge. doi:10.4324/9781315700328

Forbes, C., Zangori, L., & Schwarz, C. (2015). Using models scientifically. *Science and Children*, *53*(2). Advance online publication. doi:10.2505/4c15_053_02_42

Fretz, E. B., Wu, H. K., Zhang, B., Davis, E. A., Krajcik, J. S., & Soloway, E. (2002). An investigation of software scaffolds supporting modeling practices. *Research in Science Education*, *32*(4), 567–589. doi:10.1023/A:1022400817926

Friedrichsen, P. J., Sadler, T. D., Graham, K., & Brown, P. (2016). Design of a socio-scientific issue curriculum unit: Antibiotic resistance, natural selection, and modeling. *International Journal of Designs for Learning*, *7*(1). Advance online publication. doi:10.14434/ijdl.v7i1.19325

Gilbert, J. K. (2004). Models and modelling: Routes to more authentic science education. *International Journal of Science and Mathematics Education*, *2*(2), 115–130. doi:10.100710763-004-3186-4

Gilbert, J. K., & Justi, R. (2016). *Modelling-based teaching in science education*. Springer International Publishing. doi:10.1007/978-3-319-29039-3

Gobert, J. D., & Buckley, B. C. (2000). Introduction to model-based teaching and learning in science education. *International Journal of Science Education*, *22*(9), 891–894. doi:10.1080/095006900416839

Grant, M., & Lapp, D. (2011). Teaching science literacy. *Educational Leadership*, *68*(6).

Harris, R., & Ratcliffe, M. (2005). Socio-scientific issues and the quality of exploratory talk—What can be learned from schools involved in a 'collapsed day'project? *Curriculum Journal*, *16*(4), 439–453. doi:10.1080/09585170500384396

Harrison, A. G., & Treagust, D. F. (1998). Modelling in science lessons: Are there better ways to learn with models? *School Science and Mathematics*, *98*(8), 420–429. doi:10.1111/j.1949-8594.1998.tb17434.x

Harrison, A. G., & Treagust, D. F. (2000). A typology of school science models. *International Journal of Science Education*, *22*(9), 1011–1026. doi:10.1080/095006900416884

Justi, R. S., & Gilbert, J. K. (2002). Modelling, teachers' views on the nature of modelling, and implications for the education of modellers. *International Journal of Science Education*, *24*(4), 369–387. doi:10.1080/09500690110110142

Kahn, S., & Zeidler, D. L. (2017). A case for the use of conceptual analysis in science education research. *Journal of Research in Science Teaching*, *54*(4), 538–551. doi:10.1002/tea.21376

Klosterman, M. L., & Sadler, T. D. (2010). Multi-level assessment of scientific content knowledge gains associated with socioscientific issues based instruction. *International Journal of Science Education, 32*(8), 1017–1043. doi:10.1080/09500690902894512

Laugksch, R. C. (2000). Scientific literacy: A conceptual overview. *Science Education, 84*(1), 71–94. doi:10.1002/(SICI)1098-237X(200001)84:1<71::AID-SCE6>3.0.CO;2-C

Louca, L. T., & Zacharia, Z. C. (2012). Modeling-based learning in science education: Cognitive, metacognitive, social, material and epistemological contributions. *Educational Review, 64*(4), 471–492. doi:10.1080/00131911.2011.628748

Ministry of National Education (MoNE). (2013). *İlköğretim Kurumları (İlkokullar ve Ortaokullar) Fen Bilimleri Dersi (3, 4, 5, 6, 7 ve 8. Sınıflar) Öğretim Programı* [Primary Education Institutions (Primary and Secondary Schools) Science Course (3, 4, 5, 6, 7 and 8th Grades) Curriculum]. Academic Press.

Ministry of National Education (MoNE). (2018). *Fen Bİlimleri Dersi Öğretim Programı (İlkokul ve Ortaokul 3, 4, 5, 6, 7 ve 8. Sınıflar)* [Science Education Curriculum (Primary and Secondary Schools 3, 4, 5, 6, 7 and 8th Grades)]. Academic Press.

MOIB Model | RI[2]. (2019, June 24). http://ri2.missouri.edu/content/MOIB-Model

Namdar, B., & Shen, J. (2015). Modeling-Oriented Assessment in K-12 Science Education: A synthesis of research from 1980 to 2013 and new directions. *International Journal of Science Education, 37*(7), 993–1023. doi:10.1080/09500693.2015.1012185

National Research Council (NRC). (2012). *A Framework for K-12 Science Education: Practices, Crosscutting Concepts, and Core Ideas*. The National Academies Press.

NGSS Lead States. (2013). *The next generation science standards: For states, by states*. National Academies Press.

Ogunkola, B. J. (2013). Scientific literacy: Conceptual overview, importance and strategies for improvement. *Journal of Educational and Social Research, 3*(1), 265–274.

Peel, A., Nguyen, H., Friedrichsen, P., Sadler, T., & Zangori, L. (2019, April 3). *A Professional Design Cycle to Support Model-Oriented Issue-Based Teaching* [Poster]. NARST- 2019 Annual International Conference, Baltimore, MD.

Peel, A., Sadler, T. D., Andrew, K., Zangori, L., & Friedrichsen, P. (2017). Climate change as an issue for socio-scientific issues teaching and learning. In D. P. Shepardson, A. Roychoudhury, & A. S. Hirsch (Eds.), *Teaching and learning about climate change. A framework for educators* (pp. 153–165). Routledge. doi:10.4324/9781315629841-11

Peel, A., Zangori, L., Friedrichsen, P., Hayes, E., & Sadler, T. (2019). Students' model-based explanations about natural selection and antibiotic resistance through socio-scientific issues-based learning. *International Journal of Science Education*, *41*(4), 510–532. doi:10.1080/09500693.2018.1564084

Presley, M. L., Sickel, A. J., Muslu, N., Merle-Johnson, D., Witzig, S. B., Izci, K., & Sadler, T. D. (2013). A Framework for socio-Scientific issues based education. *Science Educator*, *22*(1), 26–32.

Roberts, D. A. (2013). Scientific literacy/science literacy. In *Handbook of research on science education* (pp. 743–794). Routledge.

Sadler, T. D. (2004). Informal reasoning regarding socioscientific issues: A critical review of research. *Journal of Research in Science Teaching*, *41*(5), 513–536. doi:10.1002/tea.20009

Sadler, T. D. (2009). Situated learning in science education: Socio-scientific issues as contexts for practice. *Studies in Science Education*, *45*(1), 1–42. doi:10.1080/03057260802681839

Sadler, T. D. (Ed.). (2011). *Socio-scientific issues in the classroom: Teaching, learning and research*. Springer. doi:10.1007/978-94-007-1159-4

Sadler, T. D., Barab, S., & Scott, B. (2007). What do students gain by engaging in socioscientific inquiry? *Research in Science Education*, *37*(4), 371–391. doi:10.100711165-006-9030-9

Sadler, T. D., Foulk, J. A., & Friedrichsen, P. J. (2017). Evolution of a Model for Socio-Scientific Issue Teaching and Learning. *International Journal of Education in Mathematics. Science and Technology*, *5*(2), 75–87.

Sadler, T. D., Friedrichsen, P., & Zangori, L. (2019). A framework for teaching for socio-scientific issue and model based learning (SIMBL). *Educação e Fronteiras*, *9*(25), 08–26.

Savin-Baden, M., & Major, C. H. (2013). *Qualitative Research: The essential guide to theory and practice*. Routledge.

Schwarz, C. V., Reiser, B. J., Davis, E. A., Kenyon, L., Achér, A., Fortus, D., Shwartz, Y., Hug, B., & Krajcik, J. (2009). Developing a learning progression for scientific modeling: Making scientific modeling accessible and meaningful for learners. *Journal of Research in Science Teaching*, *46*(6), 632–654. doi:10.1002/tea.20311

Shen, J., Lei, J., Chang, H.-Y., & Namdar, B. (2014). Technology-enhanced, modeling-based instruction (TMBI) in science education. In J. M. Spector, M. D. Merrill, J. Elen, & M. J. Bishop (Eds.), *Handbook of research on educational communications and technology* (pp. 529–540). Springer. doi:10.1007/978-1-4614-3185-5_41

Treagust, D., & Harrison, A. (1999). The genesis of effective scientific explanations for the classroom. In J. Loughran (Ed.), *researching teaching methodologies and practices for understanding pedagogy* (pp. 28–43). Routledge.

Treagust, D. F., Chittleborough, G., & Mamiala, T. L. (2002). Students' understanding of the role of scientific models in learning science. *International Journal of Science Education*, *24*(4), 357–368. doi:10.1080/09500690110066485

Van Driel, J. H., & Verloop, N. (2002). Experienced teachers' knowledge of teaching and learning of models and modelling in science education. *International Journal of Science Education*, *24*(12), 1255–1272. doi:10.1080/09500690210126711

Yager, R. E. (1991). The constructivist learning model. *Science Teacher (Normal, Ill.)*, *58*(6), 52.

Zangori, L., Foulk, J., Sadler, T. D., & Peel, A. (2018). Exploring Elementary Teachers' Perceptions and Characterizations of Model-Oriented Issue-Based Teaching. *Journal of Science Teacher Education*, *29*(7), 555–577. doi:10.1080/1046560X.2018.1482173

Zangori, L., Peel, A., Kinslow, A., Friedrichsen, P., & Sadler, T. D. (2017). Student development of model-based reasoning about carbon cycling and climate change in a socio-scientific issues unit. *Journal of Research in Science Teaching*, *54*(10), 1249–1273. doi:10.1002/tea.21404

Section 6
Concluding Remarks

Chapter 11
Unifying Themes in Socioscientific Issues-Based Instruction for Scientific Literacy Development

Wardell A. Powell
https://orcid.org/0000-0002-3486-2077
Framingham State University, USA

ABSTRACT

This chapter presents the unifying themes in socioscientific issues-based instruction for scientific literacy development. Section 1 presented an overview of how to effectively implement socioscientific issues in the elementary grades to provide students with opportunities to apply science to their everyday lives. Section 2 built upon where the authors left off in section one. In this section, the authors used real-world scientific context to provide opportunities to use character and values and moral reasoning as they think about finding solutions to real-world scientific problems. Section 3 showed the continued use of socioscientific issues with an upward trajectory to enhance scientific literacy at the college level. Section 4 demonstrated socioscientific issues being successfully implemented at the core of the P-12 educational system. In Section 5, the authors revealed the integrative nature among STEM, model-based learning, and socioscientific issues in achieving scientific literacy.

DOI: 10.4018/978-1-7998-4558-4.ch011

INTRODUCTION

The COVID-19 Pandemic highlights the urgent need to enhance scientific literacy among the American public in general and school-age children in particular. As thousands of Americans succumbed to contracting the Coronavirus, many policymakers in the United States federal and state governments seem to disregard known scientific principles, practices, and findings regarding the type of medication to use and the scientific processes required to produce vaccines. Some of these policymakers even advocated for widespread use of antimalarial medicines such as Hydroxychloroquine to treat COVID-19, asking the people what do they have to lose by not taking Hydroxychloroquine, when leading scientists are saying potentially one's life because of the harmful side effects of this medication. The Center for Disease Control and Prevention had to put out bulletins to warn the public against ingesting household cleaners and disinfectants to treat the COVID-19. Many in government even disregard the warnings of scientists on the wearing of face masks and social distancing as the United States move towards re-opening its economy. These dangerous suggestions by policymakers highlight the urgency needed to develop scientific literacy among the public in general, and school-age children in particular. The most recent PISA report defined scientific literacy as follows:

Scientific literacy is the ability to engage with science-related issues, and with the ideas of science, as a reflective citizen. A scientifically literate person is willing to engage in reasoned discourse about science and technology, which requires the competencies to explain phenomena scientifically, evaluate and design scientific enquiry, and interpret data and evidence scientifically. (OECD, 2018, p. 75)

PISA's definition of scientific literacy features three competencies. These are:

1. Ability to explain phenomena scientifically
2. The ability to interpret data and evidence scientifically
3. Ability to evaluate and design scientific enquiry

The socioscientific issues framework is critical to helping students achieve the identified scientific literacy competencies. Socioscientific issues ill-structured nature enables students to engage in discourse on real-world scientific problems that are socially-related (Zeidler, 2014). Studies have shown that socioscientific issues enhance students' abilities to explain phenomena scientifically (Kosterman & Sadler, 2009; Nielson, 2012; Sadler, Klosterman, & Topcu, 2011; Zeidler, Sadler, Simmons, & Howes, 2005). Additionally, socioscientific issues have shown to impact students' ability to evaluate and design scientific inquiry (Zeidler, Applebaum, &

Sadler, 2011; Bencze, Sperling, & Carter, 2011), in addition to enhancing students' ability to interpret data and evidence scientifically (Powell, 2019 & 2020: Zeidler, Sadler, Applebaum, & Callahan, 2009). Furthermore, socioscientific issues have the potential to enhance students' moral and character development (Owens, Sadler, & Zeidler, 2017; Zeidler, 2014), which are essential blueprints in helping students to make sound decisions on issues requiring morality. Additionally, it is important to keep in mind that socioscientific issues goes beyond the classroom, because by its very nature it enables *all* to consider virtue as they engage in real-world scientific problems. The COVID-19 Pandemic is a case in point that requires our elected officials to keep virtue at the forefront of their decisions. The general public needs our elected officials to be trustworthy, caring, courageous, honest, and fair, as they use wisdom to solve this pandemic. The socioscientific issues framework is effectively poised to fulfill the scientific literacy advocated by PISA and provide more for the students, whether we are talking about the COVID-19 Pandemic, in whose communities should natural gas pipelines run, where to build landfills, should we consume genetically modified foods and climate change among many other issues. The authors of this volume are aware of the relationships between socioscientific issues and scientific literacy and have all contributed to sharing the many ways in which they used socioscientific issues to enhance scientific literacy among their students.

Socioscientific Issues and Achieving Scientific Literacy

PISA's approach to scientific literacy has a lot of commonalities to the socioscientific issues framework. According to Sadler and Zeidler (2009), both "PISA's approach and the socioscientific issues framework to scientific literacy emphasize preparing students for life and citizenship, complex reasoning and reflective practices, and robust understandings of the nature of science, particularly as it is practiced in society" (p. 1). However, the socioscientific issues framework for achieving scientific literacy is much more robust as it considers moral development (Zeidler, 2014; Sadler, 2011; Zeidler et al., 2005), affective emotions (Hodson, 2006; Kolsto, 2001b; Saunders and Rennie, 2013; Wu & Tsai, 2011), character education and development (Zeidler, Sadler, Simmons, & Howes, 2005), and sociomoral discourse (Zeidler, 2014). This view encapsulates functional scientific literacy (Zeidler & Keefer, 2003), which serves to connect real-world scientific problems to the science education curriculum. What good would it be to determine that someone is scientifically literate (i.e., abilities to explain phenomena scientifically, to interpret data and evidence scientifically, and to evaluate and design scientific inquiry), if that person isn't equipped to consider the moral atrocities that are committed against marginalized population who are forced to drink polluted water? The proposed notion of scientific literacy that is advocated by the socioscientific issues framework includes opportunities for students to consider moral and ethical factors as they make decisions about the preservation of the health of the general public and the environment (Sadler, 2011; Zeidler, Herman, & Sadler, 2019). Developing scientific literacy among students will require teachers to provide opportunities for their students to engage in sociomoral discourses that enhance their students' abilities to empathize and consider others' perspectives as they develop their character.

Clearly, our responses to climate change, pollution, ecosystem destructions, the current COVID-19 Pandemic, to name a few real-world scientific issues, are not void of morality, sociomoral discourse, character development, cognitive development, affective emotions and the nature of science. These are all needed to achieve the level of scientific literacy need to solve problems. Figure 1 below provides a visual of factors to advance the level of scientific literacy that is needed to solve real world problems.

Figure 1. Socioscientific issues framework for advancing PISA's scientific literacy

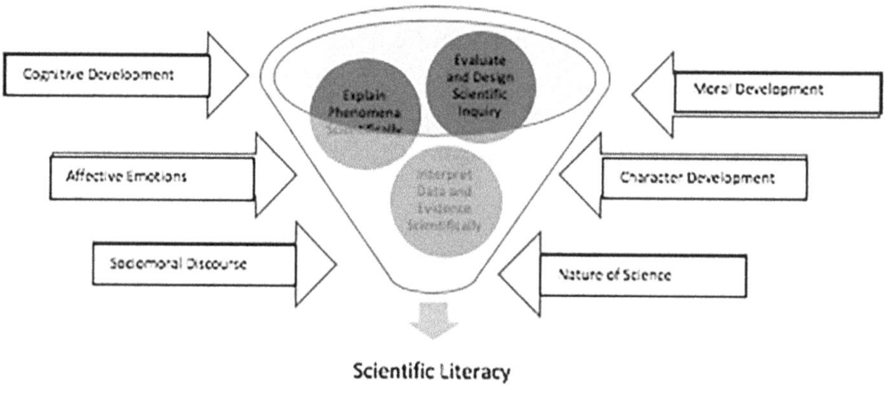

Unity in Using Socioscientific Issues to Achieving Scientific Literacy

The authors of this volume have all shared their expertise and creativity in advancing scientific literacy through the socioscientific issues framework. Consequently, they are united in their beliefs that moral development, character education, nature of science, sociomoral discourse, affective emotions and cognitive development are essential to developing all students' abilities to explain phenomena scientifically, interpret data and evidence scientifically, and evaluate and design scientific inquiry.

Section I presented an overview of how to effectively implement socioscientific issues in the elementary grades to provide students with opportunities to apply science to their everyday lives.

Additionally, science is often taken a back seat to subjects such as reading and math. This resulted in many elementary students to not experience much science until the upper primary level when it's time to take high stakes states' elementary science examination. In fact, in some regions of the United States, science is not given equality values as mathematics and English Language Arts in the schools' accountability reports. This does not advance much hope of laying a firm foundation on which to develop scientific literacy among elementary students. The authors in this section indeed show that no child is too young to do science.

Section II built upon where the authors left off in section one. In this section, the authors used real-world scientific context to provide opportunities to use character and values and moral reasoning as they think about finding solutions to real-world scientific problems. The authors in this section provided some blueprints of how to develop and implement effective socioscientific issues instruction with middle grades students.

Section III showed us the continued use of socioscientific issues with upward trajectory to enhance scientific literacy at the college level. In this section, the authors showed the relationships between socioscientific reasoning and scientific literacy in the undergraduate classroom. Additionally, the authors showed the close association between socioscientific issues and the nature of technology, and how these can be used in the undergraduate classroom to leverage students in finding solutions for real-world problems.

Section IV demonstrated socioscientific issues being successfully implemented at the core of the P-12 educational system. If preservice teachers have a positive perception of socioscientific issues and its effect in advancing scientific literacy among their students, then they will be more open to provide instructional opportunities to engage in argumentation and sociomoral discourse activities on real world community based scientific issues. This will certainly advance the level of scientific literacy advocated throughout the text.

In Section V, the authors revealed the integrative nature among STEM, Model-Based Learning, and socioscientific issues in achieving scientific literacy. In order to prepare students as future global citizens, advocates for STEM have called on all who are involved in research, policy development, and teaching to make STEM education a priority (ICASE, 2013). However, we have seen that minority groups in general, and black female, in particular, are marginalized at the predominantly white institutions (Dorch & Patel, 2017), and are underrepresented in STEM (McGee & Bentley, 2017). Model-based learning within the socioscientific issues framework facilitates the constructivist learning approaches for women and people of color to enter the academy.

In Closing

We have argued the merits of using socioscientific issues to enhance scientific literacy among all students. As we conclude this volume, I thought it is fitting to end with the following:

The main criterion for evaluating age, youth, and old age cannot be that of the calendar.

No one is old just because he or she was born a long time ago or young just because he or she was born a short time ago. People are old or young much more a function of how they think of the world, the availability they have for curiously giving themselves to knowledge (Freire, 2004a, p, 72).

Freire's quote reinforces what this volume is about as we have seen the contributing authors artfully using their creativity and skills to engage ***all*** students in socioscientific issues instruction. These instructional practices enable students to think of the world, the affinity they have for nature, and giving themselves knowledge through the lens of moral reasoning, character development, emotive reasoning, cognitive development, and the nature of science as they ponder solutions to real-world scientific problems. We have seen students express their willingness to consider the perspectives of others as they decide on scientific issues that go against their beliefs. For example, when determining if scientists should clone animals for food to prevent malnutrition and deaths, students make comments such as, "*My religion teaches that God is the creator of life. Although cloning goes against my religion, I imagine how they feel. As a Christian, I believe it is my job to serve and help people when they are in need. Others have helped us, so we should help now in whatever ways we can.*"

This is an example of the type of virtue, a sense of caring that enables one to be courageous to go against entrenched beliefs that effective socioscientific issues instruction develops among students.

REFERENCES

Bencze, L. L., Sperling, E., & Carter, L. (2011). Students' research informed socio-scientific activism: Re/Visions for a sustainable future. *Research in Science Education*, *42*(1), 129–148. doi:10.100711165-011-9260-3

Dortch, D., & Patel, C. (2017). Black undergraduate women and their sense of belonging in STEM at predominantly White institutions. *NASPA Journal About Women in Higher Education*, *10*(2), 202–215. doi:10.1080/19407882.2017.1331854

Freire, P. (2004). *Pedagogy of indignation*. Paradigm Publishers.

Hodson, D. (2006). Why we should prioritize learning about science. *Canadian Journal of Science. Mathematics and Technology Education*, *6*(3), 293–311. doi:10.1080/14926150609556703

ICASE. (2013). The Kuching Declaration. *Proceeding of the World Conference on Science and Technology Education (WorldSTE2013)*. Available online at: http:// www.icaseonline.net/ICASE%20Kuching%20Declaration_Final. pdf

Kolstø, S. D. (2001a). Scientific literacy for citizenship: Tools for dealing with the science dimension of controversial socioscientific issues. *Science Education*, *85*(3), 291–310. doi:10.1002ce.1011

Kosterman, M. L., & Sadler, T. D. (2010). Multi-level assessment of scientific content knowledge gains associated with socioscientific issues-based instruction. *International Journal of Science Education*, *32*(8), 1017–1043. doi:10.1080/09500690902894512

McGee, E. O., & Bentley, L. (2017). The Troubled Success of Black Women in STEM. *Cognition and Instruction*, *35*(4), 265–289. doi:10.1080/07370008.2017.1355211

Nielson, J. A. (2012). Science in discussion: An analysis of the use of science content in socioscientific discussion. *Science Education*, *96*(3), 428–456. doi:10.1002ce.21001

OECD. (2018). *PISA for development assessment and analytical framework: Reading, Mathematics and Science*. OECD.

Owens, D. C., Sadler, T. D., & Zeidler, D. L. (2017). Controversial issues in the science classroom. *Phi Delta Kappan*, *99*(4), 45–49. doi:10.1177/0031721717745544

Powell, W. A. (2020). Framework for Using Geospatial Technology to Promote Middle School Students' Critical Thinking on Socioscientific Issues. In *Next Generation Digital Tools and Applications for Teaching and Learning Enhancement* (pp. 47–78). IGI Global. doi:10.4018/978-1-7998-1770-3.ch004

Powell, W. A., & Fuchs, D. (2019). Using Socioscientific Issues to Enhance Evidence-Based Reasoning Among Middle School Students. In S. Robinson & V. Knight (Eds.), *Handbook of Research on Critical Thinking and Teacher Education Pedagogy* (pp. 150–176). IGI Global. doi:10.4018/978-1-5225-7829-1.ch009

Sadler, T., Klosterman, M., & Topcu, M. (2011). Learning science content and socio-scientific reasoning through classroom explorations of global climate change. In T. D. Sadler (Ed.), *Socio-scientific issues in the classroom: Teaching, learning and research* (pp. 45–77). Springer. doi:10.1007/978-94-007-1159-4_4

Sadler, T. D. (2011). Situating socio-scientific issues in classrooms as a means of achieving goals of science education. In *Socio-scientific issues in science classrooms: Teaching, learning and research* (pp. 1–9). Springer. doi:10.1007/978-94-007-1159-4_1

Sadler, T. D., & Zeidler, D. L. (2009). Scientific literacy, PISA, and socioscientific discourse: Assessment for progressive aims of science education. *Journal of Research in Science Teaching*, *46*(8), 909–921. doi:10.1002/tea.20327

Saunders, K. J., & Rennie, L. J. (2013). A pedagogical model for ethical inquiry into socioscientific issues in science. *Research in Science Education, 43*(1), 253–274. doi:10.100711165-011-9248-z

Wu, Y. T., & Tsai, C. C. (2011). High school students' informed reasoning regarding a socio-scientific issue, with relation to scientific epistemological beliefs and cognitive structures. *International Journal of Science Education, 33*(3), 371–400. doi:10.1080/09500690903505661

Zeidler, D. L. (2014). Socioscientific Issues as a Curriculum Emphasis: Theory, Research, and Practice. In N. G. Lederman & S. K. Abell (Eds.), *Handbook of Research on Science Education* (Vol. 2, pp. 697–726). Routledge.

Zeidler, D. L., Applebaum, S. M., & Sadler, T. D. (2011). Enacting a socioscientific issues classroom: Transformative transformations. In T. D. Sadler (Ed.), *Socio-scientific issues in science classrooms: Teaching, learning and research* (pp. 277–306). Springer. doi:10.1007/978-94-007-1159-4_16

Zeidler, D. L., Herman, B. C., & Sadler, T. D. (2019). New directions in socioscientific issues research. *Disciplinary and Interdisciplinary Science Education Research, 1*(11), 1–9. doi:10.118643031-019-0008-7

Zeidler, D. L., & Keefer, M. (2003). The role of moral reasoning and the status of socio-scientific issues in science education. In D. L. Zeidler (Ed.), *The role of moral reasoning on socio- scientific issues and discourse in science education* (pp. 7–38). Kluwer Academic Publishers.

Zeidler, D. L., Sadler, T. D., Applebaum, S., & Callahan, B. E. (2009). Advancing reflective judgment through socioscientific issues. *Journal of Research in Science Teaching*, *46*(1), 74–101. doi:10.1002/tea.20281

Zeidler, D. L., Sadler, T. D., Simmons, M. L., & Howes, E. V. (2005). Beyond STS: A research-based framework for socioscientific issues education. *Science education*, *89*(3), 357–377. doi:10.1002ce.20048

Compilation of References

Abd-El-Khalick, F. (2003). Socioscientific issues in pre-college science classrooms. In D. L. Zeidler (Ed.), *The role of moral reasoning on socioscientific issues and discourse in science education* (pp. 41–61). Springer. doi:10.1007/1-4020-4996-X_3

Abd-El-Khalick, F. (2006). Socioscientific issues in pre-college science classrooms. In D. L. Zeidler (Ed.), *The role of moral reasoning and discourse on socioscientific issues in science education* (pp. 41–61). Springer.

Abell, S. K., Appleton, K., & Hanuscin, D. L. (2010). *Designing and Teaching the Elementary Science Methods Course*. Routledge. doi:10.4324/9780203859131

ABET. (2019). *Criteria for accrediting engineering programs*. Retrieved from https://www.abet.org/accreditation/accreditation-criteria/accreditation-policy-and-procedure-manual-appm-2019-2020/

Adıgüzel, A. (2015). *Eğitimde yaratıcı drama* [Creative drama in education]. Pegem Akademi.

Adıgüzel, H. Ö. (2006). Yaratıcı drama kavramı, bileşenleri ve aşamaları [The concept, components and stages of creative drama]. *Yaratıcı Drama Dergisi, 1*(1), 17–27.

Agell, L., Soria, V., & Carrio, M. (2015). Using role play to debate animal testing. *Journal of Biological Education, 49*(3), 309–321. doi:10.1080/00219266.2014.943788

Aikenhead, G. S. (2007). Expanding the research agenda for scientific literacy. In *Promoting scientific literacy: Science education research in transaction, Proceedings of the Linnaeus Tercentenary Symposium held at Uppsala University*, (pp. 64–71). Uppsala: Geotryckeriet.

Aikenhead, G., Orpwood, G., & Fensham, P. (2011). Scientific literacy for a knowledge society. In Exploring the landscape of scientific literacy. New York, NY: Routledge / Taylor & Francis Group.

Aikenhead, G. S. (2006). *Science education for everyday life: Evidence-based practice*. Teachers College Press.

Ainsworth, S. (2008). The educational value of multiple-representations when learning complex scientific concepts. In J. K. Gilbert, M. Reiner, & M. Nakhleh (Eds.), *Visualization: Theory and practice in science education* (pp. 191–208). Springer. doi:10.1007/978-1-4020-5267-5_9

Akkocaoğlu Çayır, N., & Erdoğan, T. (2017). Dramada planlama [Planning in drama]. In T. Erdoğan (Ed.), *Okul öncesinden ilköğretime kuramdan uygulamaya drama* [Drama from preschool to elementary, theory to practice]. Eğiten Kitap.

Albe, V. (2008). Students' positions and considerations of scientific evidence about a controversial socioscientific issue. *Science and Education, 17*(8-9), 805–827. doi:10.100711191-007-9086-6

Albe, V. (2008). When scientific knowledge, daily life experience, epistemological and social considerations Intersect: Students' argumentation in group discussions on a socio-scientific issue. *Research in Science Education, 38*(1), 67–90. doi:10.100711165-007-9040-2

American Association for the Advancement of Science (AAAS). (1990). *Science for all Americans.* New York, NY: Oxford University Press.

American Association for the Advancement of Science (AAAS). (1993). *Benchmarks for science literacy.* New York, NY: Oxford University Press.

American Association for the Advancement of Science. (1989). *Science for all Americans: a Project 2061 Report on Literacy Goals in Science, Mathematics and Technology.* AAS Publishing. www.project2061.org/publications/sfaa/online/sfaatoc.htm

Anderson, D., & Zhang, Z. (2003). Teacher perceptions of field trip planning and implementation. *Visitor Studies Today, 6*(3), 6–11.

Appleton, K., & Kindt, I. (1999). Why teach primary science? Influences on beginning teachers' practices. *International Journal of Science Education, 21*(2), 155–168. doi:10.1080/095006999290769

Archila, P. (2017). Using drama to promote argumentation in science education. *Science & Education, 26*(3–4), 345–375. doi:10.100711191-017-9901-7

Arık, M., & Topçu, M.S. (2020). Implementation of engineering design process in the K-12 science classrooms: Trends and issues. *Research in Science Education.* doi:10.100711165-019-09912-x

Armstrong, K., & Weber, K. (1991). Genetic engineering: A lesson on bioethics for the classroom. *The American Biology Teacher, 53*(5), 294–297. doi:10.2307/4449297 PMID:11655276

Armstrong, T. (2012). *Neurodiversity in the classroom.* Alexandria, VA: Association for Supervision and Curriculum Development.

Arnold, Cohen, & Warner. (2009). Youth and Environmental Action: Perspectives of Young Environmental Leaders on Their Formative Influences. *Journal of Environmental Education, 40*(3), 27–36. doi:.3.27-36. doi:10.3200/JOEE.40

Ary, D., Jacobs, L. C., & Sorensen, C. (2010). *Introduction to research in education.* Belmont, CA: Wadsworth.

Assael, H. (1995). *Consumer behavior & marketing action* (5th ed.). PWS-Kent Publishing Company.

Atasoy, Ş. (2018). Öğretmen adaylarının yaşam alanlarına göre yerel sosyobilimsel konularla ilgili informal muhakemeleri [Student teachers' informal reasoning of local socioscientific issues according to living places]. *Fen Bilimleri Öğretimi Dergisi*, *6*(1), 60–72.

Aufderheide, P. (1993). *Media Literacy: A report of the national leadership conference on media literacy*. Aspen Institute.

Australian Curriculum, Assessment, and Reporting Authority. (2015). Retrieved on October 14, 2019, from https://australiancurriculum.edu.au/media/5558/science_-_sequence_of_content.pdf

Bairaktarova, D., & Pilotte, M. (2020). Person or thing oriented: A comparative study of individual differences of first-year engineering students and practitioners. *Journal of Engineering Education*, *109*(2), 1–13. doi:10.1002/jee.20309

Bandura, A. (1997). *Self-efficacy: The exercise of control*. W. H. Freeman.

Barton, A. C. (2001). Science education in urban settings: Seeking new ways of praxis through critical ethnography. *Journal of Research in Science Teaching*, *38*(8), 899–917. doi:10.1002/tea.1038

Batson, C. D., Early, S., & Salvarani, G. (1997). Perspective taking: Imagining how another feels versus imagining how you would feel. *Personality and Social Psychology Bulletin*, *68*, 619–631. doi:10.1037/0022-3514.68.4.619

Bauer, M. W. (2009). The evolution of public understanding of science - discourse and comparative evidence. *Science, Technology & Society*, *14*(2), 221–240. doi:10.1177/097172180901400202

Bell, R. L., & Lederman, N. G. (2003). Understandings of the nature of science and decision making on science and technology based issues. *Science Education*, *87*(3), 352–377. doi:10.1002ce.10063

Belova, N., Feierabend, T., & Eilks, I. (2015). The evaluation of role-playing in the context of teaching climate change. *International Journal of Science and Mathematics Education*, *13*(S1), 165–190. doi:10.100710763-013-9477-x

Bencze, L. L., Sperling, E., & Carter, L. (2011). Students' research informed socio-scientific activism: Re/Visions for a sustainable future. *Research in Science Education*, *42*(1), 129–148. doi:10.100711165-011-9260-3

Berkowitz, M. W., & Simmons, P. (2003). Integrating science education and character education. In D. L. Zeidler (Ed.), *The role of moral reasoning on socioscientific issues and discourse in science education* (pp. 117–138). Kluwer. doi:10.1007/1-4020-4996-X_7

Binder, A. R., Cacciatore, M. A., Scheufele, D. A., Shaw, B. R., & Corley, E. A. (2012). Measuring risk/benefit perceptions of emerging technologies and their potential impact on communication of public opinion toward science. *Public Understanding of Science (Bristol, England)*, *21*(7), 830–847. doi:10.1177/0963662510390159 PMID:23832561

Blank, R. K. (2013). Science instructional time is declining in elementary schools: What are the implications for student achievement and closing the gap? *Science Education*, *97*(6), 830–847. doi:10.1002ce.21078

Bofferding, L., & Kloser, M. (2015). Middle and High School Students Conceptions of Climate Change Mitigation and Adaptation Strategies. *Environmental Education Research*, *21*(2), 275–294. doi:10.1080/13504622.2014.888401

Boone, H. N., & Boone, D. A. (2012). Analyzing Likert Data. *Journal of Extension*, *50*(2), 2–5.

Borgerding, L., & Dagistan, M. (2018). Preservice science teachers' concerns and approaches for teaching socioscientific and controversial issues. *Journal of Science Teacher Education*, *29*(4), 283–306. doi:10.1080/1046560X.2018.1440860

Braund, M. (2015). No Drama and learning science: An empty space? *British Educational Research Journal*, *41*(1), 102–121. doi:10.1002/berj.3130

Brito, A. J., de Almedia, A. T., & Mota, M. M. C. (2009). A multicriteria model for risk sorting of natural gas pipelines based on ELECTRE TRI integrating Utility Theory. *European Journal of Operational Research*, *200*(3), 812–821. doi:10.1016/j.ejor.2009.01.016

Brophy, S., Klein, S., Portsmore, M., & Rogers, C. (2008). Advancing engineering education in P-12 classrooms. *Journal of Engineering Education*, *97*(3), 369–387. doi:10.1002/j.2168-9830.2008.tb00985.x

Burek, K., & Zeidler, D. L. (2015). Seeing the forest for the trees! Conservation and activism through socioscientific issues for young students. In M. Mueller & D. J. Tippins (Eds.), *Ecojustice, citizen science & youth activism: Situated tensions for science education* (pp. 425–441). Springer.

Burns, J. (1992). Student perceptions of technology and implications for an empowering curriculum. *Research in Science Education*, *22*(1), 72–80. doi:10.1007/BF02356881

Bursal, M., & Paznokas, L. (2006). Mathematics anxiety and preservice elementary teachers' confidence to teach mathematics and science. *School Science and Mathematics*, *106*(4), 173–180. doi:10.1111/j.1949-8594.2006.tb18073.x

Buser, M., & Koch, C. (2012). Eyes wide shut? Loyalty, and practical morality in Engineering Education. In S. Christensen, C. Mitcham, L. Bocong, & Y. An (Eds.), *Engineering, Development, and Philosophy: American, Chinese, and European Perspectives* (pp. 253–267). Springer. doi:10.1007/978-94-007-5282-5_15

Buxton, C. A. (2010). Social problem solving through science: An approach to critical, place-based, science teaching and learning. *Equity & Excellence in Education*, *43*(1), 120–135. doi:10.1080/10665680903408932

Bybee, R. W. (1997). *Achieving Scientific Literacy: From Purposes to Practices*. Heinemann.

Bybee, R. W. (1997). Towards an understanding of scientific literacy. In W. Gräber & C. Bolte (Eds.), Scientific literacy. An international symposium (pp. 37-68). Institut für die Pädagogikder Naturwissenschaften (IPN).

Bybee, R. W. (2015). *The BSCS 5E Instructional Model: Creating Teachable Moments*. doi:10.2505/9781941316009

Bybee, R. W. (1995). Achieving scientific literacy. *Science Teacher (Normal, Ill.)*, *62*(7), 28.

Bybee, R. W. (2010). Advancing STEM education: A 2020 vision. *Technology and Engineering Teacher*, *70*(1), 30–35.

Bybee, R. W. (2013). *The case for STEM education: Challenges and opportunities*. NSTA Press.

Bybee, R. W. (2014). NGSS and the next generation of science teachers. *Journal of Science Teacher Education*, *25*(2), 211–221. doi:10.100710972-014-9381-4

Carlone, H. B., Haun-Frank, J., & Webb, A. (2011). Assessing equity beyond knowledge- and skills-based outcomes: A comparative ethnography of two fourth-grade reform-based science classrooms. *Journal of Research in Science Teaching*, *48*(5), 459–485. doi:10.1002/tea.20413

Cebesoy, U. B., & Oztekin, C. (2016). Relationships among Turkish pre-service science teachers' genetics literacy levels and their attitudes towards issues in genetics literacy. *Journal of Baltic Science Education*, *15*(2), 159–172.

Cebesoy, U. B., & Oztekin, C. (2018). Genetics literacy: Insights from science teachers' knowledge, attitude, and teaching perceptions. *International Journal of Science and Mathematics Education*, *16*(7), 1247–1268. doi:10.100710763-017-9840-4

Chang Rundgren, S. N., & Rundgren, C. J. (2010). SEE-SEP: From a separate to a holistic view of socioscientific issues. *Asia-Pacific Forum on Science Learning & Teaching*, *11*(1), 1–24.

Chang, H. S., & Lee, H. J. (2010). College students' decision-making tendencies in the context of socioscientific issues (SSI). *Journal of the Korean Association for Science Education*, *30*(7), 887–900.

Chang, H. Y., Wu, H. K., & Hsu, Y. S. (2013). Integrating a mobile augmented reality activity to contextualize student learning of a socioscientific issue. *British Journal of Educational Technology*, *44*(3), E95–E99. doi:10.1111/j.1467-8535.2012.01379.x

Chang-Rundgren, S. N., & Rundgren, C.-J. (2010). SEE-SEP: From a separate to a holistic view of socioscientific issues. *Asia-Pacific Forum on Science Learning and Teaching*, *11*(1), Article 2.

Chang, S. N., & Chiu, M. H. (2008). Lakatos' Scientific Research Programmes as a Framework for Analysing Informal Argumentation about Socioscientific Issues. *International Journal of Science Education*, *30*(13), 1753–1773. doi:10.1080/09500690701534582

Chinn, C. A., & Brewer, W. F. (1993). The role of anomalous data in knowledge acquisition: A theoretical framework and implications for science instruction. *Review of Educational Research, 63*(1), 1–49. doi:10.3102/00346543063001001

Choi, S., Niyogi, D., Shepardson, D. P., & Charusombat, U. (2010). Do earth and environmental science textbooks promote middle and high school students' conceptual development about climate change? Textbooks' consideration of students' misconceptions. *Bulletin of the American Meteorological Society, 91*, 889-898. . doi:10.1175/2009BAMS2625.1

Choi, K., Lee, H., Shin, N., Kim, S., & Krajcik, J. (2011). Re-conceptualization ofscientific literacy in South Korea for the 21st Century. *Journal of Research in Science Teaching, 48*(6), 670–679. doi:10.1002/tea.20424

Choi, S., & Ramsey, J. (2009). Constructing elementary teachers' beliefs, attitudes, and practical knowledge through an inquiry-based elementary science course. *School Science and Mathematics, 109*(6), 313–324. doi:10.1111/j.1949-8594.2009.tb18101.x

Chokker, K., Dua, S., Taylor, N., Boyes, E., & Stanisstreet, M. (2011). Indian secondary students' views about global warming: Beliefs about the usefulness of actions and willingness to act. *International Journal of Science and Mathematics Education, 9*(5), 1167–1188. doi:10.100710763-010-9254-z

Christensen, R., & Knezek, G. (2015). The climate change attitude survey: Measuring middle school student beliefs and intentions to enact positive environmental change. *International Journal of Environmental and Science Education, 10*(5), 773–788.

Christenson, N. (2015). *Socioscientific argumentation: Aspects of content and structure* (Doctoral dissertation, Karlstads University). http://www.diva-portal.org/smash/record.jsf?pid=diva2%3A806023&dswid=-8412

Christenson, N., Chang Rundgren, S. N., & Höglund, H. O. (2012). Using the SEE-SEP model to analyze upper secondary students' use of supporting reasons in arguing socioscientific issues. *Journal of Science Education and Technology, 21*(3), 342–352. doi:10.100710956-011-9328-x

Christenson, N., Chang Rundgren, S. N., & Zeidler, D. L. (2014). The relationship of discipline background to upper secondary students' argumentation on socioscientific issues. *Research in Science Education, 44*(4), 581–601. doi:10.100711165-013-9394-6

Clarson, D. L., & Dormody, T. J. (1994). Analyzing data measured by individual Likert-type items. *Journal of Agricultural Education, 35*(4), 31–35.

Clement, J. (2000). Model based learning as a key research area for science education. *International Journal of Science Education, 22*(9), 1041–1053. doi:10.1080/095006900416901

Clement, J. (2008). The role of explanatory models in teaching for conceptual change. In S. Vosniadou (Ed.), *International handbook of research on conceptual change* (pp. 417–452). Routledge.

Clough, M. P. (2011). Teaching and assessing the nature of science. *Science Teacher (Normal, Ill.)*, *78*(6), 56–60.

Clough, M. P. (2013). Teaching about the nature of technology: Issues and pedagogical practices. In M. P. Clough, J. K. Olson, & D. S. Niederhauser (Eds.), *The nature of technology: Implications for learning and teaching* (pp. 345–369). Sense. doi:10.1007/978-94-6209-269-3_19

Cochran-Smith, M., & Lytle, S. (1999). The teacher research movement: A decade later. *Educational Researcher*, *28*(7), 15–25. doi:10.3102/0013189X028007015

Cohen, J. (2006). Social, emotional, ethical, and academic education: Creating a climate for learning, participation in democracy, and well-being. *Harvard Educational Review*, *76*(2), 201–237. doi:10.17763/haer.76.2.j44854x1524644vn

Cole, P. M., Martin, S. E., & Dennis, T. A. (2004). Emotion regulation as a scientific construct: Methodological challenges and directions for child development research. *Child Development*, *75*(2), 317–333. doi:10.1111/j.1467-8624.2004.00673.x PMID:15056186

Collaborative for Academic, Social, and Emotional Learning. (2017). *Social and emotional learning (SEL) competencies*. Retrieved March 2, 2020, from https://casel.org/wp-content/uploads/2019/12/CASEL-Competencies.pdf

Constantinou, C., Hadjilouca, R., & Papadouris, N. (2010). Students' epistemological awareness concerning the distinction between science and technology. *International Journal of Science Education*, *32*(2), 143–172. doi:10.1080/09500690903229296

Cook, J. (2019). Understanding and countering misinformation about climate change. In I. Chiluwa & S. Samoilenko (Eds.), *Handbook of Research on Deception, Fake News, and Misinformation Online* (pp. 281–306). IGI-Global. doi:10.4018/978-1-5225-8535-0.ch016

Copple, C., Bredekamp, S., Koralek, D., & Charner, K. (2014). *Developmentally appropriate practice: Focus on children in first, second, and third grades*. National Association for the Education of Young Children.

Corbin, J., & Strauss, A. (2008). *Basics of qualitative research: Techniques and procedures for developing grounded theory* (3rd ed.). Sage Publications. doi:10.4135/9781452230153

Cress, U., & Kimmerle, J. (2008). A systemic and cognitive view on collaborative knowledge building with wikis. *International Journal of Computer-Supported Collaborative Learning*, *3*(2), 105–122. doi:10.100711412-007-9035-z

Creswell, J. W., & Miller, D. L. (2000). Determining validity in qualitative inquiry. *Theory into Practice*, *39*(3), 124–130. doi:10.120715430421tip3903_2

Creswell, J. W., & Plano Clark, V. L. (2011). *Designing and conducting mixed methods research* (2nd ed.). Sage Publications.

Cunningham, C. M., & Carlsen, W. S. (2014). Teaching engineering practices. *Journal of Science Teacher Education*, *25*(2), 197–210. doi:10.100710972-014-9380-5

Curran, F. C., & Kitchin, J. (2019). Early elementary science instruction: Does more time on science or science topics/skills predict science achievement in the early grades? *AERA Open*, *5*(3). Advance online publication. doi:10.1177/2332858419861081

Custer, R. L. (1995). Examining the dimensions of technology. *International Journal of Technology and Design Education*, *5*(3), 219–244. doi:10.1007/BF00769905

Czerniak, C. M., & Lumpe, A. T. (1996). Relationship between teacher beliefs and science education reform. *Journal of Science Teacher Education*, *7*(4), 247–266. doi:10.1007/BF00058659

Dani, D., Wan, G., & Henning, J. E. (2010). A case for media literacy in the context of socioscientific issues. *New Horizons in Education*, *58*(3), 85.

Danusso, L., Testa, I., & Vicentini, M. (2010). Improving Prospective Teachers' Knowledge about Scientific Models and Modelling: Design and evaluation of a teacher education intervention. *International Journal of Science Education*, *32*(7), 871–905. doi:10.1080/09500690902833221

Davenport, C., & Landler, M. (2019). Trump Administration Hardens its Attack on Climate Science. *New York Times*. Available from: https://www.nytimes.com/2019/05/27/us/politics/trump-climate-science.html

Davies, I. (2004). Science and citizenship education. *International Journal of Science Education*, *26*(14), 1751–1763. doi:10.1080/0950069042000230785

Dawson, V. M., & Venville, G. (2010). Teaching strategies for developing students' argumentation skills about socioscientific issues in high school genetics. *Research in Science Education*, *40*(2), 133–148. doi:10.100711165-008-9104-y

Dawson, V., & Venville, G. (2013). Introducing high school biology students to argumentation about socioscientific issues. *Canadian Journal of Science. Mathematics and Technology Education*, *13*(4), 356–372. doi:10.1080/14926156.2013.845322

Day, S. P., & Bryce, T. G. (2013). The benefits of cooperative learning to socio-scientific discussion in secondary school science. *International Journal of Science Education*, *35*(9), 1533–1560. do i:10.1080/09500693.2011.642324

De Klerk Wolters, F. (1989). A PATT study among 10 to 12-year-old students in the Netherlands. *Journal of Technology Education*, *1*(1), 21–31.

De Vere, I., Johnson, K. B., & Thong, C. (2009). *Educating the responsible engineer: Socially responsible design and sustainability in the curriculum*. International conference on engineering and product design education, University of Brighton, UK.

De Vries, M. J. (2005). *Teaching about technology: An introduction to the philosophy of technology for non-philosophers* (Vol. 27). Springer.

De Vries, M. J., & Tamir, A. (1997). Shaping concepts of technology: What concepts and how to shape them. In M. J. De Vries & A. Tarmir (Eds.), *Shaping concepts of technology* (pp. 3–10). Springer. doi:10.1007/978-94-011-5598-4_1

DeBoer, G. E. (2000). Scientific literacy: Another look at its historical and contemporary meanings and its relationship to science education reform. *Journal of Research in Science Teaching*, *37*(6), 582–601. doi:10.1002/1098-2736(200008)37:6<582::AID-TEA5>3.0.CO;2-L

Demir, A., & Namdar, B. (2019). The effect of modeling activities on grade 5 students' informal reasoning about a real-life issue. *Research in Science Education*. Advance online publication. doi:10.100711165-019-09896-8

Denessen, E., Vos, N., Hasselman, F., & Monika Louws, M. (2015). The relationship between primary school teacher and student attitudes towards science and technology. *Education Research International*, *2015*, 534690. Advance online publication. doi:10.1155/2015/534690

Denham, S. A. (2015). Assessment of SEL in educational contexts. In J. A. Durlak, C. E. Domitrovich, R. P. Weissberg, & T. P. Gullotta (Eds.), *Handbook of social and emotional learning: Research and practice* (pp. 285–300). New York: Guilford.

Denton, P. (2007). *The power of our words: Teacher language that helps children learn*. Turners Fall, MA: Northeast Foundation for Children.

DeVries, R., & Zan, B. (2012). *Moral classrooms, moral children: Creating a constructivist atmosphere in early education*. Teachers College Press.

Dewey, J. (1975). Moral principles in education. Carbondale: Southern Illinois University Press. (Originally published 1909)

Diekman, A. B., Brown, E. R., Johnston, A. M., & Clark, E. K. (2010). Seeking congruity between goals and roles: A new look at why women opt out of science, technology, engineering, and mathematics careers. *Psychological Science*, *21*(8), 1051–1057. doi:10.1177/0956797610377342 PMID:20631322

DiGironimo, N. (2011). What is technology? Investigating student conceptions about the nature of technology. *International Journal of Science Education*, *33*(10), 1337–1352. doi:10.1080/09500693.2010.495400

DiSessa, A. (2004). Metarepresentation: Native competence and targets for instruction. *Cognition and Instruction*, *22*(3), 293–331. doi:10.12071532690xci2203_2

Dolan, T. J., Nichols, B. H., & Zeidler, D. L. (2009). Using socioscientific issues in primary classrooms. *Journal of Elementary Science Teacher Education*, *21*(3), 1–12. doi:10.1007/BF03174719

Dong, T., Anderson, R. C., Kim, I., & Li, Y. (2008). Collaborative reasoning in China and Korea. *Reading Research Quarterly*, *43*(4), 400–424. doi:10.1598/RRQ.43.4.5

Dong, T., Anderson, R. C., Lin, T., & Wu, X. (2009). Concurrent student-managed discussions in a large class. *International Journal of Educational Research*, *48*(5), 352–367. doi:10.1016/j.ijer.2010.03.005

Dorion, K. R. (2009). Science through Drama: A multiple case exploration of the characteristics of drama activities used in secondary science lessons. *International Journal of Science Education, 31*(16), 2247–2270. doi:10.1080/09500690802712699

Dortch, D., & Patel, C. (2017). Black undergraduate women and their sense of belonging in STEM at predominantly White institutions. *NASPA Journal About Women in Higher Education, 10*(2), 202–215. doi:10.1080/19407882.2017.1331854

Driver, R., Leach, J., Millar, R., & Scott, P. (1996). *Young people's images of science*. Open University Press.

Du Plessis, C. (2009). *An approach to studying urban sustainability from within an ecological worldview* (Unpublished doctoral thesis). University of Salford.

Du Plessis, C., & Brandon, P. (2015). An ecological worldview as basis for a regenerative sustainability paradigm for the built environment. *Journal of Cleaner Production, 109*, 53–61. doi:10.1016/j.jclepro.2014.09.098

Dunlap, R. E., Van Liere, K. D., Mertig, A. G., & Jones, R. E. (2000). Measuring endorsement of the new ecological paradigm: A revised NEP scale. *The Journal of Social Issues, 56*(3), 425–442. doi:10.1111/0022-4537.00176

Durlak, J. A., Weissberg, R. P., Dymnicki, A. B., Taylor, R. D., & Schellinger, K. B. (2011). The impact of enhancing students' social and emotional learning: A meta-analysis of school-based universal interventions. *Child Development, 82*(1), 405–432. doi:10.1111/j.1467-8624.2010.01564.x PMID:21291449

Du, S., Fu, Z., & Wang, Y. (2014). The flipped classroom – advantages and challenges. In *International Conference on Economic Management and Trade Cooperation* (pp. 17–20). Atlantis Press.

Duschl, R. A., Bismack, A. S., Greeno, J., Gitomer, D. H., Bismack, A. S., Greeno, J., & Gitomer, D. H. (2016). Coordinating preK-16 STEM education research and practices for advancing and refining reform agendas. In R. A. Duschl & A. S. Bismack (Eds.), *Reconceptualizing STEM education: The central role of practices* (pp. 1–32). Routledge. doi:10.4324/9781315700328

Duveen, J., & Solomon, J. (1994). The great evolution trial: Use of role play in the classroom. *Journal of Research in Science Teaching, 31*(5), 575–582. doi:10.1002/tea.3660310510

Dym, C. L. (1999). Learning engineering: Design, learning, and experiences. *Journal of Engineering Education, 88*(2), 145–148. doi:10.1002/j.2168-9830.1999.tb00425.x

Eastwood, J. L., Sadler, T. D., Zeidler, D. L., Lewis, A., Amiri, L., & Applebaum, S. (2012). Contextualizing nature of science instruction in socioscientific issues. *International Journal of Science Education, 34*(15), 2289–2315. doi:10.1080/09500693.2012.667582

Eggert, S., & Bögeholz, S. (2010). Students' use of decision-making strategies with regard to socioscientific issues: An application of the Rasch partial credit model. *Science Education, 94*(2), 230–258.

Eggert, S., & Bögeholz, S. (2010). Students' use of decision-making strategies with regard to socioscientific issues: An application of the Rasch partial credit model. *Science Education, 94*(2), 230–258. doi:10.1002ce.20358

Eisenberg, N., Fabes, R., Murphy, B., Karbon, M., Maszk, P., Smith, M., O'Boyle, C., & Suh, K. (1994). The relations of emotionality and regulation to dispositional and situational empathy-related responding. *Journal of Personality and Social Psychology, 66*(4), 776–797. doi:10.1037/0022-3514.66.4.776 PMID:8189352

Ekborg, M., Ottander, C., Silfver, E., & Simon, S. (2013). Teachers' Experience of Working with Socio-scientific Issues: A Large Scale and in Depth Study. *Research in Science Education, 43*(2), 599–617. doi:10.100711165-011-9279-5

English, L. D. (2016). STEM education K-12: Perspectives on integration. *International Journal of STEM Education, 3*(1), 1–8. doi:10.118640594-016-0036-1

Erikson, E. H. (1959). Identity and the life cycle. *Psychological Issues, 1*, 1–71.

Erikson, E. H. (1963). *Childhood and society*. Norton.

Es, H., & Ozturk, N. (2019). An Activity for transferring the multidimensional structure of SSI to middle school science courses: I discover myself in the decision-making process with SEE-STEP! *Research in Science Education*, 1–22. doi:10.100711165-019-09865-1

Eşkin, H., & Ogan-Bekiroğlu, F. (2009). Investigation of a pattern between students' engagement in argumentation and their science content knowledge: A Case Study. *Eurasia Journal of Mathematics. Science and Technology Education, 5*(1), 63–70.

Espeja, A. G., & Largarón, D. C. (2015). Socio-scientific issues (SSI) in initial training of primary school teachers: Pre-service teachers' conceptualization of SSI and appreciation of the value of teaching SSI. *Procedia: Social and Behavioral Sciences, 196*, 80–88. Advance online publication. doi:10.1016/j.sbspro.2015.07.015

European Commission. (1995). *Teaching and learning: Towards the learning society*. White Paper on Education and Training, Office for Official Publications in European Countries. https://europa.eu/documents/comm/white_papers/pdf/com95_590_en.pdf

Evagorou, M. (2011). Discussing a socioscientific issue in a primary school classroom: The case of using a technology-supported environment in formal and nonformal settings. In T. D. Sadler (Ed.), *Socio-scientific issues in science classrooms: Teaching, learning and research* (pp. 133–159). Springer. doi:10.1007/978-94-007-1159-4_8

Evagorou, M., Guven, D., & Mugaloglu, E. (2014). Preparing elementary and secondary pre-service teachers for everyday science. *Science Education International, 25*(1), 68–78.

Evagorou, M., & Osborne, J. (2013). Exploring young students' collaborative argumentation within a socioscientific issue. *Journal of Research in Science Teaching, 50*(2), 209–237. doi:10.1002/tea.21076

Feldman, A. (1996). Enhancing the practice of physics teachers: Mechanisms for the generation and sharing of knowledge and understanding in collaborative action research. *Journal of Research in Science Teaching, 33*(5), 513–540. doi:10.1002/(SICI)1098-2736(199605)33:5<513::AID-TEA4>3.0.CO;2-U

Feldman, L., Nisbet, M., Leiserowitz, A., & Maibach, E. W. (2010). *The Climate Change Generation? Survey Analysis of the Perceptions and Beliefs of Young Americans.* Yale Project on Climate Change Communication.

Feldman, M. P., Morris, M. L., & Hoisington, D. (2000). Why All the Controversy? *Choice (Chicago, Ill.), 15*(1), 8–12.

Fisher, D. R., Waggle, J., & Leifeld, P. (2012). Where Does Political Polarization Come From? Locating Polarization Within the U.S. Climate Change Debate. *The American Behavioral Scientist*, 1–23.

Flora, J. A., Saphir, M., Lappé, M., Roser-Renouf, C., Maibach, E. W., & Leiserowitz, A. A. (2014). Evaluation of a National High School Entertainment Education Program: The Alliance for Climate Education. *Climatic Change, 127*(3–4), 419–434. doi:10.100710584-014-1274-1

Foor, C. E., & Walden, S. E. (2009). "Imaginary engineering" or "re-imagined engineering": Negotiating gendered identities in the borderland of a college of engineering. *NWSA Journal, 21*(2), 41–64.

Forbes, C., & Davis, E. (2008). Exploring preservice elementary teachers' critique and adaptation of science curriculum materials in respect to socioscientific issues. *Science & Education, 17*(8-9), 829–854. doi:10.100711191-007-9080-z

Forbes, C., Zangori, L., & Schwarz, C. (2015). Using models scientifically. *Science and Children, 53*(2). Advance online publication. doi:10.2505/4c15_053_02_42

Ford, M. J., & Forman, E. A. (2006). Refining disciplinary learning in classroom contexts. *Review of Research in Education, 30*(1), 1–32. doi:10.3102/0091732X030001001

Foster, S. J. (2001). Historical empathy in theory and practice: Some final thoughts. In O. L. Davis, E. A. Yeager, & S. J. Foster (Eds.), *Historical empathy and perspective taking in the social studies*. Rowman and Littlefield.

Fowler, S. R., Zeidler, D. L., & Sadler, T. D. (2009). Moral sensitivity in the context of socioscientific issues in high school science students. *International Journal of Science Teacher Education, 31*(2), 279–296. doi:10.1080/09500690701787909

Freeman, B., Marginson, S., & Tytler, R. (2014). *The age of STEM: Educational policy and practice across the world in science, technology, engineering and mathematics*. Routledge. doi:10.4324/9781315767512

Freire, P. (2004). *Pedagogy of indignation.* Paradigm Publishers.

Fretz, E. B., Wu, H. K., Zhang, B., Davis, E. A., Krajcik, J. S., & Soloway, E. (2002). An investigation of software scaffolds supporting modeling practices. *Research in Science Education, 32*(4), 567–589. doi:10.1023/A:1022400817926

Frey, R. E. (1991). Another look at technology and science. *Journal of Technology Education, 3*(1), 16–29. doi:10.21061/jte.v3i1.a.2

Friedel, R. (2010). *A culture of improvement: Technology and the western millennium.* MIT Press.

Friedrichsen, P. J., Sadler, T. D., Graham, K., & Brown, P. (2016). Design of a Socio-scientific Issue Curriculum Unit: Antibiotic Resistance, Natural Selection, and Modeling. *International Journal of Designs for Learning, 7*(1). Advance online publication. doi:10.14434/ijdl.v7i1.19325

Garcia-Mila, M., & Andersen, C. (2007). Cognitive foundations of learning argumentation. In S. Erduran & M. P. Jimenez-Aleixandre (Eds.), *Argumentation in science education: Perspectives from classroom-based research* (pp. 29–45). Springer. doi:10.1007/978-1-4020-6670-2_2

Gardner, P. L. (1992). The application of science to technology. *Research in Science Education, 22*(1), 140–148. doi:10.1007/BF02356889

Gardner, P. L. (1994). The relationship between technology and science: Some historical and philosophical reflections. Part I. *International Journal of Technology and Design Education, 4*(2), 123–153. doi:10.1007/BF01204544

Gardner, P. L. (1995). The relationship between technology and science: Some historical and philosophical reflections. Part II. *International Journal of Technology and Design Education, 5*(1), 1–33. doi:10.1007/BF00763650

Garner, P. W. (2010). Emotional competence and its influences on teaching and learning. *Educational Psychology Review, 22*(3), 297–321. doi:10.100710648-010-9129-4

Gay, G. (2013). Teaching to and through cultural diversity. *Curriculum Inquiry, 43*(1), 48–70. doi:10.1111/curi.12002

Gelman, R., & Baillargeon, R. (1983). A review of some Piagetian concepts. In Handbook of child psychology (4th ed., pp. 167–230). New York, NY: Wiley.

Genel, A., & Topcu, M. S. (2016). Turkish preservice science teachers' socioscientific issues-based teaching practices in middle school science classrooms. *Research in Science & Technological Education, 34*(1), 105–123. doi:10.1080/02635143.2015.1124847

Gilbert, J. (2005). Visualization: A metacognitive skill in science and science education. In J. Gilbert (Ed.), *Visualization in science education. Models and modeling in science education* (pp. 9–27). Springer. doi:10.1007/1-4020-3613-2_2

Gilbert, J. K. (2004). Models and modelling: Routes to more authentic science education. *International Journal of Science and Mathematics Education, 2*(2), 115–130. doi:10.100710763-004-3186-4

Gilbert, J. K., & Justi, R. (2016). *Modelling-based teaching in science education*. Springer International Publishing. doi:10.1007/978-3-319-29039-3

Gobert, J. D., & Buckley, B. C. (2000). Introduction to model-based teaching and learning in science education. *International Journal of Science Education*, *22*(9), 891–894. doi:10.1080/095006900416839

Goodnight, G. T. (2005). Science and technology controversy: A rationale for inquiry. *Argumentation and Advocacy*, *42*(1), 26–29. doi:10.1080/00028533.2005.11821636

Grace, M. M., & Ratcliffe, M. (2002). The science and values that young people draw upon to make decisions about biological conservation issues. *International Journal of Science Education*, *24*(11), 1157–1169. doi:10.1080/09500690210134848

Grandin, J. M., & Hirleman, E. D. (2009). Educating engineers as global citizens: A call for action: A report of the national summit meeting on the globalization of engineering education. *Online Journal of Global Engineering Education*, *4*(1), 10–25.

Grant, M., & Lapp, D. (2011). Teaching science literacy. *Educational Leadership*, *68*(6).

Green, T. F. (1985). The formation of conscience in an age of technology. *American Journal of Education*, *94*(1), 1–32. doi:10.1086/443829

Grimson, J. (2010). Re-engineering the curriculum for the 21st century. *European Journal of Engineering Education*, *27*(1), 31–37. doi:10.1080/03043790110100803

Guzey, S. S., Moore, T., & Harwell, M. (2014). Development of an instrument to measure students' attitudes toward STEM. *School Science and Mathematics*, *114*(6), 271–279. doi:10.1111sm.12077

Guzey, S. S., Moore, T., & Morse, G. (2016). Student interest in engineering design-based science. *School Science and Mathematics*, *116*(8), 411–419. doi:10.1111sm.12198

Haban, T. (2004). *Public Attitude towards Agricultural Biotechnology*. ESA Working Paper.

Halverson, K. L., Siegel, M. A., & Freyermuth, S. K. (2009). Lenses for framing decisions: Undergraduates' decision making about stem cell research. *International Journal of Science Education*, *31*(9), 1249–1268. doi:10.1080/09500690802178123

Hancock, E. S., & Gallard, A. J. (2004). Preservice science teachers' beliefs about teaching and learning: The influence of K-12 field experiences. *Journal of Science Teacher Education*, *15*(4), 281–291. doi:10.1023/B:JSTE.0000048331.17407.f5

Hancock, T. S., Friedrichsen, P. J., Kinslow, A. T., & Sadler, T. D. (2019). Selecting socio-scientific issues for teaching. *Science & Education*, *28*(6), 639–667. doi:10.100711191-019-00065-x

Hand, B., & Choi, A. (2010). Examining the impact of student use of multiple modal representations in constructing arguments in organic chemistry laboratory classes. *Research in Science Education*, *40*(1), 29–44. doi:10.100711165-009-9155-8

Haney, J. J., Lumpe, A. T., Czerniak, C. M., & Egan, V. (2002). From beliefs to actions: The beliefs and actions of teachers implementing change. *Journal of Science Teacher Education, 13*(3), 171–187. doi:10.1023/A:1016565016116

Harrison, A. G., & Treagust, D. F. (1998). Modelling in science lessons: Are there better ways to learn with models? *School Science and Mathematics, 98*(8), 420–429. doi:10.1111/j.1949-8594.1998.tb17434.x

Harrison, A. G., & Treagust, D. F. (2000). A typology of school science models. *International Journal of Science Education, 22*(9), 1011–1026. doi:10.1080/095006900416884

Harris, R., & Ratcliffe, M. (2005). Socio-scientific issues and the quality of exploratory talk—What can be learned from schools involved in a 'collapsed day'project? *Curriculum Journal, 16*(4), 439–453. doi:10.1080/09585170500384396

Herman, B. C., Zeidler, D. L., & Newton, M. (2018). Students' emotive reasoning though place-based environmental socioscientific issues. *Research in Science Education.* https://doi-doi:10.100711165-018-9764-1

Herman, B. C., & Clough, M. P. (2017). The role of history and nature of science in climate change teaching and learing. In D. Shepardson, R. Roychoudhury, & A. Hirsch (Eds.), *Teaching and learning about climate change: A framework of educators* (pp. 31–44). Routledge.

Herman, B. C., Sadler, T. D., Zeidler, D. L., & Newton, M. (2018). A socioscientific issues approach to environmental education. In G. Reis & J. Scott (Eds.), *International perspectives on the theory and practice of environmental education: A reader* (pp. 145–161). Springer International Publishers. doi:10.1007/978-3-319-67732-3_11

Hermans, M., & Korhonen, J. (2017). Ninth graders and climate change: Attitudes towards consequences, views on mitigation, and predictors of willingness to act. *International Research in Geographical and Environmental Education, 26*(3), 223–239. doi:10.1080/10382046.2017.1330035

Herschbach, D. R. (2011). The STEM initiative: Constraints and challenges. *Journal of STEM Teacher Education, 48*(1), 96–112. doi:10.30707/JSTE48.1Herschbach

Higher Education Council (HEC). (2007). *Öğretmen Yetiştirme ve Eğitim Fakülteleri (1982-2007)* [Teacher training and faculties of education (1982-2007)]. https://www.yok.gov.tr/Documents/Yayinlar/Yayinlarimiz/ogretmen-yetistirme-ve-egitim-fakulteleri.pdf

Hildebrandt, C., & Zan, B. (2014). Constructivist approaches to moral education in early childhood. In L. Nucci, T. Krettenauer, & D. Narvaez (Eds.), *Handbook of moral and character education* (pp. 180–197). Routledge.

Hmelo, C. E., Holton, D., & Kolodner, J. L. (2000). Designing to learn about complex systems. *Journal of the Learning Sciences, 9*(3), 247–298. doi:10.1207/S15327809JLS0903_2

Hodson, D. (2006). Why we should prioritize learning about science. *Canadian Journal of Science. Mathematics and Technology Education, 6*(3), 293–311. doi:10.1080/14926150609556703

Hofstein, A., Eilks, I., & Bybee, R. (2011). Societal issues and their importance for contemporary science education—A pedagogical justification and the state-of-the-art in Israel, Germany, and the USA. *International Journal of Science and Mathematics Education*, *9*(6), 1459–1483. doi:10.100710763-010-9273-9

Holbrook, J., & Rannikmae, M. (Eds.). (1997). Supplementary teaching materials promoting scientific and technological literacy. Tartu, Estonia: ICASE (International Council of Associations for Science Education).

Holbrook, J., & Rannikmae, M. (2007). The nature of science education for enhancing scientific literacy. *International Journal of Science Education*, *29*(11), 1347–1362. doi:10.1080/09500690601007549

Honey, M., Pearson, G., & Schweingruber, H. A. (Eds.). (2014). *STEM integration in K-12 education: Status, prospectus, and an agenda for research*. The National Academies Press.

Hong, S. (1999). *Science and technology as productive power and culture*. Moonji Press.

Hopwood, B., Mellor, M., & O'Brien, G. (2005). Sustainable Development: Mapping different approaches. *Sustainable Development*, *13*(1), 38–52. doi:10.1002d.244

Hughes, T. P. (1983). *Networks of power: Electrification in western society*. Johns Hopkins University Press.

Hughes, T. P. (1986). The seamless web: Technology, science, etcetera, etcetera. *Social Studies of Science*, *16*(2), 281–292. doi:10.1177/0306312786016002004

Hughes, T. P. (2004). *Human-built world: How to think about technology and culture*. University of Chicago Press. doi:10.7208/chicago/9780226120669.001.0001

Hughes, T. P. (2012). The evolution of large technological systems. In W. E. Bijker, T. P. Hughes, T. J. Pinch, & D. G. Douglas (Eds.), *The social construction of technological systems: New directions in the sociology and history of technology* (pp. 45–74). MIT Press.

Hung, W., Jonassen, D. H., & Liu, R. (2008). Problem-based learning. In J. M. Spector, J. G. van Merrienboer, M. D. Merrill, & M. Driscoll (Eds.), *Handbook of Research on Educational Communications and Technology* (3rd ed., pp. 485–506). Erlbaum.

Hurd, P. D. (1958). Science literacy: Its meaning for American schools. *Educational Leadership*, *16*(1), 13–16.

ICASE. (2013). The Kuching Declaration. *Proceeding of the World Conference on Science and Technology Education (WorldSTE2013)*. Available online at: http://www.icaseonline.net/ICASE%20Kuching%20Declaration_Final. pdf

International Technology Education Association (ITEA). (2007). *Standards for technological literacy: Content for the study of technology*. International Technology Education Association.

Jarvis, T., & Rennie, L. J. (1996). Understanding technology: The development of a concept. *International Journal of Science Education, 18*(8), 977–992. doi:10.1080/0950069960180809

Jensen, B. B., & Schnack, K. (1997). The action competence approach in environmental education. *Environmental Education Research, 3*(2), 163–178. doi:10.1080/1350462970030205

Jho, H., Yoon, H. G., & Kim, M. (2014). The relationship of science knowledge, attitude and decision making on socio-scientific issues: The case study of students' debates on a nuclear power plant in Korea. *Science & Education, 23*(5), 1131–1151. doi:10.100711191-013-9652-z

Jiménez-Aleixandre, M. P., Bugallo Rodríguez, A., & Duschl, R. A. (2000). "Doing the lesson" or "doing science": Argument in high school genetics. *Science Education, 84*(6), 757–792. doi:10.1002/1098-237X(200011)84:6<757::AID-SCE5>3.0.CO;2-F

Jiménez-Aleixandre, M. P., & Erduran, S. (2008). Argumentation in science education: An overview. In S. Erduran & M. P. Jimenez-Aleixandre (Eds.), *Argumentation in science education: Perspectives from classroom-based research* (pp. 3–27). Springer.

Justi, R. S., & Gilbert, J. K. (2002). Modelling, teachers' views on the nature of modelling, and implications for the education of modellers. *International Journal of Science Education, 24*(4), 369–387. doi:10.1080/09500690110110142

Justi, R., & Gilbert, J. (2003). Teachers' views on the nature of models. *International Journal of Science Education, 25*(11), 1369–1386. doi:10.1080/0950069032000070324

Kahn. (2015). Another "M" for STEM? Moral Considerations for Advancing STEM Literacy. *K-12 STEM Education, 1*(4), 149-156.

Kahn, S. (2018). From access to assets: Strength-based visions for inclusive science education. In M. Koomen, S. Kahn, C. L. Atchison, & T. A. Wild (Eds.), *Towards inclusion of all learners through science teacher education* (pp. 105–114). Brill. doi:10.1163/9789004368422_012

Kahn, S. (2019). *It's still debatable! Using socioscientific issues to develop scientific literacy, K-5*. NSTA Press.

Kahn, S., & Zeidler, D. L. (2016). Using our heads and HARTSS*: Developing perspective-taking skills for socioscientific reasoning (*Humanities, ARTs, and Social Sciences). *Journal of Science Teacher Education, 27*(3), 261–281. doi:10.100710972-016-9458-3

Kahn, S., & Zeidler, D. L. (2017). A case for the use of conceptual analysis in science education research. *Journal of Research in Science Teaching, 54*(4), 538–551. doi:10.1002/tea.21376

Kahn, S., & Zeidler, D. L. (2019). A conceptual analysis of perspective taking in support of socioscientific reasoning. *Science & Education, 28*(6-7), 605–638. doi:10.100711191-019-00044-2

Karisan, D., Tüzün, Ö. Y., & Zeidler, D. L. (2017). Quality of preservice teachers argumentation in socioscientific issues context. *Journal of Human Sciences, 14*(4), 3504–3520. doi:10.14687/jhs.v14i4.4949

Karpudewan, M., & Roth, W. M. (2018). Changes in primary students' informal reasoning during an environment-related curriculum on socio-scientific issues. *International Journal of Science and Mathematics Education, 16*(3), 401–419. doi:10.100710763-016-9787-x

Ke, L., Sadler, T., Zangori, L., & Friedrichsen, P. (2020). Students' perceptions of socio-scientific issue-based learning and their appropriation of epistemic tools for systems thinking. *International Journal of Science Education*, 1–23. Advance online publication. doi:10.1080/09500693.2020.1759843

Ke, L., & Schwarz, C. (2020). Using epistemic considerations in teaching: Fostering students' meaningful engagement in scientific modeling. In A. Upmeier zu Belzen, D. Krüger, & J. Van Driel (Eds.), *Towards a Competence-based View on Models and Modeling in Science Education* (pp. 181–199). Springer International Publishing.

Kelley, T. R., & Knowles, J. G. (2016). A conceptual framework for integrated STEM education. *International Journal of STEM Education, 3*(11), 1–11. doi:10.118640594-016-0046-z

Khishfe, R. (2012). Nature of science and decision-making. *International Journal of Science Education, 34*(1), 67–100. doi:10.1080/09500693.2011.559490

Khishfe, R. (2014). Explicit Nature of Science and Argumentation Instruction in the Context of Socioscientific Issues: An effect on student learning and transfer. *International Journal of Science Education, 36*(6), 974–1016. doi:10.1080/09500693.2013.832004

Khishfe, R., Alshaya, F. S., BouJaoude, S., Mansour, N., & Alrudiyan, K. I. (2017). Students' understandings of nature of science and their arguments in the context of four socio-scientific issues. *International Journal of Science Education, 39*(3), 299–334. doi:10.1080/09500693.2017.1280741

Kilinc, A., Demiral, U., & Kartal, T. (2017). Resistance to dialogic discourse in SSI teaching: The effects of an argumentation-based workshop, teaching practicum, and induction on a preservice science teacher. *Journal of Research in Science Teaching, 54*(6), 764–789. doi:10.1002/tea.21385

Kılınç, A., Kartal, T., Eroğlu, B., Demiral, Ü., Afacan, Ö., Polat, D., Guler, M. P., & Görgülü, Ö. (2013). Preservice science teachers' efficacy regarding a socioscientific issue: A belief system approach. *Research in Science Education, 43*(6), 2455–2475. doi:10.100711165-013-9368-8

Kim, D., Moon, J., Lee, J., Song, C., & Park, J. (2013). *The birth of modern engineering*. Ecolivres Press.

Kim, G., Ko, Y., & Lee, H. (2020). The effects of community-based socioscientific issues program (SSI-COMM) on promoting students' sense of place and character as citizens. *International Journal of Science and Mathematics Education, 18*(3), 399–418. doi:10.100710763-019-09976-1

King, A. (1993). From sage on the stage to guide on the side. *College Teaching, 41*(1), 30–35. doi:10.1080/87567555.1993.9926781

Kinslow, A. T. (2018). *The development and implementation of a heuristic for teaching reflective scientific skepticism within a socio-scientific issue instructional framework* (Unpublished dissertation). University of Missouri.

Kinslow, A. T., Sadler, T. D., & Nguyen, H. T. (2019). Socio-scientific reasoning and environmental literacy in a field-based ecology class. *Environmental Education Research, 25*(3), 388–410. do i:10.1080/13504622.2018.1442418

Klosterman, M. L., Sadler, T. D., & Brown, J. (2012). Science teachers' use of mass media to address socio-scientific and sustainability issues. *Research in Science Education, 42*(1), 51–74. doi:10.100711165-011-9256-z

Klosterman, M., & Sadler, T. (2010). Multi-level assessment of scientific content knowledge gains associated with socioscientific issues-based instruction. *International Journal of Science Education, 32*(8), 1017–1043. doi:10.1080/09500690902894512

Knight, A. J. (2006). Does application matter? An examination of public perception of agricultural biotechnology applications. *AgBioForum, 9*(2), 121–128.

Kollmuss, A., & Agyeman, J. (2002). Mind the gap: Why do people act environmentally and what are the barriers to pro-environmental behaviour? *Environmental Education Research, 8*(3), 239–260. doi:10.1080/13504620220145401

Kolstø, S. D. (2001a). Scientific literacy for citizenship: Tools for dealing with the science dimension of controversial socioscientific issues. *Science Education, 85*(3), 291–310. doi:10.1002ce.1011

Korpan, C., Bisanz, G., Bisanz, J., & Henderson, J. (1997). Assessing literacy in science: Evaluation of scientific news briefs. *Science Education, 81*(5), 515–532. doi:10.1002/(SICI)1098-237X(199709)81:5<515::AID-SCE2>3.0.CO;2-D

Krajcik, J. S., & Sutherland, L. (2010). Supporting students in developing literacy in science. *Science, 328*(5977), 456–459. doi:10.1126cience.1182593 PMID:20413490

Krippendorff, K. (2006). *The semantic turn: A new foundation for design.* CRC Press Taylor & Francis Group.

Kroes, P. (2012). *Technical artefacts: Creations of mind and matter: A philosophy of engineering design.* Springer. doi:10.1007/978-94-007-3940-6

Lachapelle, C., & Cunningham, C. (2014). Engineering in elementary schools. In S. Purzer, J. Strobel, & M. Cardella (Eds.), *Engineering in pre-college settings: Synthesizing research, policy, and practices* (pp. 61–88). Purdue University Press. doi:10.2307/j.ctt6wq7bh.8

Lasry, N., Dugdale, M., & Charles, E. (2014). Just in time to flip your classroom. *The Physics Teacher, 52*(1), 34–37. doi:10.1119/1.4849151

Laugksch, R. C. (2000). Scientific Literacy: A conceptual overview. *Science Education, 84*(1), 71–94. doi:10.1002/(SICI)1098-237X(200001)84:1<71::AID-SCE6>3.0.CO;2-C

Lawson, D. F., Stevenson, K. T., Peterson, M. N., Carrier, S. J., Seekamp, E., & Strnad, R. (2019). Evaluating climate change behaviors and concern in the family context. *Environmental Education Research, 25*(5), 1–13. doi:10.1080/13504622.2018.1564248

Layton, D. (1993). *Technology's challenge to science education.* Open University Press.

Layton, E. T. (1974). Technology as knowledge. *Technology and Culture, 15*(1), 31–41. doi:10.2307/3102759

Lazarowitz, R., & Bloch, I. (2005). Awareness of societal issues among high school biology teachers teaching. *Journal of Science Education and Technology, 14*(5), 437–457. doi:10.100710956-005-0220-4

Leammukda, F. D., & Roehrig, G. H. (2020). *Community-Based Conceptual Framework for STEM Integration.* Paper presented at the annual meeting of the Association for Science Teacher Education, San Antonio, TX.

Lederman, N. G. (1992). Students' and teachers' conceptions of the nature of science: A review of the research. *Journal of Research in Science Teaching, 29*(4), 331–359. doi:10.1002/tea.3660290404

Lederman, N. G. (2003). Introduction. In D. Zeidler (Ed.), *The role of Moral Reasoning on Socioscientific Issues and Discourse in Science* (pp. 1–4). Kluwer Academic Publishers. doi:10.1007/1-4020-4996-X_1

Lederman, N. G. (2007). Nature of science: Past, present, and future. In S. K. Abell & N. G. Lederman (Eds.), *Handbook of research on science education* (pp. 831–879). Lawrence Erlbaum Associates.

Lederman, N. G., Abd-El-Khalick, F., Bell, R. L., & Schwartz, R. S. (2002). Views of nature of science questionnaire: Toward valid and meaningful assessment of learners' conceptions of nature of science. *Journal of Research in Science Teaching, 39*(6), 497–521. doi:10.1002/tea.10034

Lederman, N. G., Antink, A., & Bartos, S. (2014). Nature of science, scientific inquiry, and socio-scientific issues arising from genetics: A pathway to developing a scientifically literate citizenry. *Science & Education, 23*(2), 285–302. doi:10.100711191-012-9503-3

Lee, H. (2015). *Construction of nature of technology framework and its utilization for investigation of changes in college students' perception of nature of technology through SSI-based program* (Unpublished doctoral dissertation). Ewha Womans University, Korea.

Lee, H., Abd-El-Khalick, F., & Choi, K. (2006). Korean science teachers' perceptions of the introduction of socio-scientific issues into the science curriculum. *Canadian Journal of Science. Mathematics and Technology Education, 6*(2), 97–117. doi:10.1080/14926150609556691

Lee, H., Chang, H., Choi, K., Kim, S. W., & Zeidler, D. L. (2012). Developing character and values for global citizens: Analysis of pre-service science teachers' moral reasoning on socioscientific issues. *International Journal of Science Education, 34*(6), 925–953. doi:10.1080/09500693.2011.625505

Lee, H., & Lee, H. (2015). Analysis of students' socioscientific decision-making from the nature of technology perspectives. *Journal of The Korean Association For Science Education, 35*(1), 169–177. doi:10.14697/jkase.2015.35.1.0169

Lee, H., & Lee, H. (2016). Contextualized nature of technology in socioscientific issues. *Journal of the Korean Association for Research in Science Education, 36*(2), 303–315. doi:10.14697/jkase.2016.36.2.0303

Lee, H., & Lee, H. (2017). Development and application of rubric for assessing nature of technology in the context of socioscientific issues. *Journal of the Korean Association for Research in Science Education, 37*(2), 323–334.

Lee, H., Yoo, J., Choi, K., Kim, S., Krajcik, J., Herman, B., & Zeidler, D. L. (2013). Socioscientific Issues as a Vehicle for Promoting Character and Values for Global Citizens. *International Journal of Science Education, 35*(12), 2079–2113. doi:10.1080/09500693.2012.749546

Lee, Y. C. (2007). Developing decision-making skills for socio-scientific issues. *Journal of Biological Education, 41*(4), 170–177. doi:10.1080/00219266.2007.9656093

Lehrer, R., & Schauble, L. (2006). Scientific thinking and science literacy: Supporting development learning in contexts. In W. Damon, R. M. Learner, K. A. Renninger, & I. E. Sigel (Eds.), Handbook of child psychology (6th ed., Vol. 4). John Wiley and Sons.

Leiserowitz, A., & Smith, N. (2011). *Knowledge of climate change among visitors to Science & Technology Museums*. Retrieved from Yale Program on Climate Change Communication website: https://environment.yale.edu/climate/fles/MuseumReport.pdf

Leiserowitz, A., Maibach, E., Roser-Renouf, C., Rosenthal, S., & Cutler, M. (2017). *Climate change in the American mind: May 2017. Yale University and George Mason University*. Yale Program on Climate Change Communication.

Lemke, J. L. (1998). Multiplying meaning: visual and verbal semiotics in scientific text. In J. R. Martin & R. Vell (Eds.), *Reading science: critical and functional perspectives on discourses of science* (pp. 87–113). Routledge.

Levinson, R. (2006). Towards a theoretical framework for teaching controversial socio-scientific issues. *International Journal of Science Education, 28*(10), 1201–1224. doi:10.1080/09500690600560753

Lewis, J., & Leach, J. (2006). Discussion of socio-scientific issue: The role of science knowledge. *International Journal of Science Education, 28*(11), 1267–1287. doi:10.1080/09500690500439348

Lindahl, M. G., Folkesson, A.-M., & Zeidler, D. L. (2019). Students ' recognition of educational demands in the context of a socioscientific issues curriculum. *Journal of Research in Science Teaching*, 1–28. doi:10.1002/tea.21548

Lindsay, P., & Norman, D. A. (1977). *Human information processing: An introduction to psychology*. Harcourt Brace Jovanovich, Inc.

Linn, S.-S., & Mitzes, J. J. (2010). Learning argumentation skills through instruction in socioscientific issues: The effect of ability level. *International Journal of Science and Mathematics Education*, *8*(6), 993–1007. doi:10.100710763-010-9215-6

Livingstone, S. (2004). Media literacy and the challenge of new information and communication technologies. *Communication Review*, *1*(7), 3–14. doi:10.1080/10714420490280152

Lombardi, D., Sinatra, G. M., & Nussbaum, E. M. (2013). Plausibility reappraisals and shifts in middle school students' climate change conceptions. *Learning and Instruction*, *27*, 50–62. doi:10.1016/j.learninstruc.2013.03.001

Louca, L. T., & Zacharia, Z. C. (2012). Modeling-based learning in science education: Cognitive, metacognitive, social, material and epistemological contributions. *Educational Review*, *64*(4), 471–492. doi:10.1080/00131911.2011.628748

Louca, L. T., Zacharia, Z. C., & Constantinou, C. P. (2011). In Quest of productive modeling-based learning discourse in elementary school science. *Journal of Research in Science Teaching*, *48*(8), 919–951. doi:10.1002/tea.20435

Loureiro, L. M., & Bugbee, M. (2005). Enhanced GM foods: Are consumers ready to pay for the potential benefits of biotechnology? *The Journal of Consumer Affairs*, *39*(1), 52–70. doi:10.1111/j.1745-6606.2005.00003.x

Magnusson, T. (2009). Of Epistemic Tools: Musical instruments as cognitive extensions. *Organised Sound*, *14*(2), 168–176. doi:10.1017/S1355771809000272

Maloney, J. (2007). Children's role and use of evidence on science: An analysis of decision-making in small groups. *British Educational Research Journal*, *33*(3), 371–401. doi:10.1080/01411920701243636

Manz, E. (2012). Understanding the codevelopment of modeling practice and ecological knowledge. *Science Education*, *96*(6), 1071–1105. doi:10.1002ce.21030

Marshall, J. C., Horton, R. M., Igo, B. L., & Switzer, D. M. (2009). K-12 science and mathematics teachers' beliefs about and use of inquiry in the classroom. *International Journal of Science and Mathematics Education*, *7*(3), 575–596. doi:10.100710763-007-9122-7

Martin, M., & Schinzinger, R. (1989). *Ethics in Engineering*. McGraw-Hill.

Matuk, C., & Linn, M. (2012). Technology integration to scaffold and assess students ' use of visual evidence in science inquiry. In American Educational Research Association (AERA) Conference (pp. 1–8). Academic Press.

Maxwell, J. A. (2009). Designing a qualitative study. In L. Bickman & D. J. Rog (Eds.), *The Handbook of Applied Social Research Methods* (pp. 214–253). Sage. doi:10.4135/9781483348858.n7

McBeth, W., & Volk, T. L. (2009). The National Environmental Literacy Project: A baseline study of middle grade students in the United States. *The Journal of Environmental Education, 41*(1), 55–67. doi:10.1080/00958960903210031

McCright, A. M., & Dunlap, R. E. (2011). Cool dudes: The denial of climate change among conservative white males in the United States. *Global Environmental Change, 21*(4), 1163–1172. doi:10.1016/j.gloenvcha.2011.06.003

McDonald, C. V. (2014). Preservice primary teachers' written arguments in a socioscientific argumentation task. *The Electronic Journal of Science Education, 18*(7), 1–20.

McGee, E. O., & Bentley, L. (2017). The Troubled Success of Black Women in STEM. *Cognition and Instruction, 35*(4), 265–289. doi:10.1080/07370008.2017.1355211

McMillan, D. W., & Chavis, D. M. (1986). Sense of community: A definition and theory. *Journal of Community Psychology, 14*(1), 6–23. doi:10.1002/1520-6629(198601)14:1<6::AID-JCOP2290140103>3.0.CO;2-I

McNaughton, M. J. (2004). Educational drama in the teaching of education for sustainability. *Environmental Education Research, 10*(2), 139–155. doi:10.1080/13504620242000198140

McSharry, G., & Jones, S. (2000). Role-play in science teaching and learning. *The School Science Review, 82*, 73–82.

Mead, E., Roser-Renouf, C., Flora, J. A., Maibach, E. W., Leiserowitz, A., & Rimal, R. N. (2012). Information Seeking about Global Climate Change among Adolescents: The Role of Risk Perceptions, Efficacy Beliefs and Parental Influences. *Atlantic Journal of Communication, 20*(1), 31–52. doi:10.1080/15456870.2012.637027 PMID:22866024

Mead, M., & Métraux, R. (1957). Image of the scientist among high-school students: A pilot study. *Science, 126*(3270), 384–390. doi:10.1126cience.126.3270.384 PMID:17774477

Means, M. L., & Voss, J. F. (1996). Who reasons well? Two studies of informal reasoning among children of different grade, ability, and knowledge levels. *Cognition and Instruction, 14*(2), 139–178. doi:10.12071532690xci1402_1

Mehalik, M., Doppelt, Y., & Schunn, C. D. (2008). Middle school science through design-based learning versus scripted inquiry: Better overall science concept learning and equity gap reduction. *Journal of Engineering Education, 97*(1), 71–85. doi:10.1002/j.2168-9830.2008.tb00955.x

Merriam, S. B. (1998). *Qualitative research and case study applications in education*. Jossey-Bass.

Miles, M. B., & Huberman, A. M. (1994). *Qualitative data analysis: An expanded sourcebook*. Sage.

Miles, M. B., & Huberman, A. (1994). *Qualitative data analysis: An expanded sourcebook*. Sage.

Miller. (1987). Scientific Literacy in the United States. In *Communicating Science to the Public*. London: Wiley.

Miller. (1995). Scientific Literacy for Effective Citizenship. In *Science/Technology/Society as Reform in Science Education*. New York: State University Press of New York.

Miller, J. (1983). Scientific Literacy: A Conceptual and Empirical Review. *Daedalus*, *112*, 29–48.

Miller, J. (1998). The measurement of civic scientific literacy. *Public Understanding of Science (Bristol, England)*, *7*(3), 203–223. doi:10.1088/0963-6625/7/3/001

Ministry of National Education (MoNE). (2013). *Fen Bilimleri dersi öğretim program (İlkokul ve ortaokul 3, 4, 5, 6, 7 ve 8. Sınıflar)* [Science teaching program (Primary and middle 3rd, 4th, 5th, 6th, 7th and 8th grades]. Board of Education. Retrieved from http://mufredat.meb.gov.tr/ProgramDetay.aspx?PID=325

Ministry of National Education (MoNE). (2013). *İlköğretim Kurumları (İlkokullar ve Ortaokullar) Fen Bilimleri Dersi (3, 4, 5, 6, 7 ve 8. Sınıflar) Öğretim Programı* [Primary Education Institutions (Primary and Secondary Schools) Science Course (3, 4, 5, 6, 7 and 8th Grades) Curriculum]. Academic Press.

Ministry of National Education (MoNE). (2018). *Fen BİlimIeri Dersi Öğretim Programı (İlkokul ve Ortaokul 3, 4, 5, 6, 7 ve 8. Sınıflar)* [Science Education Curriculum (Primary and Secondary Schools 3, 4, 5, 6, 7 and 8th Grades)]. Academic Press.

Ministry of National Education. (2018). *Fen bilimleri dersi öğretim programı (İlkokul ve Ortaokul 3,4,5,6,7 ve 8. sınıflar)* [Science course instruction program (Elementary and middle school 3,4,5,6,7, and 8 grades]. Ankara, Turkey: Talim Terbiye Kurulu Başkanlığı.

Mitcham, C. (1994). *Thinking through technology: The path between engineering and philosophy*. University of Chicago Press.

Mogensen, F., & Schnack, K. (2010). The action competence approach and the 'new' discourses of education for sustainable development, competence and quality criteria. *Environmental Education Research*, *16*(1), 59–74. doi:10.1080/13504620903504032

MOIB Model | RI[2]. (2019, June 24). http://ri2.missouri.edu/content/MOIB-Model

Moll, L., Amanti, C., Neff, D., & Gonzalez, N. (1992). Funds of knowledge for teaching: Using a qualitative approach to connect homes and classrooms. *Theory into Practice*, *31*(2), 132–141. doi:10.1080/00405849209543534

Monson, D., & Besser, D. (2015). Smashing milk cartons: Third-grade students solve a real-world problem using the engineering design process, collaborative group work, and integrated STEM education. *Science and Children*, *52*(9), 38–43. doi:10.2505/4c15_052_09_38

Montes, L. D., & Rockley, M. G. (2002). Teacher perceptions in the selection of experiments. *Journal of Chemical Education*, *79*(2), 244–247. doi:10.1021/ed079p244

Moore, T. J., Glancy, A. W., Tank, K. M., Kersten, J. A., Smith, K. A., & Stohlmann, M. S. (2014). A framework for quality K-12 engineering education: Research and development. *Journal of Pre-College Engineering Education Research*, *4*(1), 1–13. doi:10.7771/2157-9288.1069

Moore, T. J., Stohlmann, M. S., Wang, H.-H., Tank, K. M., & Roehrig, G. H. (2014). Implementation and integration of engineering in K-12 STEM education. In J. Strobel, S. Purzer, & M. Cardella (Eds.), *Engineering in Pre-College Settings: Research into Practice*. Sense Publishers.

Moore, T. J., Tank, K. M., Glancy, A. W., & Kersten, J. A. (2015). NGSS and the landscape of engineering in K-12 state science standards. *Journal of Research in Science Teaching, 52*(3), 296–318. doi:10.1002/tea.21199

Morris, H. (2014). Socioscientific issues and multidisciplinarity in school science textbooks. *International Journal of Science Education, 36*(7), 1137–1158. doi:10.1080/09500693.2013.848493

Namdar, B. (2015). An examination of preservice science teachers' representational modality preferences during computer-supported knowledge organization. *Journal of Theory and Practice in Education, 11*(3), 949–970.

Namdar, B., & Shen, J. (2015). Modeling-Oriented Assessment in K-12 Science Education: A synthesis of research from 1980 to 2013 and new directions. *International Journal of Science Education, 37*(7), 993–1023. doi:10.1080/09500693.2015.1012185

Namdar, B., & Shen, J. (2016). Intersection of argumentation and the use of multiple representations in the context of socioscientific issues. *International Journal of Science Education, 38*(7), 1100–1132. doi:10.1080/09500693.2016.1183265

Namdar, B., & Shen, J. (2018). Knowledge organization through multiple representations in a computer-supported collaborative learning environment. *Interactive Learning Environments, 26*(5), 638–653. doi:10.1080/10494820.2017.1376337

National Academies of Sciences, Engineering, and Medicine. (2018). *How people learn II: Learners, contexts, and cultures*. National Academies Press.

National Academy of Sciences. (2004). *The engineer of 2020: Visions of engineering in the new century*. National Academies Press.

National Governors Association. (2010). *Common core state standards*.

National Research Council (NRC). (1996). *National science education standards*. Washington, DC: National Academy Press.

National Research Council (NRC). (2006). *Tech tally: Approaches to assessing technological literacy*. Washington, DC: National Academy Press.

National Research Council (NRC). (2012). *A Framework for K-12 Science Education: Practices, Crosscutting Concepts, and Core Ideas*. The National Academies Press.

National Research Council (NRC). (2012). *A framework for k-12 science education: Practices, crosscutting concepts, and core ideas*. The National Academy Press.

National Research Council (NRC). (2012). *A framework for K-12 science education: Practices, crosscutting concepts, and core ideas*. Washington, DC: National Academies Press.

National Research Council. (2000). *How people learn: Brain, mind, experience, and school: Expanded edition*. The National Academies Press.

National Research Council. (2011). *Successful K-12 STEM education: Identifying effective approaches in science, technology, engineering, and mathematics*. National Academies Press.

National Research Council. (2012). *A framework for K-12 science education: Practices, crosscutting concepts, and core ideas. Committee on conceptual framework for the new K-12 science education standards*. The National Academies Press.

National Research Council. (2012). *A framework for K-12 science education: Practices, crosscutting concepts, and core ideas*. National Academies Press.

National Research Council. (2012). *A Framework for K-12 Science Education: Practices, Crosscutting Concepts, and Core Ideas*. The National Academies Press., doi:10.17226/13165

National Science Education Standards. (1996). *National Academy of Sciences*. National Academy Press.

National Science Teachers Association (NSTA). (1982). Science/technology/society (STS). Washington, DC: NSTA.

National Society of Professional Engineers. (2015). *Code of ethics for engineers*. NSPE Press.

Nersessian, N. J. (2002). The cognitive basis of model-based reasoning in science. In P. Carruthers, S. Stich, & M. Siegal (Eds.), *The Cognitive Basis of Science* (pp. 133–153). Cambridge University Press. doi:10.1017/CBO9780511613517.008

Newton, M. H. (2016). *A longitudinal examination of a SSI-embedded experiential environmental education course and environmental behaviors* (Order No. 10244480). Available from Agricultural & Environmental Science Collection.

Next Generation Science Standards Lead States. (2013). *Next Generation Science Standards: For States, By States*. The National Academies Press.

NGSS Lead States. (2013). *Next Generation Science Standards: For states, by states*. National Academies Press.

NGSS Lead States. (2013). *Next generation science standards: For states, by states*. The National Academy Press.

NGSS Lead States. (2013). *The next generation science standards: For states, by states*. National Academies Press.

Nielson, J. A. (2012). Science in discussion: An analysis of the use of science content in socioscientific discussion. *Science Education*, *96*(3), 428–456. doi:10.1002ce.21001

Nussbaum, E. M., & Kardash, C. M. (2005). The effects of goal instructions and text on the genera- tion of counterarguments during writing. *Journal of Educational Psychology, 97*(2), 157–169. doi:10.1037/0022-0663.97.2.157

Nussbaum, E. M., & Schraw, G. (2007). Promoting argument-counterargument integration in students' writing. *Journal of Experimental Education, 76*(1), 59–92. doi:10.3200/JEXE.76.1.59-92

Nye, D. E. (2006). *Technology matters: Questions to live with.* MIT Press.

Odegaard, M. (2003). Dramatic science: A critical review of drama in science education. *Studies in Science Education, 39*(1), 75–102. doi:10.1080/03057260308560196

OECD. (2003). *The PISA 2003 assessment framework.* Retrieved November 2008 from: http://www.pisa.oecd.org/dataoecd/46/14/33694881.pdf

OECD. (2006). *Assessing Scientific, Reading and Mathematical Literacy: A Framework for PISA 2006, PISA.* OECD Publishing. doi:10.1787/9789264026407-

OECD. (2007). PISA 2006: Science Competencies for Tomorrow's World: Volume 1: Analysis, PISA. OECD Publishing. doi:10.1787/9789264040014-en

OECD. (2009). *PISA 2006 Technical Report, PISA.* OECD Publishing. doi:10.1787/9789264048096-

OECD. (2012). What kinds of careers do boys and girls expect for themselves? In *PISA in focus, No 14.* OECD Publishing. doi:10.1787/5k9d417g2933-

OECD. (2017). *PISA 2015 assessment and analytical framework: Science, reading, mathematics, financial literacy and collaborative problem solving* (Rev. ed.). Program for International Student Assessment, Organisation for Economic Co-operation and Development. doi:10.1787/9789264281820-

OECD. (2018). *PISA for development assessment and analytical framework: Reading, Mathematics and Science.* OECD.

Ogunkola, B. J. (2013). Scientific literacy: Conceptual overview, importance and strategies for improvement. *Journal of Educational and Social Research, 3*(1), 265–274.

Olafson, L. J., & Schraw, G. (2002). Some final thoughts on the epistemological melting pot. *Issues in Education, 8*(2), 233–247.

Olson, J. K. (2013). The purposes of schooling and the nature of technology: The end of education? In M. P. Clough, J. K. Olson, & D. S. Niederhauser (Eds.), *The nature of technology: Implications for learning and teaching.* Sense Publishers. doi:10.1007/978-94-6209-269-3_13

Olson, J., Tippett, C., Milford, T., Ohana, C., & Clough, M. (2015). Science teacher preparation in a north American context. *Journal of Science Teacher Education, 26*(1), 7–28. doi:10.100710972-014-9417-9

Ong, M. (2005). Body projects of young women of color in physics: Intersections of gender, race, and science. *Social Problems, 52*(4), 593–617. doi:10.1525p.2005.52.4.593

Osborne, J., Erduran, S., & Simon, S. (2004). Enhancing the quality of argumentation in school science. *Journal of Research in Science Teaching, 41*(10), 994–1020. doi:10.1002/tea.20035

Österlind, K. (2005). Concept formation in environmental education: 14-year olds' work on the intensified greenhouse effect and the depletion of the ozone layer. *International Journal of Science Education, 27*, 891-908. . doi:10.1080/09500690500038264

Owens, D. C., Sadler, T. D., & Zeidler, D. L. (2017). Controversial issues in the science classroom. *Phi Delta Kappan, 99*(4), 45–49. doi:10.1177/0031721717745544

Ozdem, Y., Dal, B., Öztürk, N., Sönmez, D., & Alper, U. (2014). What is that thing called climate change? An investigation into the understanding of climate change by seventh-grade students. *International Research in Geographical and Environmental Education, 23*(4), 294–313. doi:10 .1080/10382046.2014.946323

Özden, M. (2015). Prospective elementary teachers' views about socioscientific issues: A concurrent parallel design study. *International Journal of Elementary Education, 7*(3), 333–354. Retrieved July 28, 2019, from

Öztürk, N., & Yılmaz-Tüzün, Ö. (2017). Preservice science teachers' epistemological beliefs and informal reasoning regarding socioscientific issues. *Research in Science Education, 47*(6), 1275–1304. doi:10.100711165-016-9548-4

Pacey, A. (1983). *The culture of technology*. MIT Press.

Pallant, A., & Lee, H.-S. (2015). Constructing scientific arguments using evidence from dynamic computational climate models. *Journal of Science Education and Technology, 24*(2), 378–395. doi:10.100710956-014-9499-3

Papadimitriou, V. (2004). Prospective primary teachers' understanding of climate change, greenhouse effect, and ozone layer depletion. *Journal of Science Education and Technology, 13*, 299-307. :JOST.0000031268.72848.6d doi:10.1023/B

Passmore, C., Gouvea, J. S., & Giere, R. (2014). Models in Science and in Learning Science: Focusing Scientific Practice on Sense-making. In M. R. Matthews (Ed.), *International Handbook of Research in History, Philosophy and Science Teaching* (pp. 1171–1202). Springer Netherlands. doi:10.1007/978-94-007-7654-8_36

Patronis, T., Potari, D., & Spiliotopoulou, V. (1999). Students' argumentation in decision-making on a socio-scientific issue: Implications for teaching. *International Journal of Science Education, 21*(7), 745–754. doi:10.1080/095006999290408

Patton, M. Q. (1990). *Qualitative evaluation and research methods*. Sage.

Patton, M. Q. (2002). *Qualitative research and evaluation methods* (3rd ed.). Sage.

Paul, R., & Elder, L. (2009). Critical thinking, creativity, ethical reasoning: A unity of opposites. In D. Ambrose & T. Cross (Eds.), *Morality, Ethics, and Gifted Minds* (pp. 117–132). Springer. doi:10.1007/978-0-387-89368-6_8

Pearson, P. D., & Hiebert, E. H. (2013). Understanding the common core state standards. *Teaching with the Common Core Standards for English language arts PreK–2*, 1-21.

Pedretti, E. (2003). Teaching science, technology, society and environment (STSE) education. In D. L. Zeidler (Ed.), *The role of moral reasoning on socioscientific issues and discourse in science education* (pp. 219–239). Springer. doi:10.1007/1-4020-4996-X_12

Peel, A., Nguyen, H., Friedrichsen, P., Sadler, T., & Zangori, L. (2019, April 3). *A Professional Design Cycle to Support Model-Oriented Issue-Based Teaching* [Poster]. NARST- 2019 Annual International Conference, Baltimore, MD.

Peel, A., Sadler, T. D., Andrew, K., Zangori, L., & Friedrichsen, P. (2017). Climate change as an issue for socio-scientific issues teaching and learning. In D. P. Shepardson, A. Roychoudhury, & A. S. Hirsch (Eds.), *Teaching and learning about climate change. A framework for educators* (pp. 153–165). Routledge. doi:10.4324/9781315629841-11

Peel, A., Zangori, L., Friedrichsen, P., Hayes, E., & Sadler, T. (2019). Students' model-based explanations about natural selection and antibiotic resistance through socio-scientific issues-based learning. *International Journal of Science Education, 41*(4), 510–532. doi:10.1080/09500693.2018.1564084

Pew Initiative on Food and Biotechnology. (2000). *Public Sentiment about Genetically Modified Food.* Available at: http://www.pewagbiotech.org

Pew Research Center. (2015). *Amid debate over labeling GM foods, most Americans believe they're unsafe*. Retrieved from http://pewrsr.ch/1MhCFSN

Pew Research Center. (2018). *Americans are narrowly divided over health effects of genetically modified foods*. Retrieved from https://pewrsr.ch/2FDG8RB

Pew Research Center. (2020). *About half of U.S. adults are wary of health effects of genetically modified foods, but many also see advantages*. Retrieved from https://pewrsr.ch/2w5UmGU

Piaget, J. (1926). *The language and thought of the child*. Harcourt Brace.

Pickens, J. (2005). Attitudes and Perceptions. In N. Borkowski (Ed.), *Organizational Behavior and Health Care* (pp. 43–76). Jones & Barlett Publishing.

Pitiporntapin, S., Yutakom, N., & Sadler, T. (2016). Thai pre-service science teachers' struggles in using socio-scientific issues (SSIs) during practicum. *Asia-Pacific Forum on Science Learning and Teaching, 17*(2), 1–21.

Plutzer, E., McCaffrey, M., Hannah, A. L., Rosenau, J., Berbeco, M., & Reid, A. H. (2016). Climate confusion among U.S. teachers. *Science, 351*(6274), 664–665. doi:10.1126cience.aab3907 PMID:26912845

Polanyi, M. (1958). *Personal knowledge: Towards a post-critical philosophy*. University of Chicago Press.

Powell, W. A. (2014). *The effects of emotive reasoning on secondary school students' decision-making in the context of socioscientific issues*. https://scholarcommons.usf.edu/etd/5385/

Powell, W. A. (2020). Framework for Using Geospatial Technology to Promote Middle School Students' Critical Thinking on Socioscientific Issues. In *Next Generation Digital Tools and Applications for Teaching and Learning Enhancement* (pp. 47–78). IGI Global. doi:10.4018/978-1-7998-1770-3.ch004

Powell, W. A., & Fuchs, D. (2019). Using Socioscientific Issues to Enhance Evidence-Based Reasoning Among Middle School Students. In S. Robinson & V. Knight (Eds.), *Handbook of Research on Critical Thinking and Teacher Education Pedagogy* (pp. 150–176). IGI Global. doi:10.4018/978-1-5225-7829-1.ch009

Prados, J. W. (1998). *Engineering education in the United States: Past, present, and future*. Paper presented at the International Conference on Engineering Education.

President's Council of Advisors on Science and Technology. (2010). *Report to the president: Prepare and inspire: K-12 education in science, technology, engineering, and mathematics (STEM) for America's future*. Washington, DC: Executive Office of the President. Retrieved from https://obamawhitehouse.archives.gov/sites/default/files/microsites/ostp/pcast-stem-ed-final.pdf

Presley, M. L., Sickel, A. J., Muslu, N., Merle-Johnson, D., Witzig, S. B., Izci, K., & Sadler, T. D. (2013). A Framework for socio-Scientific issues based education. *Science Educator*, *22*(1), 26–32.

Presley, M. L., Sickel, A. J., Muslu, N., Merle-Johnson, D., Witzig, S. B., Izci, K., & Sadler, T. D. (2013). A framework for socio-scientific issues based education. *Science Educator*, *22*, 26.

Ramsey, J. (1993). The science education reform movement: Implications for social responsibility. *Science Education*, *77*(2), 235–258. doi:10.1002ce.3730770210

Ratcliffe, M. (1999). Evaluation of abilities in interpreting media reports of scientific Research. *International Journal of Science Education*, *21*(10), 1085–1099. doi:10.1080/095006999290200

Redclift, M. (2005). Sustainable Development (1987–2005): An Oxymoron Comes of Age. *Sustainable Development*, *13*(4), 212–227. doi:10.1002d.281

Reinfried, S., Aeschbacher, U., & Rottermann, B. (2012). Improving Students' conceptual Understanding of the Greenhouse Effect Using Theory-based Learning Materials That Promote Deep Learning. *International Research in Geographical and Environmental Education*, *21*(2), 155–178. doi:10.1080/10382046.2012.672685

Remillard, J. T., & Bryans, M. B. (2004). Teachers' orientations toward mathematics curriculum materials: Implications for teacher learning. *Journal for Research in Mathematics Education*, *35*(5), 352–358. doi:10.2307/30034820

Responsive Classroom. (2020). *Responsive Classroom: Principals and Practices*. Retrieved from https://www.responsiveclassroom.org/about/principles-practices/

Rimm-Kaufman, S., & Hulleman, C. S. (2015). SEL in elementary school settings: Identifying mechanisms that matter. In J. A. Durlak, C. E. Domitrovich, R. P. Weissberg, & T. P. Gullotta (Eds.), *Handbook of social and emotional learning: Research and practice* (pp. 151–166). Guilford.

Roberts, D. A. (2007). Scientific literacy/Science literacy. In S. K. Abell & N. G. Lederman (Eds.), *Handbook of research on science education* (pp. 729–780). Lawrence Erlbaum Associates.

Roberts, D. A. (2007). Scientific Literacy/Science Literacy. In S. K. Abell & N. G. Lederman (Eds.), *Handbook of Research on Science Education*. Lawrence Erlbaum Associates., doi:10.4324/9780203824696-32

Roberts, D. A. (2011). Competing visions of scientific literacy: The influence of a science curriculum policy image. In C. Linder, L. Ostman, D. A. Roberts, P. Wickman, G. Erickson, & A. MacKinnon (Eds.), *Promoting scientific literacy: Science education research in transaction* (pp. 11–27). Routledge/Taylor & Francis Group.

Roberts, D. A. (2013). Scientific literacy/science literacy. In *Handbook of research on science education* (pp. 743–794). Routledge.

Roberts, D. A., & Bybee, R. (2014). Scientific literacy, science literacy, and science education. In N. G. Lederman & S. K. Abell (Eds.), *Handbook of research on science education* (Vol. 2, pp. 545–558). Routledge.

Roberts, D. A., & Bybee, R. W. (2014). Scientific literacy, science literacy, and science education. In N. G. Lederman & S. K. Abell (Eds.), *Handbook of research in science education* (Vol. 2, pp. 545–558). Routledge.

Roberts, D. A., & Bybee, R. W. (2014). Scientific Literacy, Science Literacy, and Science Education. In N. G. Lederman & S. K. Abell (Eds.), *Handbook of Research on Science Education* (Vol. 2, pp. 545–558). Routledge. doi:10.4324/9780203097267-38

Roeser, S. (2012). Emotional Engineers: Toward Morally Responsible Design. *Science and Engineering Ethics*, *18*, 103–115.

Romine, W., Sadler, T. D., & Dauer, J. (in review). *Measurement of socio-scientific reasoning (SSR) and exploration of how students' SSR competencies improve over time.*

Romine, W. L., Sadler, T. D., & Kinslow, A. T. (2017). Assessment of scientific literacy: Development and validation of the Quantitative Assessment of Socio-Scientific Reasoning (QuASSR). *Journal of Research in Science Teaching*, *54*(2), 274–295. doi:10.1002/tea.21368

Roth, K. (2014). Elementary science teaching. In N. Lederman & S. Abell (Eds.), *Handbook of research on science education* (Vol. 2, pp. 361–394). Routledge.

Russell, J. S., Hanna, A., Bank, L. C., & Shapira, A. (2007). Education in construction engineering and management built on tradition: Blueprint for tomorrow. *Journal of Construction Engineering and Management*, *133*(9), 661–668.

Rust, F. O'C. (2009). Teacher research and the problem of practice. *Teachers College Record, 111*(8), 1882–1893.

Rust, F., & Meyers, E. (2006). The bright side: Teacher research in the context of educational reform and policy-making. *Teachers and Teaching, 12*(1), 69–86. doi:10.1080/13450600500365452

Sadler, T. (2007). *The aims of science education: Unifying the fundamental and derived senses of scientific literacy*. Paper presented at the Linnaeus Tercentenary 2007 Symposium "Promoting Scientific Literacy," Uppsala University, Uppsala, Sweden.

Sadler, T. D., Friedrichsen, P., & Zangori, L. (2019). A framework for teaching for socio-scientific issue and model based learning (SIMBL). *Educação e Fronteiras, 9*(25), 08–26.

Sadler, T. D., Friedrichsen, P., & Zangori, L. (2019). A framework for teaching for socio-scientific issue and model based learning (SIMBL). *Educação e Fronteiras/Education and Borders, 9*(25), 8-26.

Sadler, P. M., Coyle, H. P., & Schwartz, M. (2000). Engineering competitions in the middle school classroom: Key elements in developing effective design challenges. *Journal of the Learning Sciences, 9*(3), 299–327. doi:10.1207/S15327809JLS0903_3

Sadler, T. (2004). Informal reasoning regarding socioscientific issues: A critical review of the research. *Journal of Research in Science Teaching, 41*(5), 513–536. doi:10.1002/tea.20009

Sadler, T. (2004). Moral sensitivity and its contribution to the resolution of socio-scientific issues. *Journal of Moral Education, 33*(3), 339–358. doi:10.1080/0305724042000733091 PMID:16317839

Sadler, T. (2011). Situating socioscientific issues in classrooms as a means of achieving goals of science education. In T. Sadler (Ed.), *Socio-scientific issues in the classroom: Teaching, learning and research* (pp. 1–9). Springer. doi:10.1007/978-94-007-1159-4_1

Sadler, T. D. (2004b). Moral and ethical dimensions of socioscientific decision-making as integral components of scientific literacy. *Science Educator, 13*(1), 39–48.

Sadler, T. D. (2009). Situated learning in science education: Socio-scientific issues as contexts for practice. *Studies in Science Education, 45*(1), 1–42. doi:10.1080/03057260802681839

Sadler, T. D. (2009). Socioscientific issues in science education: Labels, reasoning, and transfer. *Cultural Studies of Science Education, 4*(3), 697–703. doi:10.100711422-008-9133-x

Sadler, T. D. (2011). Socio-scientific Issues-Based Education: What We Know About Science Education in the Context of SSI. In T. D. Sadler (Ed.), *Socio-scientific Issues in the Classroom: Teaching, Learning and Research* (pp. 355–369). Springer Netherlands. doi:10.1007/978-94-007-1159-4_20

Sadler, T. D. (Ed.). (2011). *Socio-scientific issues in the classroom: Teaching, learning and research*. Springer. doi:10.1007/978-94-007-1159-4

Sadler, T. D., Amirshokoohi, A., Kazempour, M., & Allspaw, K. M. (2006). Socioscience and ethics in science classrooms: Teacher perspectives and strategies. *Journal of Research in Science Teaching*, *43*(4), 353–376. doi:10.1002/tea.20142

Sadler, T. D., Barab, S. A., & Scott, B. (2007). What do students gain by engaging in socioscientific inquiry? *Research in Science Education*, *37*(4), 371–391. doi:10.100711165-006-9030-9

Sadler, T. D., Chambers, F. W., & Zeidler, D. L. (2004). Student conceptualizations of the nature of science in response to a socioscientific issue. *International Journal of Science Education*, *26*(4), 387–409. doi:10.1080/0950069032000119456

Sadler, T. D., & Donnelly, L. A. (2006). Socioscientific Argumentation: The effects of content knowledge and morality. *International Journal of Science Education*, *28*(12), 1463–1488. doi:10.1080/09500690600708717

Sadler, T. D., Foulk, J. A., & Friedrichsen, P. J. (2017). Evolution of a Model for Socio-Scientific Issue Teaching and Learning. *International Journal of Education in Mathematics. Science and Technology*, *5*(2), 75–87.

Sadler, T. D., Romine, W. L., & Topcu, M. S. (2016). Learning science content through socio-scientific issues based instruction: A multi-level assessment study. *International Journal of Science Education*, *38*(10), 1622–1635. doi:10.1080/09500693.2016.1204481

Sadler, T. D., & Zeidler, D. (2009). Scientific literacy, PISA, and socioscientific discourse: Assessment for progressive aims of science education. *Journal of Research in Science Teaching*, *46*(8), 909–921. doi:10.1002/tea.20327

Sadler, T. D., & Zeidler, D. L. (2004). The morality of socioscientific issues: Construal and resolution of genetic engineering dilemmas. *Science Education*, *88*(1), 4–27. doi:10.1002ce.10101

Sadler, T., Klosterman, M., & Topcu, M. (2011). Learning science content and socio-scientific reasoning through classroom explorations of global climate change. In T. D. Sadler (Ed.), *Socio-scientific issues in the classroom: Teaching, learning and research* (pp. 45–77). Springer. doi:10.1007/978-94-007-1159-4_4

Sadler, T., & Zeidler, D. (2005). The significance of content knowledge for informal reasoning regarding socioscientific issues: Applying genetics knowledge to genetic engineering issues. *Science Education*, *89*(1), 71–93. doi:10.1002ce.20023

Saldaña, J. (2009). *The coding manual for qualitative researchers*. Sage Publications Ltd.

Sampson, V., & Clark, D. B. (2008). Assessment of the ways students generate arguments in science education: Current perspectives and recommendations for future directions. *Science Education*, *92*(3), 447–472. doi:10.1002ce.20276

Sampson, V., Enderle, P., & Grooms, J. (2013). Argumentation in science education: Helping students understand the nature of scientific argumentation so they can meet the new science standards. *Science Teacher (Normal, Ill.)*, *80*(5), 30–33. doi:10.2505/4/tst13_080_05_30

Sampson, V., Enderle, P., Grooms, J., & Witte, S. (2013). Writing to learn by learning to write during the school science laboratory: Helping middle and high school students develop argumentative writing skills as they learn core ideas. *Science Education, 97*(5), 643–670. doi:10.1002ce.21069

Sanders, M. (2009). STEM, STEM education, STEM mania. *Technology Teacher, 68*(4), 20–26.

San, İ. (1990). Eğitimde yaratıcı drama [Creative drama in education]. *Ankara Üniversitesi Eğitim Bilimleri Fakültesi Dergisi, 23*(2), 573–583.

Sauer, G. (2012). *Redesigning engineering curricula for the 21st century*. Iowa State University.

Saunders, K. J., & Rennie, L. J. (2013). A pedagogical model for ethical inquiry into socioscientific issues in science. *Research in Science Education, 43*(1), 253–274. doi:10.100711165-011-9248-z

Savin-Baden, M., & Major, C. H. (2013). *Qualitative Research: The essential guide to theory and practice*. Routledge.

Scherz, Z., & Oren, M. (2006). How to change students' images of science and technology. *Science Education, 90*(6), 965–985. doi:10.1002ce.20159

Schoenewolf, G. (1990). Emotional contagion: Behavioral induction in individuals and groups. *Modern Psychoanalysis, 15*(1), 49–61.

Schwarz, C. V., Reiser, B. J., Davis, E. A., Kenyon, L., Achér, A., Fortus, D., Shwartz, Y., Hug, B., & Krajcik, J. (2009). Developing a learning progression for scientific modeling: Making scientific modeling accessible and meaningful for learners. *Journal of Research in Science Teaching, 46*(6), 632–654. doi:10.1002/tea.20311

Schwarz, C. V., & White, B. Y. (2005). Metamodeling Knowledge: Developing Students' Understanding of Scientific Modeling. *Cognition and Instruction, 23*(2), 165–205. doi:10.12071532690xci2302_1

Segal, E. A. (2011). Social empathy: A model built on empathy, contextual understanding, and social responsibility that promotes social justice. *Journal of Social Service Research, 37*(3), 266–277. doi:10.1080/01488376.2011.564040

Shamos, M. (1995). *The myth of scientific literacy*. Rutgers University Press.

Shaw, J. A. (2012). Using small group debates to actively engage students in an introductory microbiology course. *Journal of Microbiology & Biology Education, 13*(2), 155–160. doi:10.1128/jmbe.v13i2.420 PMID:23653803

Shen, J., Lei, J., Chang, H.-Y., & Namdar, B. (2014). Technology-enhanced, modeling-based instruction (TMBI) in science education. In J. M. Spector, M. D. Merrill, J. Elen, & M. J. Bishop (Eds.), *Handbook of research on educational communications and technology* (pp. 529–540). Springer. doi:10.1007/978-1-4614-3185-5_41

Shepardson, D. P., Choi, S., Niyogi, D., & Charusombat, U. (2011). Seventh grade students' mental models of the greenhouse effect. *Environmental Education Research, 17*, 1-17. . doi:10.1080/13504620903564549

Sherif, M., & Cantril, H. (1945). The psychology of attitudes: I. *Psychological Review, 52*(6), 295–319. doi:10.1037/h0062252 PMID:21012631

Shulman, L. (1987). Knowledge and teaching: Foundations of the new reform. *Harvard Educational Review, 57*(1), 1–22. doi:10.17763/haer.57.1.j463w79r56455411

Siebert, E. D., & McIntosh, W. J. (Eds.). (2001). *College pathways to the science education standards*. NSTA Press.

Simmons, M. L., & Zeidler, D. L. (2003). Beliefs in the nature of science and responses to socioscientific issues. In D. L. Zeidler (Ed.), *The Role of Moral Reasoning on Socioscientific Issues and Discourse in Science Education* (pp. 81–94). Kluwer Academic Publishers. doi:10.1007/1-4020-4996-X_5

Simonneaux, L. (2001). Role-play or debate to promote students' argumentation and justification on an issue in animal transgenesis. *International Journal of Science Education, 23*(9), 903–927. doi:10.1080/09500690010016076

Simonneaux, L. (2002). Analysis of classroom debating strategies in the field of biotechnology. *Journal of Biological Education, 37*(1), 9–12. doi:10.1080/00219266.2002.9655839

Simonneaux, L., & Simonneaux, J. (2009). Students' socio-scientific reasoning on controversies from the viewpoint of education for sustainable development. *Cultural Studies of Science Education, 4*(3), 657–687. doi:10.100711422-008-9141-x

Skamp, K., Boyes, E., & Stanisstreet, M. (2004). Students' ideas about air quality. *Research in Science Education, 34*(3), 313–342. doi:10.1023/B:RISE.0000044643.24770.5c

Skamp, K., Boyes, E., & Stannistreet, M. (2009). Global warming responses at the primary secondary interface 1. students' beliefs and willingness to act. *Australian Journal of Environmental Education, 25*, 15–30. doi:10.1017/S0814062600000379

Sklavounos, S., & Rigas, F. (2006). Estimation of safety distances in the vicinity of fuel gas pipeline. *Journal of Loss Prevention in the Process Industries, 19*(1), 24–31. doi:10.1016/j.jlp.2005.05.002

Slavin, R. E. (1980). Cooperative learning. *Review of Educational Research, 50*(2), 315–342. doi:10.3102/00346543050002315 PMID:6698102

Smith, G. A., & Williams, D. R. (1999). *Ecological education in action: On weaving education, culture, and the environment*. State University of New York Press.

Solomon, J., & Aikenhead, G. (1994). *STS Education: International Perspectives on Reform. Ways of Knowing Science Series*. Teachers College Press.

Somers, J. (1994). *Drama in the curriculum.* Cassell Education.

Son, W. C. (2011). Engineering design and philosophy of technology. *Journal of the Society of Philosophical Studies, 94*, 107–136.

Stahl, G., Ludvigsen, S., Law, N., & Cress, U. (2014). CSCL artifacts. *International Journal of Computer-Supported Collaborative Learning, 9*(3), 237–245. doi:10.100711412-014-9200-0

Stahl, S., Hynd, C., Britton, B., McNish, M., & Bosquet, D. (1996). What happens when students read multiple source documents in history? *Reading Research Quarterly, 31*(4), 430–456. doi:10.1598/RRQ.31.4.5

Staker, B. H., & Horn, M. B. (2012). *Classifying K – 12 Blended Learning.* Academic Press.

Stevenson, K. T., Peterson, M. N., & Bondell, H. D. (2016). The Influence of Personal Beliefs, friends, and Family in Building Climate Change Concern among Adolescents. *Environmental Education Research, 4622*, 1–14. doi:10.1080/13504622.2016.1177712

Stevenson, K. T., Peterson, M. N., Carrier, S. J., Strnad, R. L., Bondell, H. D., Kirby-Hathaway, T., & Moore, S. E. (2014). Role of Significant Life Experiences in Building Environmental Knowledge and Behavior Among Middle School Students. *The Journal of Environmental Education, 45*(3), 163–177. doi:10.1080/00958964.2014.901935

Stolz, M., Witteck, T., Marks, R., & Eilks, I. (2013). Reflecting socio-scientific issues for science education coming from the case of curriculum development on doping in chemistry education. *Eurasia Journal of Mathematics, Science and Technology Education, 9*(4), 361–370.

Strobel, J., Hess, J., Pan, R., & Wachter Morris, C. A. (2013). Empathy and care within engineering: Qualitative perspectives from engineering faculty and practicing engineers. *Engineering Studies, 5*(2), 137–159. doi:10.1080/19378629.2013.814136

Sumida, M. (2010). Identifying twice-exceptional children and three gifted styles in the Japanese primary science classroom. *International Journal of Science Education, 32*(15), 2097–2111. doi:10.1080/09500690903402018

Tal, R. T., & Kedmi, Y. (2006). Teaching socioscientific issues: Classroom culture and students' performances. *Cultural Studies of Science Education, 1*(4), 615–644. doi:10.100711422-006-9026-9

Tashakkori, A., & Teddlie, C. (1998). *Mixed methodology: Combining qualitative and quantitative approaches.* Sage.

Taylor, R. D., Oberle, E., Durlak, J. A., & Weissberg, R. P. (2017). Promoting positive youth development through school-based social and emotional learning interventions: A meta-analysis of follow-up effects. *Child Development, 88*(4), 1156–1171. doi:10.1111/cdev.12864 PMID:28685826

Teddlie, C., & Tashakkori, A. (2006). A general typology of research designs featuring mixed methods. *Research in the Schools, 13*(1), 12–28.

Teo, T. W., Tan, K. C. D., Yan, Y. K., Teo, Y. C., & Yeo, L. W. (2014). How flip teaching supports undergraduate chemistry laboratory learning. *Chemistry Education Research and Practice*, *15*(4), 550–567. doi:10.1039/C4RP00003J

Tidemand, S., & Nielsen, J. A. (2017). The role of socioscientific issues in biology teaching: From the perspective of teachers. *International Journal of Science Education*, *39*(1), 44–61. do i:10.1080/09500693.2016.1264644

Topçu, M. S., Sadler, T. D., & Yilmaz-Tuzun, O. (2010). Preservice science teachers' informal reasoning about socioscientific issues: The influence of issue of context. *International Journal of Science Education*, *32*(18), 2475–2495. doi:10.1080/09500690903524779

Topçu, M. S., Yılmaz-Tüzün, Ö., & Sadler, T. D. (2011). Turkish preservice science teachers' informal reasoning regarding socioscientific issues and the factors influencing their informal reasoning. *Journal of Science Teacher Education*, *22*(4), 313–332. doi:10.100710972-010-9221-0

Tosun, T. (2000). The beliefs of preservice elementary teachers toward science and science teaching. *School Science and Mathematics*, *100*(7), 374–379. doi:10.1111/j.1949-8594.2000.tb18179.x

Toulmin, S. (1958). *The uses of argument*. Cambridge University Press.

Treagust, D. F., Chittleborough, G., & Mamiala, T. L. (2002). Students' understanding of the role of scientific models in learning science. *International Journal of Science Education*, *24*(4), 357–368. doi:10.1080/09500690110066485

Treagust, D., & Harrison, A. (1999). The genesis of effective scientific explanations for the classroom. In J. Loughran (Ed.), *researching teaching methodologies and practices for understanding pedagogy* (pp. 28–43). Routledge.

Trochim, W. (2020). *Research methods data base*. Retrieved from https://conjointly.com/kb/

Tsai, C. Y. (2018). The effect of online argumentation of socio-scientific issues on students' scientific competencies and sustainability attitudes. *Computers & Education*, *116*, 14–27. doi:10.1016/j.compedu.2017.08.009

Uluçınar-Sağır, Ş., & Kılıç, Z. (2013). İlköğretim öğrencilerinin bilimin doğasını anlama düzeylerine bilimsel tartışma odaklı öğretimin etkisi. *Hacettepe Üniversitesi Eğitim Fakültesi Dergisi*, *44*, 308–318.

UNEP. (2012). *21 Issues for the 21st Century: Result of the UNEP Foresight Process on Emerging Environmental Issues*. United Nations Environment Programme (UNEP). www.unep.org/pdf/Foresight_Report-21_Issues_for_the_21st_Century.pdf

UNESCO. (2019). *Science for a sustainable future*. Retrieved from https://en.unesco.org/themes/science-sustainable-future

Upadhyay, B. R. (2005). Using students' lived experiences in an urban science classroom: An elementary school teacher's thinking. *Science Education*, *90*(1), 94–110. doi:10.1002ce.20095

USGCRP. (2017). Climate Science Special Report: Fourth National Climate Assessment, Volume I. U.S. Global Change Research Program. doi:10.7930/J0J964J6

Vakil, S., & Ayers, R. (2019). The racial politics of STEM education in the USA: Interrogations and explorations. *Race, Ethnicity and Education, 22*(4), 449–458. doi:10.1080/13613324.2019.1592831

Valli, L. (1997). Listening to other voices: A description of teacher reflection in the United States. *Peabody Journal of Education, 72*(1), 67–88. doi:10.120715327930pje7201_4

Van Driel, J. H., & Verloop, N. (2002). Experienced teachers' knowledge of teaching and learning of models and modelling in science education. *International Journal of Science Education, 24*(12), 1255–1272. doi:10.1080/09500690210126711

Vesterinen, V.-M., Tolppanen, S., & Aksela, M. (2016). Towards Citizenship Science Education: What Students do to make the World a Better Place? *International Journal of Science Education, 38*(1), 30–50. doi:10.1080/09500693.2015.1125035

Vieira, R. M., & Tenreiro-Vieira, C. (2016). Fostering scientific literacy and critical thinking in elementary science education. *International Journal of Science and Mathematics Education, 14*(4), 659–680. doi:10.100710763-014-9605-2

Vilorio, D. (2014). STEM 101: Intro to tomorrow's jobs. *Occupational Outlook Quarterly*, 2–12. https://www.bls.gov/careeroutlook/

Vincenti, W. G. (1990). *What engineers know and how they know it*. John Hopkins University Press.

Volti, R. (2010). *Society and technological change*. Worth.

Vreede, C. D., Warner, A., & Pitter, R. (2017). Facilitating Youth to Take Sustainability Actions: The Potential of Peer Education. *The Journal of Environmental Education, 45*(1), 37–46. doi:10.1080/00958964.2013.805710

Waight, N., & Abd-El-Khalick, F. (2012). Nature of technology: Implications for design, development, and enactment of technological tools in school science classrooms. *International Journal of Science Education, 34*(18), 2875–2905. doi:10.1080/09500693.2012.698763

Walker, K. A., & Zeidler, D. L. (2007). Promoting discourse about socioscientific issues through scaffolded inquiry. *International Journal of Science Education, 29*(11), 1387–1410. doi:10.1080/09500690601068095

Walther, J., Miller, S. E., & Sochacka, N. W. (2017). A model of empathy in engineering as a core skill, practice orientation, and professional way of being. *Journal of Engineering Education, 106*(1), 23–148. doi:10.1002/jee.20159

Watkins, J., Spencer, K., & Hammer, D. (2014) Examining young students' problem scoping in engineering design. *Journal of Pre-College Engineering Education Research, 4*(1), Article 5.

Wilkerson, S. (2008). Application of the Paired t-test. *Xavier University Of Louisiana's Undergraduate Research Journal, 5*(1).

Wilson, K. M. (2000). Drought, debate, and uncertainty: Measuring reporters' knowledge and ignorance about climate change. *Public Understanding of Science (Bristol, England), 9*(1), 1–13. doi:10.1088/0963-6625/9/1/301

Wilson-Lopez, A., & Minichiello, A. (2017). Disciplinary literacy in engineering. *Journal of Adolescent & Adult Literacy, 61*(1), 7–14. doi:10.1002/jaal.658

World Health Organization (WHO). (2010b). *The World Health Report - Health Systems Financing: The Path to Universal Coverage*. Retrieved from http://www.who.int/entity/whr/2010/whr10_en.pdf

World Health Organization. (2000). *Turning the tide of malnutrition: responding to the challenge of the 21st century*. WHO.

Wu, Y. T., & Tsai, C. C. (2011). High school students' informed reasoning regarding a socio-scientific issue, with relation to scientific epistemological beliefs and cognitive structures. *International Journal of Science Education, 33*(3), 371–400. doi:10.1080/09500690903505661

Wu, Y.-T., & Tsai, C.-C. (2007). High school students' informal reasoning on a socio-scientific issue: Qualitative and quantitative analyses. *International Journal of Science Education, 29*(9), 1163–1187. doi:10.1080/09500690601083375

Yager, R. E. (1991). The constructivist learning model. *Science Teacher (Normal, Ill.), 58*(6), 52.

Yapıcıoğlu, A., & Kaptan, F. (2017). A mixed method research study on the effectiveness of socioscientific issue based instruction. *Education in Science, 42*(192), 113–137.

Yerdelen, S., Cansiz, M., Cansiz, N., & Akcay, H. (2018). Promoting preservice teachers' attitudes toward socioscientific issues. *Journal of Education in Science Environment and Health, 4*(1), 1–11.

Yoon, J., & Onchwari, J. A. (2006). Teaching young children science: Three key points. *Journal of Early Childhood Education, 33*(6), 419–423. doi:10.100710643-006-0064-4

Zangori, L., Ke, L., Sadler, T. D., & Peel, A. (2020). Exploring primary students causal reasoning about ecosystems. *International Journal of Science Education*. DOI: 10.1080/09500693.2020.1783718

Zangori, L., & Forbes, C. T. (2014). Scientific Practices in Elementary Classrooms: Third-Grade Students' Scientific Explanations for Seed Structure and Function. *Science Education, 98*(4), 614–639. doi:10.1002ce.21121

Zangori, L., Foulk, J., Sadler, T. D., & Peel, A. (2018). Exploring Elementary Teachers' Perceptions and Characterizations of Model-Oriented Issue-Based Teaching. *Journal of Science Teacher Education, 29*(7), 555–577. doi:10.1080/1046560X.2018.1482173

Zangori, L., Peel, A., Kinslow, A., Friedrichsen, P., & Sadler, T. D. (2017). Student development of model-based reasoning about carbon cycling and climate change in a socio-scientific issues unit. *Journal of Research in Science Teaching*, *54*(10), 1249–1273. doi:10.1002/tea.21404

Zeidler, D. L. (2007). *An inclusive view of scientific literacy: Core issues and future directions.* Paper presented at "Promoting Scientific Literacy: Science Education Research and Practice in Transaction," LSL Symposium, Uppsala, Sweden.

Zeidler, D. L., & Sadler, D. L. (2011). An inclusive view of scientific literacy: Core issues and future directions of socioscientific reasoning. In C. Linder, L. Ostman, D. A. Roberts, P. Wickman, G. Erickson & A. MacKinnon (Eds.), Promoting scientific literacy: Science education research in transaction (pp. 176-192). New York, NY: Routledge / Taylor & Francis Group.

Zeidler, D. L. (2014). Socioscientific issues as a curriculum emphasis: Theory, research and practice. In N. G. Lederman & S. K. Abell (Eds.), *Handbook of research in science education* (Vol. 2, pp. 697–726). Routledge.

Zeidler, D. L. (2014). Socioscientific issues as a curriculum emphasis: Theory, research, and practice. In N. G. Lederman & S. K. Abell (Eds.), *Handbook of research on science education* (Vol. 2, pp. 697–726). Routledge.

Zeidler, D. L. (2014). Socioscientific Issues as a Curriculum Emphasis: Theory, Research, and Practice. In N. G. Lederman & S. K. Abell (Eds.), *Handbook of Research on Science Education* (Vol. 2, pp. 711–740)., doi:10.4324/9780203097267-45

Zeidler, D. L. (2014). Socioscientific issues as a curriculum emphasis: Theory, research, and practice. In N. G. Lederman & S. K. Abell (Eds.), *Handbook of Research on Science Education, 2* (pp. 711–740). Routledge.

Zeidler, D. L. (2014). STEM education: A deficit framework for the twenty first century? A sociocultural socio-scientific response. *Cultural Studies of Science Education*, *11*(1), 11–26. doi:10.100711422-014-9578-z

Zeidler, D. L., Applebaum, S. M., & Sadler, T. D. (2011). Enacting a socioscientific issues classroom: Transformative transformations. In T. D. Sadler (Ed.), *Socio-scientific issues in science classrooms: Teaching, learning and research* (pp. 277–306). Springer. doi:10.1007/978-94-007-1159-4_16

Zeidler, D. L., Berkowitz, M. W., & Bennett, K. (2013). Thinking (scientifically) responsibly: the cultivation of character in a global science education community. In *Assessing schools for generation R (responsibility), contemporary trends and issues in science education 41* (pp. 83–99). Springer.

Zeidler, D. L., Herman, B. C., Clough, M. P., Olson, J. K., Kahn, S., & Newton, M. (2016). Humanitas emptor: Reconsidering recent trends and policy in science teacher education. *Journal of Science Teacher Education*, *27*(5), 465–476. doi:10.100710972-016-9481-4

Zeidler, D. L., Herman, B. C., & Sadler, T. D. (2019). New directions in socioscientific issues research. *Disciplinary and Interdisciplinary Science Education Research*, *1*(11), 1–9. doi:10.118643031-019-0008-7

Zeidler, D. L., & Kahn, S. (2014). *It's debatable! Using socioscientific issues to develop scientific literacy, K-12*. NSTA Press.

Zeidler, D. L., & Kahn, S. (2014). *It's debatable: using socioscientific issues to develop scientific literacy, K-12*. NSTA Press.

Zeidler, D. L., & Keefer, M. (2003). The role of moral reasoning and the status of socio-scientific issues in science education. In D. L. Zeidler (Ed.), *The role of moral reasoning on socio- scientific issues and discourse in science education* (pp. 7–38). Kluwer Academic Publishers.

Zeidler, D. L., & Keefer, M. (2003). The role of moral reasoning and the status of socioscientific issues in science education. In D. L. Zeidler (Ed.), *The role of moral reasoning on socioscientific issues and discourse in science education* (pp. 7–38). Springer. doi:10.1007/1-4020-4996-X_2

Zeidler, D. L., & Newton, M. H. (2017). Using a socioscientific issues framework for climate change education: An ecojustice approach. In D. Shepardson, A. Roychoudhury, & A. Hirsch (Eds.), *Teaching and learning about climate change: A framework for educators* (pp. 56–65). Routledge. doi:10.4324/9781315629841-5

Zeidler, D. L., & Nichols, B. H. (2009). Socioscientific issues: Theory and practice. *Journal of Elementary Science Education*, *21*(2), 49–58. doi:10.1007/BF03173684

Zeidler, D. L., & Sadler, T. D. (2008). The role of moral reasoning in argumentation: conscience, character and care. In S. Erduran & M. Pilar Jimenez-Aleixandre (Eds.), *Argumentation in science education: perspectives from classroom-based research* (pp. 201–216). Springer Press.

Zeidler, D. L., & Sadler, T. D. (2008). The role of moral reasoning in argumentation: Conscience, character and care. In S. Erduran & M. Pilar Jimenez-Aleixandre (Eds.), *Argumentation in science education: Perspectives from classroom-based research* (pp. 201–216). Springer Press.

Zeidler, D. L., Sadler, T. D., Applebaum, S., & Callahan, B. E. (2009). Advancing reflective judgment through socioscientific issues. *Journal of Research in Science Teaching*, *46*(1), 74–101. doi:10.1002/tea.20281

Zeidler, D. L., Sadler, T. D., Simmons, M. L., & Howes, E. V. (2005). Beyond STS: A research-based framework for socioscientific issues education. *Science education*, *89*(3), 357–377. doi:10.1002ce.20048

Zeidler, D. L., & Schafer, L. E. (1984). Identifying mediating factors of moral reasoning in science education. *Journal of Research in Science Teaching*, *21*(1), 1–15. doi:10.1002/tea.3660210102

Zeidler, D., Walker, K., Ackett, W., & Simmons, M. (2002). Tangled up in views: Beliefs in the nature of science and responses to socioscientific dilemmas. *Science Education*, *3*(86), 343–367. doi:10.1002ce.10025

Zohar, A., & Nemet, F. (2002). Fostering students' knowledge and argumentation skills through dilemmas in human genetics. *Journal of Research in Science Teaching*, *39*(1), 35–62. doi:10.1002/tea.10008

Zoltowski, C. B., Oakes, W. C., & Cardella, M. E. (2012). Students' ways of experiencing human-centered design. *Journal of Engineering Education*, *101*(1), 28–59. doi:10.1002/j.2168-9830.2012.tb00040.x

About the Contributors

Wardell Powell is an assistant professor of education at Framingham State University. He began his teaching career in the Bronx, New York, where he taught life and health science with the New York City Department of Education before relocating to Florida in 2003. Dr. Powell is a certified biology teacher in grades 6-12 in the state of Florida and has taught life science courses in the K-12 system for ten years before becoming a faculty member at Framingham State University. His scholarly interest centers on effective science instruction for students with and without specific learning disabilities and the development of accessible science curriculum for such students. Specifically, he is interested in the use of socioscientific issues as a critical pedagogical strategy to enhance scientific literacy among marginalized students.

* * *

Umran Betul Cebesoy completed her PhD at Middle East Technical University in 2014. She explored science teachers' genetics literacy levels and their decision-making skills in genetics related socioscientific issues in her dissertation. She was a guest researcher in Stockholm University during 2019 and investigated pre-service science teachers' decision-making skills using the SEE-SEP model and developed a professional development program for enriching science teachers' pedagogical content knowledge in teaching SSI. She has authored numerous international and national peer-reviewed journal articles in various platforms. She has been working in Usak University as Associate Professor since 2018. Her research areas include socioscientific issues, environmental education and STEM.

Benzegül Durak completed her undergraduate education at Boğaziçi University in the field of Science Education in 2016. She got her master's degree in 2019 at Yıldız Technical University with her thesis on socio-scientific issues and modeling. She continues her PhD education at Yıldız Technical University. She has been working as a research assistant at Düzce University since 2017. She still continues her studies on socio-scientific issues and scientific modeling.

Patricia Friedrichsen is a Professor of Science Education in the Department of Learning, Teaching & Curriculum at the University of Missouri (U.S.A.). Her research focuses on science teacher learning across the professional continuum from pre-service teacher education to professional development for practicing teachers. She uses a variety of frameworks to study teacher learning, including teacher beliefs, pedagogical content knowledge, and core teaching practices. Friedrichsen's current research focus is on supporting secondary science teachers' use of the Socio-scientific Issues and Model-Based Learning (SIMBL) approach, wherein students engage in modeling as they make sense of complex socio-scientific issues. Her research is funded through the National Science Foundation. At the University of Missouri, Friedrichsen directs the Middle and Secondary Science Teacher Education Program and is the department Director of Graduate Studies. She is a former Executive Board member of NARST – a global organization for improving science education through research, and a former Associate Editor of the Journal for Research in Science Teaching.

Benny Mart Hiwatig is a PhD student in the STEM Education program at the University of Minnesota. He earned his bachelor's degree in Secondary Education and took graduate coursework in Chemistry Education in the Philippines, where he also taught high school chemistry. He is currently involved in an NSF-grant funded research project that develops an observation protocol for STEM lessons and relevant training materials that are all directed towards improving the quality of STEM instruction in K-12 spaces. His primary research interests include assessment of student learning in integrated STEM contexts, exploration of how integrated STEM is enacted in the secondary level (especially in chemistry classes), and assessment and promotion of students' conceptual understanding of chemical concepts.

Sami Kahn serves as Executive Director of the Council on Science and Technology at Princeton University where she works to promote scientific literacy for all through interdisciplinary course development, robust STEM education research, and creative programming. An award-winning STEM educator, teacher educator, and author, she uses her background in science education and law to inform her research on inclusive science practices, socioscientific issues (SSI), argumentation, and social justice. She holds an M.S. in Ecology and Evolutionary Biology from Rutgers University, a J.D. in Law from Rutgers School of Law, and a Ph.D. in Curriculum and Instruction with specialization in Science Education from the University of South Florida where she served as a Presidential Doctoral Fellow. Prior to arriving at Princeton, she served as Assistant Professor of Science Education at Ohio University (promoted to Associate Professor with tenure), with earlier appointments at Rutgers University and Collegiate School in New York City.

Dilek Karisan holds a PhD in Science Education from Middle East Technical University. She is currently the Department Chair and an Associate Professor with tenure in the Department of Science Education at the University of Adnan Menderes. Her career in science education has been devoted to research related to socioscientific issues (SSI), including elements of the nature of science, epistemological reasoning, argumentation and reflective judgment. She subsequently was awarded what is a highly competitive fellowship from Higher Education Council to become a visiting scholar at University of South Florida. She has published many articles in national and international journals.

Li Ke is a postdoctoral Research Associate at the University of North Carolina at Chapel Hill. Ke's current research centers around engaging students in meaningful science learning through scientific modeling in the context of socio-scientific issues.

Khomson Keratithamkul is a PhD candidate in the STEM Education program at the University of Minnesota. His research interests primarily revolve around K-12 STEM education. He is interested in using integrated STEM curriculum as a vehicle for students to acquire necessary skills and knowledge to function in the 21st century. Khomson is also passionate about learning and incorporating different cultural stories, experiences, and narratives into STEM classrooms. Specifically, he is looking for ways to bring in global perspectives into integrated STEM curricula to encourage more global awareness among students and teachers.

Melanie Kinskey is an Assistant Professor of Science Education at Sam Houston State University. She earned her Ph.D. from the University of South Florida in Tampa. Her research focuses on elementary science instruction and preparing preservice and inservice teachers to enact socioscientific issues in the elementary context.

Hyunju Lee is a professor in the department of science education at Ewha Womans University in Seoul, South Korea. She has recieved her Ph.D. degree from University of Illinois at Urbana-Champaign. She has published various research on SSI teaching and SSI teacher educaton. Currently she has been conducting research on how teachers develop their understanding on SSI and teaching strategies for SSI.

Bahadir Namdar received his bachelor's and master's degrees in biology education from Dokuz Eylul University, Izmir, Turkey. Upon receiving a full scholarship for his post-graduate education from the Ministry of National Education, he completed his Ph.D. in science education from the University of Georgia, the United States in 2014. He is currently an associate professor of science education at the Department of Mathematics and Science Education at Ege University, Izmir, Turkey. His research interests include argumentation, computer-supported collaborative learning, and modeling in science education. Dr. Namdar has developed several professional development workshops for in-service and pre-service teachers on argumentation, socioscientific argumentation, and technology-enhanced argumentation, which were supported by The Scientific and Technological Research Council of Turkey.

Mark H. Newton is an Assistant Professor of Science Education at East Carolina University. His research focuses on the association between various learning contexts and how people conceptualize socioscientific issues.

Aysegul Oguz Namdar received her bachelor's, master's, and Ph.D. degrees in primary education from Hacettepe University, Ankara, Turkey. She currently works as an associate professor of elementary education at the Department of Elementary Education at Recep Tayyip Erdogan University, Rize, Turkey. Her research interests include drama, arts education, and teacher education in elementary education. She has developed several professional development workshops for both in-service and pre-service teachers. The workshops aimed at increasing teachers' pedagogical sills on teaching and creating values education and art education courses through drama. The professional development workshops were funded by The Scientific and Technological Research Council of Turkey.

Gillian H. Roehrig is a professor of STEM Education at the University of Minnesota. Her research explores issues of professional development for K-12 science teachers, with a focus on implementation of integrated STEM learning environments and induction and mentoring of beginning secondary science teachers. Her work in integrated STEM explores teachers' conceptions and implementation of STEM, curriculum development, and student learning in small groups during STEM lessons. She has received over $30 million in federal and state grants and published over 100 peer-reviewed journal articles and book chapters. She is a former board member of the National Association of Research in Science Teaching and former president of the Association for Science Teacher Education.

Troy Sadler is a professor of science education at the University of North Carolina at Chapel Hill School of Education and serves at the Thomas James Distinguished Professor of Experiential Learning. Sadler's research focuses on how students negotiate complex socio-scientific issues and how these issues may be used as contexts for science learning. He is interested in how issues-based learning experiences can support student learning of science and development of practices essential for full participation in modern democratic societies. He has also explored ways in which innovative technologies including virtual environments and gaming can support student learning. Sadler's work has been funded by the National Science Foundation, the Institute of Education Sciences, the US Department of Education, the Howard Hughes Medical Institute as well as local foundations and state agencies. He serves as the Co-editor for the Journal of Research in Science Teaching.

Mustafa Sami Topçu completed his doctorate education in 2008 at the Middle East Technical University Elementary Science Education Program. In 2007, he was a visiting researcher at the University of Florida, the United States (USA), where he completed an important part of his doctoral thesis. His doctoral dissertation is about socio-scientific issues and argumentation. He was a visiting faculty member at the University of Missouri in the USA (University of Missouri-Columbia) between 2014 and 2015 to conduct research on socio-scientific issues and STEM Education. He has been working as a Professor of Science Education at Yıldız Technical University since 2013.

Laura Zangori is an Assistant Professor of Science Education at the University of Missouri. Over the past 6 years, Zangori has taught and worked with students ranging from elementary through undergraduate classrooms on using the NGSS practices to support causal reasoning about biological systems. Currently, her research focuses on supporting preservice and inservice elementary teachers' engagement in the practices of modeling and how their students use models to understand and causally reason about biological systems.

Dana L. Zeidler (PhD, MS, BS, AAS) is the Distinguished University Professor of Science Education in the College of Education at the University of South Florida, Tampa, Florida. He has developed an international research program in the field of Science Education centered on Socioscientific Issues -- taking a sociocultural approach to teaching and learning about how moral and ethical issues can be a means to foster the formation of epistemological sophistication and character in the pursuit of scientific literacy. Professor Zeidler's recent distinctions include: serving as the Co-Editor of the Journal of Research in Science Teaching, receiving the Distinguished Career Award from NARST: A Worldwide Organization for Improving Science Teaching and Learning through Research; Past President of NARST; Distinguished Visiting Professor of Science Education, Ewha Womans University, Seoul, South Korea; and, Honorary Professor of Science and Environmental Studies, The Hong Kong Institute of Education, The University of Hong Kong and an Honorary Professor of Science Education at Beijing Normal University. Dr. Zeidler works closely with doctoral students, international scholars and other leaders in the field.

Index

G

I

K

M

N

O

P

R

S

T

W

Ensure Quality Research is Introduced to the Academic Community

Become an IGI Global Reviewer for Authored Book Projects

The overall success of an authored book project is dependent on quality and timely reviews.

In this competitive age of scholarly publishing, constructive and timely feedback significantly expedites the turnaround time of manuscripts from submission to acceptance, allowing the publication and discovery of forward-thinking research at a much more expeditious rate. Several IGI Global authored book projects are currently seeking highly-qualified experts in the field to fill vacancies on their respective editorial review boards:

Applications and Inquiries may be sent to:
development@igi-global.com

Applicants must have a doctorate (or an equivalent degree) as well as publishing and reviewing experience. Reviewers are asked to complete the open-ended evaluation questions with as much detail as possible in a timely, collegial, and constructive manner. All reviewers' tenures run for one-year terms on the editorial review boards and are expected to complete at least three reviews per term. Upon successful completion of this term, reviewers can be considered for an additional term.

If you have a colleague that may be interested in this opportunity, we encourage you to share this information with them.

www.ingramcontent.com/pod-product-compliance
Lightning Source LLC
LaVergne TN
LVHW081350300926
849161LV00008B/2122